A Practical Guide to the NEC4 Engineering and Construction Contract

A Practical Guide to the NEC4 Engineering and Construction Contract

Michael Rowlinson

WILEY Blackwell

This edition first published 2019
© 2019 John Wiley & Sons Ltd

All rights reserved. No part of this publication may be reproduced, stored in a retrieval system, or transmitted, in any form or by any means, electronic, mechanical, photocopying, recording or otherwise, except as permitted by law. Advice on how to obtain permission to reuse material from this title is available at http://www.wiley.com/go/permissions.

The right of Michael Rowlinson to be identified as the author of this work has been asserted in accordance with law.

Registered Offices
John Wiley & Sons, Inc., 111 River Street, Hoboken, NJ 07030, USA
John Wiley & Sons Ltd, The Atrium, Southern Gate, Chichester, West Sussex, PO19 8SQ, UK

Editorial Office
9600 Garsington Road, Oxford, OX4 2DQ, UK

For details of our global editorial offices, customer services, and more information about Wiley products visit us at www.wiley.com.

Wiley also publishes its books in a variety of electronic formats and by print-on-demand. Some content that appears in standard print versions of this book may not be available in other formats.

Limit of Liability/Disclaimer of Warranty
While the publisher and authors have used their best efforts in preparing this work, they make no representations or warranties with respect to the accuracy or completeness of the contents of this work and specifically disclaim all warranties, including without limitation any implied warranties of merchantability or fitness for a particular purpose. No warranty may be created or extended by sales representatives, written sales materials or promotional statements for this work. The fact that an organization, website, or product is referred to in this work as a citation and/or potential source of further information does not mean that the publisher and authors endorse the information or services the organization, website, or product may provide or recommendations it may make. This work is sold with the understanding that the publisher is not engaged in rendering professional services. The advice and strategies contained herein may not be suitable for your situation. You should consult with a specialist where appropriate. Further, readers should be aware that websites listed in this work may have changed or disappeared between when this work was written and when it is read. Neither the publisher nor authors shall be liable for any loss of profit or any other commercial damages, including but not limited to special, incidental, consequential, or other damages.

Library of Congress Cataloging-in-Publication Data

Names: Rowlinson, Michael (Quantity Surveyor), author.
Title: A practical guide to the NEC4 Engineering and Construction Contract /
   Michael Rowlinson, Michael Rowlinson Associates Limited, Keswick, UK.
Description: First edition. | Hoboken, NJ ; Chichester, West Sussex, United
   Kingdom : John Wiley & Sons Ltd., [2019] | Includes bibliographical
   references. |
Identifiers: LCCN 2018029649 (print) | LCCN 2018035489 (ebook) | ISBN
   9781119522522 (Adobe PDF) | ISBN 9781119522546 (ePub) | ISBN 9781119522515
   (hardcover)
Subjects: LCSH: NEC4. Engineering and construction contract. | Civil
   engineering contracts–Great Britain. | Construction contracts–Great
   Britain.
Classification: LCC KD1641 (ebook) | LCC KD1641 .R696 2018 (print) | DDC
   343.4107/8–dc23
LC record available at https://lccn.loc.gov/2018029649

Cover Design: Wiley
Cover Image: © iStock.com/ivanastar

Contents

1 Introduction   1
    1.1   General   1
    1.2   Mechanics not law   2
    1.3   A simple formula for understanding a contract   3
    1.4   Mandatory or discretionary   4
    1.5   Conditions precedent   4
    1.6   Note on use of uppercase in key words and phrases   5

2 Background to the NECECC   7
    2.1   The background: First edition   7
    2.2   The second edition   8
    2.3   The third edition   9
    2.4   The third edition (reprinted)   9
    2.5   The fourth edition   9
    2.6   Endorsement of NEC3 by the Office of Government Commerce   10
    2.7   Endorsement by the Development Bureau, HKSAR Government   11
    2.8   General philosophy: Aims and objectives   12
    2.9   Flexibility   12
    2.10   Clarity and simplicity   13
    2.11   Stimulus to good management   14
    2.12   Other characteristics   15

3 The Options: An Overview   17
    3.1   General arrangement of the ECC   17
    3.2   Other documents referred to   19
    3.3   Contract Data   20
    3.4   The published documents   20
    3.5   Main Options: General outline   21

4 Spirit of Mutual Trust and Cooperation   25
    4.1   Introduction   25
    4.2   The clauses   25
    4.3   What does it mean?   27
    4.4   Practical issues   28

5 The Cast of Characters — 33
- 5.1 Introduction — 33
- 5.2 The Client — 33
- 5.3 The Project Manager — 35
- 5.4 The Supervisor — 38
- 5.5 The Contractor — 38
- 5.6 The Senior Representatives — 39
- 5.7 The Adjudicator — 40
- 5.8 The Tribunal — 41
- 5.9 The Dispute Avoidance Board — 41
- 5.10 Subcontractors — 42
- 5.11 'Others' — 42
- 5.12 Named Suppliers — 43
- 5.13 Designers — 43
- 5.14 Principal Designer — 44
- 5.15 Principal Contractor — 45
- 5.16 Practical issues — 45

6 Communications, Early Warnings and other General Matters — 47
- 6.1 Introduction — 47
- 6.2 Communications: The clause — 47
- 6.3 Communications: Practical issues — 49
- 6.4 Early warnings: The clause — 51
- 6.5 Early warnings: Practical issues — 53
- 6.6 Other matters: The clauses — 55
- 6.7 Other matters: Practical issues — 59

7 The Contractor's Main Responsibilities — 61
- 7.1 Introduction — 61
- 7.2 Providing the Works — 61
- 7.3 Contractor's design — 62
- 7.4 Information modelling — 66
- 7.5 Other matters — 68
- 7.6 Practical issues — 72

8 Subcontracting — 75
- 8.1 Introduction — 75
- 8.2 Definition of a Subcontractor — 75
- 8.3 The core clauses — 76
- 8.4 Provisions in the Main Options — 76
- 8.5 Practical issues — 77
- 8.6 Options for forms of subcontract in the NEC4 family — 78

9	**Quality Management**		**81**
	9.1	Introduction	81
	9.2	Quality management system	81
	9.3	Tests and inspections	82
	9.4	What is a Defect?	84
	9.5	The Defect procedure	84
	9.6	The Defects Certificate	86
	9.7	Uncorrected Defects	87
	9.8	Practical issues	87
10	**Title**		**91**
	10.1	Introduction	91
	10.2	The core clauses	91
	10.3	Practical issues	92
11	**Liabilities and Insurance**		**95**
	11.1	Introduction	95
	11.2	The core clauses	95
	11.3	Secondary options	98
	11.4	Practical issues	99
12	**Time**		**101**
	12.1	Introduction	101
	12.2	The programme: Contents	102
	12.3	The programme: Submitting, accepting and revising	107
	12.4	The programme: Practical issues	110
	12.5	Starting and finishing	118
	12.6	Other matters	121
	12.7	Secondary Options related to Time	124
	12.8	Practical issues	126
13	**Payment**		**131**
	13.1	Introduction	131
	13.2	The payment process	131
	13.3	Payments in multiple currencies	134
	13.4	Interim payments – The amount due and the Price for Work Done to Date	135
	13.5	Supporting documents and records	145
	13.6	Final assessment	148
	13.7	The Contractor's share	150
	13.8	The Contractor's share: Practical issues	152
	13.9	Special provisions for the United Kingdom	153

	13.10	Related Secondary Options	157
	13.11	Practical issues	162

14 The Schedules of Cost Components — 169

	14.1	Introduction	169
	14.2	The Schedule of Cost Components	169
	14.3	The Short Schedule of Cost Components	174
	14.4	Application to Subcontractors	175
	14.5	Practical issues	176

15 Compensation Events: Theory and Events — 179

	15.1	Introduction	179
	15.2	The theory	179
	15.3	The events	181
	15.4	Practical issues	196

16 Compensation Events: Procedures — 199

	16.1	Introduction	199
	16.2	Notification by the Project Manager	200
	16.3	Notification by the Contractor and the Project Manager's reply	203
	16.4	Other matters associated with notifying compensation events	206
	16.5	Quotations: Substance	208
	16.6	Quotations: Submission and reply	210
	16.7	Assessments by the Project Manager	215
	16.8	Proposed instructions	217
	16.9	Implementing compensation events	218
	16.10	Practical issues	219

17 Compensation Events: Assessment — 227

	17.1	Introduction	227
	17.2	Changes to the Prices	228
	17.3	Changes to the Completion Date and Any Key Dates	232
	17.4	Project Manager's assumptions	234
	17.5	Other related matters	236
	17.6	Practical issues	238

18 Termination — 243

	18.1	Introduction	243
	18.2	Reasons for termination	243
	18.3	Secondary Option X11	247
	18.4	Implementing termination	248
	18.5	Procedures after termination	248
	18.6	Assessing the amount due after termination	250
	18.7	Practical issues	252

19	**Resolving and Avoiding Disputes**	**255**
	19.1 Introduction	255
	19.2 Option W1	256
	19.3 Option W2	261
	19.4 Option W3	267
	19.5 Practical issues	270
20	**Secondary Options**	**273**
	20.1 Introduction	273
	20.2 X2: Changes in the law	273
	20.3 X4: Ultimate holding company guarantee	274
	20.4 X12: Multiparty Collaboration	274
	20.5 X13: Performance bond	279
	20.6 X17: Low performance damages	280
	20.7 X18: Limitation of liability	280
	20.8 X20: Key Performance Indicators	281
	20.9 X21: Whole Life Cost	282
	20.10 X22: Early Contractor Involvement	283
	20.11 Y(UK)3: The Contracts (Rights of Third Parties) Act 1999	288
	20.12 Z: Additional conditions of contract	288
	20.13 Practical issues	289
21	**Completing the Contract Data**	**291**
	21.1 Introduction	291
	21.2 Purpose and form of the Contract Data	291
	21.3 Contract Data Part One	292
	21.4 Contract Data Part Two	304
	21.5 Practical issues	309
22	**The Supporting Documents: Need and Content**	**311**
	22.1 Introduction	311
	22.2 Scope	312
	22.3 Site Information	324
	22.4 Practical issues	325

Bibliography		329
Appendix 1	Tables of Clause Numbers, Case Law and Statutes	331
Appendix 2	Tables of Client's, Project Manager's, Supervisor's, Contractor's, Senior Representatives, Adjudicator's, Dispute Avoidance Board and Tribunals Actions	341
Appendix 3	Tables of Communication Forms and Their Uses	379

Chapter 1
Introduction

1.1 General

In writing this guide I have set out to provide a view, much of it personal, as to how to get the most out of the 4th Edition of the New Engineering Contract Engineering and Construction Contract (ECC). It is no secret that I am a fan of this suite of contracts and, as a result, may be willing to overlook what many perceive as it faults or weaknesses. In this guide I have tried to identify and suggest ways in which the procedures and aims of the contract can be simplified so that users do not become unnecessarily bogged down in procedure, but instead concentrate on achieving the goals of the ECC. This guide therefore goes through the procedure in detail as intended by the relevant clauses, but concentrates on practical issues to provide suggestions which the parties can use to achieve the overall intent and spirit of ECC and to reach the common goal.

With this guide, you get what it says on the cover: A Practical Guide to the NEC4 ECC Form of Contract. It is a guide to provide users of the ECC, both novice and experienced, with a view of all its various philosophies, principles, mechanisms and vagaries. The reader will be guided through the contract in a manner that will enable him or her to use this guide for reference without necessarily having to read it all: in other words, a practical guide rather than a stuffy textbook. That said, there will be an amount of cross-referencing between sections in order to avoid repetition, so users will need to follow these references to find more detailed supporting guidance to particular issues. One area that is not cross-referenced is the term 'spirit of mutual trust and cooperation' as found in clause 10.1 of the ECC, although used extensively throughout the guide. If users are uncertain of the meaning of this phrase, then they need to re-read Chapter 4.

This version of the Guide follows on from the two editions of the same title but for the NEC3 edition of the suite of contracts. Many of the clauses remain the same in NEC4 as they were in NEC3, so both the processes and the practical guidance remain the same from one edition to the next. Nonetheless the publication of NEC4 introduced a significant amount of change to the ECC, all of which is considered and included in this version of the Guide. What this Guide does not do is identify or consider the changes from NEC3 to NEC4. It is written purely about NEC4.

To assist the reader in finding where any particular clause, related legal case or UK statute is referred to in the text, a comprehensive index of such references is included in Tables A1.1–A1.3 in Appendix 1.

A Practical Guide to the NEC4 Engineering and Construction Contract, First Edition. Michael Rowlinson.
© 2019 John Wiley & Sons Ltd. Published 2019 by John Wiley & Sons Ltd.

The more I have worked with this suite of contracts over the years, the more I have come to think of it not as a contract but as a Project Management Procedures Manual. This should not be a surprise, as the original contract was drafted by project managers for construction professionals (and not by lawyers for other lawyers and judges).

Nevertheless, we must not lose sight of the fact that the ECC is a contract and, as such, legally binds those parties that enter into a contract incorporating these standard terms.

What I have also included in this Guide is advice and practical issues with regard to the NEC4 Engineering and Construction Subcontract. Except for some of the time periods being different and the absence of one main option, the Subcontract is truly back to back with the Contract. The differences are identified and additional practical issues are identified which apply to the Subcontractor/Contractor relationship. What users of the Guide will need to do in order to apply the majority of the text to the Subcontract is substitute the names of the parties and agents in the Contract with the equivalent names in the Subcontract. So 'Client', 'Project Manager' and 'Supervisor' in the Contract become 'Contractor' in the Subcontract. Similarly 'Contractor' in the Contract becomes 'Subcontractor' in the Subcontract, and 'Subcontractor' in the Contract becomes 'Subsubcontractor' in the Subcontract.

1.2 Mechanics not law

Being a practical guide, this book considers the mechanics of the contract and not the law. As a practicing construction professional, I am interested in the successful outcome of the project for all parties involved. From my point of view, the employing organisation should get what it wants in terms of a project finished on time, to the required quality and within budget (providing, of course, that the budget was reasonable in the first place). The consultants should be recognised for their contribution, whether it be design, management or commercially orientated, and be paid a reasonable fee for the service they provide. The contractors and subcontractors who carry out the work should be allowed to work efficiently, be recognised as having contributed to the project and make a profit.

Only those projects that satisfy all of the above criteria should be considered as being successful. Every organisation, whether it be a company, partnership or individual who is involved in a project, has its own needs and goals from that project. A good project will recognise this simple fact of business. It is when all the parties involved recognise each other's business goals (see Section 4.4.3) from the project, and work to align these goals, that success is achieved for all. As soon as one of the organisations involved feels dissatisfied, then the seeds of a dispute have been sown. As the industry knows, such seeds germinate easily and freely; once they appear on a project they can spread faster than any invasive weed.

Following on from the earlier editions, the ECC is drafted to impose the best practices within project management on the parties with the goal of avoiding disputes. It is the mechanics of these procedures and how to make them work effectively that is the focus of this guide.

As a consequence, the guide does not consider the law in relation to the ECC, except where reference is needed to explain why something is included or to confirm that, in relation to the law in the United Kingdom, those requirements have been complied with by the ECC (or not as the case may be).

1.3 A simple formula for understanding a contract

Let's face it: all contracts are confusing when you first try to work out what it all means. I picked up a simple formula for considering contracts many years ago from an experienced Chief Quantity Surveyor of a contracting organisation, who came to my then local centre of The Chartered Institute of Building to give an evening talk on Joint Contracts Tribunal (JCT) Contracts. It didn't matter that he was talking about JCT Contracts. What I took away from that talk was a formula that I still use today in relation to any contract or procedural document that I encounter; this formula holds good in all such situations. I still have the piece of paper on which I noted the few words I needed to remind me of what to do. I rarely look at that piece of paper now, as the formula has become second nature to me in relation to every contract or set of procedures I read.

The formula is in two parts. The first part can be remembered by four words: WHO, WHAT, WHEN and HOW.

To expand, a contract is a document that sets out the rights and obligations of the parties to that contract, no matter what the contract is for. In the construction and related industries such contracts cover (usually by necessity) a range of extensive rights and obligations for both parties, how such rights and obligations are to be administered, and the involvement of agents to carry out specified duties for one or both of the parties. WHO, the first of our four key words, relates to the administration of these rights and obligations. The WHO in the ECC will be one of the eight named persons including the Employer, the Project Manager, the Supervisor, the Contractor, the Senior Representatives, the Adjudicator, the Dispute Avoidance Board or the Tribunal. The specific roles of these individuals are covered in detail in Sections 5.2–5.9.

By its processes and procedures, the ECC sets out WHAT must or may be done in the event that a certain circumstance arises. The WHAT will involve the WHO doing something as set out in the contract.

WHEN that something is to be done is also set out by the contract. In the case of ECC, the timetable for WHEN these things shall be done is clear and forms a key part of the processes and procedures under the contract. Failure to comply with these processes and procedures in accordance with the requirements specified by WHEN can result in a right being forfeited because of this failure.

Finally, ECC sets out HOW the process or procedure shall be carried out. Again ECC is prescriptive as to the HOW, although much of the HOW is set out in general terms that apply across all of the subsequent detailed processes and procedures.

To summarise, the first part of the formula (which holds good for all contracts and not just the ECC) is to consider WHO does WHAT, WHEN they do it and HOW it is to be done. Understanding these things is important as ECC creates what are known in legal circles as conditions precedent. Although the English Courts do not like such provisions, they can be effective if drafted in certain terms (for further comment on conditions precedent see Section 1.5).

When dealing with specific processes and procedures in this guide, the WHO, WHAT, WHEN and HOW will be summarised as appropriate in each case.

These four key words were included by Rudyard Kipling in his short story 'The Elephant's Child', part of the *Just So Stories*, (1902) when he wrote:

> "I keep six honest serving-men
> (They taught me all I knew);
> Their names are What and Why and When
> And How and Where and Who."

1.4 Mandatory or discretionary

The second part of the formula I learnt that evening was to consider whether an obligation, requirement or procedure is mandatory or discretionary. The distinction is quite clear: if something is mandatory, then it must be done in order to create a right for you and/or an obligation on someone else. If something is discretionary, then the party concerned can do it if the party feels it is appropriate but loses nothing otherwise.

The key to whether something is mandatory or discretionary is in the little words. If a provision says that a party 'shall', 'must' or 'will' do something, then the requirement to do that something is mandatory; that key little word leaves that party with no other option.

On the other hand, if the provision in question says that the party 'may' or 'can' do something, then that requirement is left to the discretion of that party, i.e. the action is discretionary.

Appreciating whether a requirement or a provision is mandatory or discretionary is key to making sure that you, as a party or agent to the contract, do what is required of you at the right time and in the right way.

In the ECC, and indeed every other contract in the NEC4 family together with all the previous editions, there is little to doubt or question as to whether things are mandatory or discretionary. The first clause in the ECC, clause 10.1, clearly states that the Employer, the Project Manager, the Supervisor and the Contractor *shall* act as stated in the contract. The meaning is plain and clear: they are all required to carry out the procedures set out in the contract at all times and in the way stated. There is no discretion about it, unless such discretion is given expressly in a particular clause (there are a small number of such instances which will be pointed out as they arise).

1.5 Conditions precedent

Put as simply as possible, a condition precedent is a condition that acts to prevent either a right or an obligation from coming about until such time as the event prescribed as the condition precedent occurs. If a time limit is attached to the occurrence of the event (which is a condition precedent to a right or an obligation) and the event has not occurred within the time limit stated, the right or obligation can never come about.

It is important for users of the ECC to understand this principle; part of a mechanism that is commonly used includes such a condition precedent with a time limit. This actual condition will be highlighted when it is commented on.

While the courts in the United Kingdom do not traditionally like or support such clauses, they have enforced numerous examples where the wording has been clear. The first and second editions of the ECC were both said to include conditions precedent but it is generally felt that those conditions were not clearly enough worded to be effective. However, it was generally considered that the wording in the third edition was almost certainly clear enough to be considered as an effective condition precedent. That being the case, the fourth edition also contains the same standard of wording.

1.6 Note on use of uppercase in key words and phrases

Capital initial letters are used to identify terms that are defined as a feature of the ECC as set out in clause 11.1. Whenever I have referred to any such term, I have maintained consistency with the ECC and followed that principle of using upper case for the first letter of defined terms throughout the text of this guide. The reader will, however, come across instances where the same terms are referred to in a general sense, when lowercase is used. I have adopted this approach in order to distinguish between specific references to procedures, rights, obligations and other such matters that are directly linked to the ECC, and more general comments about good practice, the construction industry and other non-contract specific items.

For example, 'Contractor' refers to a specific issue that concerns the Contractor under the ECC and 'contractor' refers to the contractor in general terms.

Chapter 2
Background to the NECECC

2.1 The background: First edition

The timescale that we are looking at starts with a consultative document published in 1991, which was followed by the first edition in 1993, the second edition in 1995, the third edition in July 2005, the reprinted third edition in April 2013 and then the fourth edition in June 2017 (NEC Panel 2017a).

Many people believe the first edition was published in response to Sir Michael Latham's (1994) report, *Constructing the Team*. This report was, however, predated by both the consultative document and First Edition.

In his report, Sir Michael identified the New Engineering Contract (NEC) (as it was then called) as being the contract, which, more than any other in general circulation at the time, contained many of the provisions, which he considered should be adopted in Construction Contracts. Out of the thirteen key issues that Latham thought should be adopted, the NEC contained eight. The full list of key issues from *Constructing the Team* is as follows:

1. A specific duty for all parties to deal fairly with each other, and with their subcontractors, specialists and suppliers, in an atmosphere of mutual cooperation.
2. Firm duties of teamwork, with shared financial motivation to pursue those objectives. These should involve a general presumption to achieve 'win-win' solutions to problems which may arise during the course of the project.
3. A wholly interrelated package of documents which clearly defines the roles and duties of all involved, and which is suitable for all types of project and for any procurement route.
4. Easily comprehensive language and with Guidance Notes attached.
5. Separation of the roles of contract administrator, project or lead manager and adjudicator. The Project or lead Manager should be clearly defined as the client's representative.
6. A choice of allocation of risks, to be decided as appropriate to each project but then allocated to the party best able to manage, estimate and carry the risk.
7. Taking all reasonable steps to avoid changes to pre-planned works information. But, where variations do occur, they should be priced in advance, with provision for independent adjudication if agreement cannot be reached.

A Practical Guide to the NEC4 Engineering and Construction Contract, First Edition. Michael Rowlinson.
© 2019 John Wiley & Sons Ltd. Published 2019 by John Wiley & Sons Ltd.

8. Express provision for assessing interim payments by methods other than monthly valuation i.e. milestones, activity schedules or payment schedules. Such arrangements must also be reflected in the related subcontract documentation. The eventual aim should be to phase out the traditional system of monthly measurement or remeasurement but meanwhile provision should still be made for it.
9. Clearly setting out the period within which interim payments must be made to all participants in the process, failing which they will have an automatic right to compensation, involving payment of interest at a sufficiently heavy rate to deter slow payment.
10. Providing for secure trust fund routes of payment.
11. While taking all possible steps to avoid conflict on site, providing for speedy dispute resolution if any conflict arises by a pre-determined impartial adjudicator/referee/expert.
12. Providing for incentives for exceptional performance.
13. Making provision where appropriate for advance mobilisation payments (if necessary, bonded) to contractors and subcontractors, including in respect of off-site prefabricated materials provided by part of the construction team.

By comparison with the other major standard forms available at the time, the NEC's score of eight was three to four times better than any of its competitors. Encouraged by this praise, the NEC Panel set about revising the contract to incorporate the balance of the ideals and to take account of other comments that had been made.

2.2 The second edition

The job of revising the first edition was completed in 1995 and, in November of that year, the second edition was published. This edition not only incorporated revisions to the provisions of the Contract but also heralded a change in name to the 'New Engineering Contract Engineering and Construction Contract', shortened in use to NECECC.

The change in name was prompted by Sir Michael's report. He had commented that the name NEC, coupled with it being published by the commercial arm of the Institution of Civil Engineers (Thomas Telford), served to suggest that it was a civil engineering contract. This impression was already restricting and would continue to restrict the use of this otherwise versatile contract to the civil and related engineering sectors; Sir Michael believed, however, that it was suitable for all types of construction, including not least building.

The change of name had the desired effect as, over the next 10 years, the use of the NECECC by employers in the building industry steadily increased.

The principle aim of the revisions that brought about the second edition was to incorporate all thirteen of the ideals (see Section 2.1) set out in *Constructing the Team*. Provision was made in the second edition to cater for the five that had not been included in the first edition. It is interesting to note, however, that the third edition only catered for twelve of the thirteen ideals. The provision for trust funds found at Secondary Option V of the second edition was dropped in the third edition for the simple reason that it was never used.

2.3 The third edition

Throughout the life of the second edition, the NEC User's Group sought and collected feedback from its members on the aspects of the contract, where it was felt that revision was required. Taking this feedback into account, the NEC Panel not only revised the NECECC but worked to consolidate the other contracts that they had drafted using the same principles and to bring them together into one unified family.

While the revision of the NECECC had at one time been expected in late 2002/early 2003, the work to the whole family delayed the publication until 14 July 2005. (Those readers with a keen eye will have noted that the NEC3 family of documents printed at that time all bear the date June 2005 on the front cover. This had been the intended month of publication; however production problems delayed the actual launch until July.)

2.4 The third edition (reprinted)

The third edition was initially reprinted in June 2006 to correct a very small number of typographical errors in the June 2005 publication. Apart from these changes no others were made.

Separate Amendments were issued in September 2011 to alter dispute resolution Option W2 and secondary Option Y(UK)2 to cater for the changes required by the amendments to the Housing Grants, Construction and Regeneration Act 1996 contained within the Local Democracy, Economic Development and Construction Act 2009, which came into effect for any new contract entered into from 1 October 2011.

In April 2013 the whole family of documents was reprinted with further amendments. These amendments made some minor changes to the compensation event clauses (core clause 6) for the sake of clarity rather than to change the meaning or application, together with a minor amendment to clause 40.1 and some more significant alterations to dispute resolution Option W2. These were accompanied by the introduction of a new UK specific secondary Option, labelled Y(UK)1 and titled 'Project Bank Account'. Detailed comment on this new secondary Option is provided at Section 13.10.4.

The April 2013 reprint also involved a change in image, in that the front covers were redesigned and the introduction of seven 'How To' guides covering the writing of the Works Information (NEC Panel 2013b), Scope (PSC) (NEC Panel 2013c) and Service Information (TSC) (NEC Panel 2013d), together with Communication Forms for each of the ECC, PSC and TSC (NEC Panel 2013e–g), with the seventh dealing with the incorporation of Building Information Modelling (BIM) (NEC Panel 2013h) into NEC3 Contracts. The total number of documents in the box set increased to thirty-nine.

2.5 The fourth edition

The fourth edition was published in June 2017 with the strap line 'evolution, not revolution'. The intent was to incorporate changing patterns in procurement and project management. The preface to the documents says that the drafting panel had three key objectives, these being:

- provide greater stimulus to good management (see Section 2.11);
- support new approaches to procurement which improve contract management; and
- inspire increased use of NEC in new markets and sectors.

Some of this was achieved by adding new contracts to the suite of contracts, in the form of a new design, build and operate form, a forthcoming alliance contract and subcontracts to support some of the other contracts which did not already have such a document to support them.

Other enhancements have included new or expanded provisions to cover finalising elements of cost as contracts proceed, providing for final assessments, expanding the available dispute resolution provisions, creating more standardisation across the suite and reworking the guidance into what is seen by the drafters as a more user friendly format.

This last element has been created by replacing the previous guidance documents with a four-volume set, which is intended to assist users through the use of the contract from conception to completion. The four volumes provided are:

- Volume 1 – Establishing a Procurement and Contract Strategy (NEC Panel 2017b)
- Volume 2 – Preparing an Engineering and Construction Contract (NEC Panel 2017c)
- Volume 3 – Selecting a Supplier (NEC Panel 2017d)
- Volume 4 – Managing an Engineering and Construction Contract (NEC Panel 2017e)

These documents are referred to at various places within this Guide.

What users familiar with the second and third editions will immediately realise is that there are no flow charts in these User Guides. There are, however, books of flowcharts available for each contract for an additional charge for each volume required.

It is not the purpose of this guide to describe the changes from the third to fourth editions.[1] This guide concentrates on administering projects using the fourth edition of the ECC. That said, the principles of the third edition are very similar; it is my belief that if you are familiar with only the second or third or fourth edition, then you should easily understand the others. Equally, anyone who can properly understand the ECC should be able, with a little thought and application, to use any of the other contracts in the NEC4 family.

2.6 Endorsement of NEC3 by the Office of Government Commerce

When the third edition of the NEC family of contracts was published in 2005, all of the 23 documents in the box set carried an endorsement from the UK's Office of Government Commerce (OGC) on their title pages and back covers: the use of the family was recommended to all public sector construction procurers. By the time of the 2013 reprint,

---

[1] For those interested in the revisions from the third to fourth editions, the author wrote a series of articles that were published in *Civil Engineering Surveyor* in 2017/2018 (Rowlinson 2017a–d, 2018a,b). All are available on the author's website (www.michael-rowlinson.co.uk).

the government body endorsing the contracts had changed its name to the Construction Clients' Board and although the wording of the endorsement has changed, the meaning and intent remained the same. The recommendation was linked to a statement that such procurers must satisfy the objectives of the government's *Achieving Excellence in Construction* (AEC) principles. These principles had been launched in March 1999 with the aim of improving the performance of central government departments, executive agencies and nondepartmental public bodies as clients in the construction industry. In the United Kingdom, depending on the state of the economy, these procurers account for between 35% and 40% of all new build, refurbishment and maintenance work carried out by the construction industry.

It is difficult to find any literature that lists the principles to which the OGC refer. What is published is a list of key factors, which in summary are:

- the establishment of integrated project teams;
- the use of short and effective lines of communication;
- the consideration of design, construction, operation and maintenance as a whole;
- effective risk management;
- effective value management;
- the use of sound project management techniques; and
- creating partnering and long-term relationships.

Many of these matters have their roots in Sir Michael Latham's 1994 report *Constructing the Team* and Sir John Egan's report *Rethinking Construction* (The Construction Task Force 1998). It is fair to say that both of these reports, and especially the former, influenced the development of the NEC family of contracts. The result was the endorsement by the OGC. Later contracts in the family published in December 2009 also carried the endorsement; the only difference is that by then it was given by the Construction Client's Board (formerly known as the Public Sector Construction Clients' Forum). The AEC principles can now only be accessed in the National Archives at the link http://webarchive.nationalarchives.gov.uk/20110622151127/www.ogc.gov.uk/ppm_documents_construction.asp.

The fourth edition carries a similar endorsement, now given by the Government Construction Board, from within the Cabinet Office UK. This endorsement is similar but now talks about efficiencies across the public sector and the promotion of behaviours in line with the principles of the Government Construction Strategy. This strategy has several faces, and rather than a single link giving access to relevant documents a general internet search on the phrase *Government Construction Strategy* will reveal numerous links to government information and scholarly articles about this latest initiative.

2.7 *Endorsement by the Development Bureau, HKSAR Government*

NEC4 also carries an endorsement from the Development Bureau of the Government of the Hong Kong Special Administrative Region (HKSAR) of the People's Republic of China. This endorsement recommends a progressive transition from NEC3 to NEC4 for the management of public works projects in Hong Kong. Whilst recognising the need for suitable amendments to cater for the Hong Kong environment, which term could include legislative, economic, social and market requirements, the endorsement foresees that the use of NEC4 should enhance collaborative partnering, unlock innovations and achieve better cost management and value for money in public works contracts.

An internet search on *Development Bureau Hong Kong* immediately reveals several statements and blogs concerning the use of NEC contracts in the HKSAR. These include a proud reference to the Development Bureau having won the NEC Client of the Year 2017 category and having been Highly Commended in the 'Contract Innovation Through Additional Clauses' category of the NEC Awards 2017. The endorsement and the awards, when taken together, show a government body that is not only making the right statements but being recognised for its achievements by the promoting organisation, for which they should be congratulated.

2.8 General philosophy: Aims and objectives

The brief leading to the preparation of the initial consultative document of the ECC was to prepare a radical new style of contract form. The ECC certainly achieved that aim in that it is intentionally different from other forms of construction contracts available at that time.

In order to comply with this desire for a contract that would be seen as radical, the drafting committee developed a number of aims and objectives that they sought to introduce into the form. These are summarised in Sections 2.9–2.12.

2.9 Flexibility

One of the principle aims was to make the ECC as flexible as possible, thereby allowing the provisions:

- to be used for engineering and construction work containing any or all of the traditional disciplines such as civil, electrical, mechanical and building work;
- to be used whether the Contractor has some design responsibility, full design responsibility or no design responsibility;
- to provide all the normal current options for types of contract such as competitive tender (where the Contractor is committed to its offered prices), target contracts, cost reimbursable contracts or management contracts; and
- to be used in any country in the world.

As this guide develops, readers will appreciate how this flexibility is provided and how numerous combinations can be used to create contracts with different risk profiles to suit the needs of an individual project or series of projects.

The objective that the contract could be used anywhere in the world is being fulfilled. The NEC3 family was adopted by governments other than that of the UK; for example, both the Australian and New Zealand authorities endorsed the contract. It is known to be used on all of the five major land masses and in the Antarctic (by the UK government on the survey stations built in that region). At the time of writing, there is no indication that these positions will change following publication of NEC4.

That said, NEC3 has been widely used in Framework agreements in the United Kingdom. That these contracts have been entered into for varying periods, typically 3 to 5 years but in some instances for 10 years or longer, will mean that NEC3 remains in use for some time to come.

2.10 Clarity and simplicity

One of the more radical aims was to produce a contract that was clear and simple in its format and readily understandable by ordinary construction professionals, as opposed to being a contract that required a degree of legal ability in order to be able to understand the rights and obligations of the parties. This aim for clarity and simplicity has been incorporated in several ways, including:

- the use of ordinary language rather than legal jargon;
- the use of short sentences at all times and by using subclauses to break up large bodies of text;
- the use of a logical structure that keeps like matters grouped together;
- the provision of flow charts for each procedure in the contract;
- a consistent approach to the management and allocation of risk across the different procurement routes;
- by limiting the extent of the text and clauses in order to provide a framework rather than by being prescriptive.

The decision to use clear and simple language with short sentences and subclauses provides the user with a contract that can be read in bite-sized pieces. The downside is that, for an industry that has been used to prescriptive rules in contracts, the lack of detail and direction regarding the next step in every resulting scenario is a concern to many. In order to overcome this concern, users must learn to appreciate the goal of a clause or subclause and adopt their working practices to achieve that goal. Practical examples of such steps are considered throughout the following chapters of this guide.

The language itself is written using the present tense, a style that has been criticised by The Hon. Mr. Justice Edwards-Stuart who, in the judgement he made in the case of *Anglian Water Services Ltd v Laing O'Rourke Utilities Ltd* [2010] EWHC 1529 (TCC), said:

> I have to confess that the task of construing the provisions in this form of contract is not made any easier by the widespread use of the present tense in its operative provisions. No doubt this approach to drafting has its adherents within the industry but, speaking for myself and from the point of view of a lawyer, it seems to me to represent a triumph of form over substance.

Notwithstanding the above statement, the good judge was still able to interpret the issues before him and give meaning to them all.

2.11 Stimulus to good management

Providing a contract that acted as a stimulus to the use of good project and commercial management techniques was central to the philosophy behind the drafting of the original contract and has been improved through the revisions. These procedures are designed to contribute to the forward-looking management philosophy, which is designed to manage problems rather than to simply allow them to degenerate into disputes.

This philosophy can be described by two basic principles, both of which impact on the objective of stimulating good management:

- foresight that is applied collaboratively serves to mitigate problems and, this, in turn, reduces risk for all those involved; and
- the clear division of function and responsibility helps accountability and motivates people to play their part in the successful management of the project.

In order for this philosophy to be successful, users of the form must adopt a cultural transition that is best described by quoting the opening paragraph of Section 1.2 in Volume 1 (establishing a procurement and contract strategy (NEC Panel 2017b) that forms part of the NEC4 family. This says:

> NEC (New Engineering Contract) is a modern day family of contracts that facilitates the implementation of sound project management principles and practices as well as defining legal relationships. Key to the successful use of NEC is users adopting the desired cultural transition. The main aspect of this transition is moving away from a reactive and hindsight-based decision-making and management approach to one that is foresight based, encouraging a creative environment with pro-active and collaborative relationships.

The philosophy and cultural transition are contributed to and achieved by several matters including but not limited to:

- the provision of express requirements requiring collaboration between the parties and other personalities involved (see clause 10.2 and Section 4.2);
- providing provisions and procedures which encourage and reward foresight (including provisions that penalise a failure to use such foresight);
- by clearly allocating risks between the parties, with differing levels of risk depending on the main option chosen but with a consistent approach to risk across those main options;
- by a clear and consistent approach to the definition and administration of compensation events;
- by providing the Project Manager with options from which the solution to suit the particular problem can selected;

- by providing procedures to obtain quotations from the Contractor in relation to problem situations or in advance of proposed change; and
- by the use of up-to-date, accurate and binding programmes which are regularly monitored and revised, thereby acting as a dynamic management tool.

2.12 Other characteristics

The family provides for different methods of exerting financial control through the selection of the preferred Main Option. The principle two methods used are by bills of quantities or an activity schedule, the latter being provided by the tenderer before the Contract is formed. In both cases the primary use of the document is only for assessing payments although, by agreement, the use of a bills of quantities can be extended.

The drafters of the NEC family have avoided the use of any cross-referencing from one clause to another. This serves to remove the need or temptation to divert from the clause the user is reading to other related clauses. Instead, the user is encouraged to simply follow the particular procedure covered by the clause they are reading. As a result, the contract seems uncluttered and comes across as very easy to read.

However, this principle has more than one downside. The lack of cross-referencing can lead users who are experienced in other contractual arrangements to become puzzled as to why the contract does not provide provisions that they would expect to see in relation to a matter they are following in a core clause. In all likelihood, such a provision is provided but in another core clause. The lack of cross-referencing in this respect means that users have to understand how the contract is laid out and learn where to go to look for the conditions they expect to find in a contract of this nature.

The lack of cross-referencing also creates situations where what could be a very severe penalty for the failure to do something is not referred to at the point where the requirement to do that something is actually set out. Instead, the penalty is set out in another core clause. It is therefore not unknown for a user not to do something that they consider is simply procedural and without any consequences should the procedure not be followed; a penalty as severe as termination of the contract could, however, lurk elsewhere in the document.

Chapter 3
The Options: An Overview

3.1 General arrangement of the ECC

The principle way in which the flexibility referred to in Section 2.9 is provided by the ECC is in the arrangement of the conditions. The drafting body developed a system whereby users of the contract could select from a menu of options to produce a version of the conditions which was suitable for the project that was being considered.

That said, the arrangement of the ECC is based around nine core clauses that must in used in every contract:

1. General
2. The *Contractor's* main responsibilities
3. Time
4. Quality management
5. Payment
6. Compensation events
7. Title
8. Liabilities and insurance
9. Termination

The first part of the flexibility comes when the user, usually the Client or the Project Manager on the Client's behalf, selects which one of the Main Options the project is going to be carried out under. The ECC offers six Main Options for the user to choose. The selection of the Main Option will determine the risk profile that the Client sets for the project subject to minor adjustments, resulting from the selection of further options, which will be discussed below.

The six Main Options that are available for selection are:

- *Option A*: Priced contract with activity schedule;
- *Option B*: Priced contract with bill of quantities;
- *Option C*: Target contract with activity schedule;
- *Option D*: Target contract with bill of quantities;
- *Option E*: Cost reimbursable contract;
- *Option F*: Management contract[1].

---

[1] The ECS does not include Option F. This procurement route is not suitable for a subcontract situation.

A Practical Guide to the NEC4 Engineering and Construction Contract, First Edition. Michael Rowlinson.
© 2019 John Wiley & Sons Ltd. Published 2019 by John Wiley & Sons Ltd.

It must be emphasised that the ECC requires that just one of the Main Options listed above is selected. One must be selected, otherwise the contract will not function. The Main Option chosen dictates which additional conditions are added to the core clauses. Should more than one Main Option be included, the resulting contract would immediately contain conflicts that it would not be possible to resolve.

Comments on these Main Options are given in Section 3.5. Details of the additional conditions added to the core clauses are given in the relevant chapter that refers to the conditions under consideration.

Once the Main Option has been selected, the user must select one of the three dispute resolution options:

- Option W1 for use where adjudication is the method of dispute resolution and where the UK Construction Act *does not apply*; or
- Option W2 where the UK Construction Act *does apply*; or
- Option W3 for use where a Dispute Avoidance Board is the method of dispute resolution and where the UK Construction Act *does not apply*.

While the ECC states that one of these options must be chosen, there would not, in the author's view, be anything to prevent users from not selecting any one of the dispute resolution options and either inserting their own procedures in Option Z or opting to rely on the relevant law. In the United Kingdom, adjudication would be implied for all contracts that fall within the definition of a construction contract under the Housing Grants, Construction and Regeneration Act 1996. In all jurisdictions that I am aware of, reference to the courts is always available.

The next part of the selection procedure that the Client must consider is to decide which, if any, of the Secondary Options will be included within the Conditions. It must be emphasised that it is not necessary to select any of the Secondary Options; the core clauses and selected Main Option, in all six cases, will provide a perfectly workable and sound contract.

The only limit on how many Secondary Options can be selected is determined by the restrictions on the combination of choices; these restrictions are stated within the notes in brackets attached to some of the Secondary Options in the Schedule of Options.

The full list of Secondary Options available for use in the ECC is:

- *Option X1*: Price adjustment for inflation (used only with Options A, B, C and D);
- *Option X2*: Changes in the law;
- *Option X3*: Multiple currencies (used only with Options A and B);
- *Option X4*: Ultimate holding company guarantee;
- *Option X5*: Sectional Completion;
- *Option X6*: Bonus for early Completion;
- *Option X7*: Delay damages;
- *Option X8*: Undertakings to the Client or Others;
- *Option X9*: Transfer of rights
- *Option X10*: Information modelling
- *Option X11*: Termination by the Client;
- *Option X12*: Multiparty collaboration (not used with Option X20);
- *Option X13*: Performance bond;

- *Option X14*: Advanced payment to the Contractor;
- *Option X15*: The Contractor's design;
- *Option X16*: Retention (not used with Option F);
- *Option X17*: Low performance damages;
- *Option X18*: Limitation of liability;
- *Option X20*: Key Performance Indicators (not used with Option X12);
- *Option X21*: Whole Life Cost
- *Option X22*: Early Contractor involvement (used only with Options C and E);
- *Option Y(UK)1*: Project Bank Account
- *Option Y(UK)2*: The Housing Grants, Construction, and Regeneration Act 1996;
- *Option Y(UK)3*: The Contracts (Rights of Third Parties) Act 1999;
- *Option Z*: Additional conditions of contract.

The observant reader will have noted that the above list does not run consecutively; Secondary Option X19 is not used within the ECC.[2] This Secondary Options, plus some others numbered from X23 onwards, are used elsewhere within the NEC4 suite of contracts, as are some of those listed above. It is a feature of the suite that the wording of the Secondary Options is consistent across all the contracts. For example, in every contract in the family where Secondary Option X16 (Retention) is available, the wording for those conditions is identical in every contract. This facility makes it relatively easy for a user to move between the different contracts in the suite and be familiar with the terms and procedures in use.

Detailed comment in relation to the Secondary Options is made either in Chapter 20 or, in certain instances, with the core clause conditions that a particular Secondary Option is linked to.

3.2 Other documents referred to

The conditions refer to other documents both in the core clauses, the Main Option conditions and in some of the Secondary Options. The documents that fall under this heading are:

- the Scope (see Section 22.2);
- the Site Information (see Section 22.3);
- the Accepted Programme (see Section 12.2);
- the Schedule of Cost Components (see Section 14.2);
- the Short Schedule of Cost Components (see Section 14.3);
- an activity schedule (Main Options A and C) (see Section 13.5.1);
- a bill of quantities (Main Options B and D) (see Section 13.5.2);
- ultimate holding company guarantee (X4) (see Section 20.3);
- undertakings to the Client and Others (X8) (see Section 7.3.2);
- transfer of rights documentation (X9) (see Section 7.3.3);
- the Information Execution Plan (X10) (see Section 7.4.2);
- the Project Information (X10) (see Section 7.4.2);

---

[2] Option X22 is not used in the ECS; otherwise the lists are the same.

- the Information Model (X10) (see Section 7.4.2);
- the Partnering Information (X12) (see Section 20.4);
- Schedule of Core Group Members (X12) (see Section 20.4.5);
- a Schedule of Partners (X12) (see Section 20.4.4);
- performance bond (X13) (see Section 20.5);
- advanced payment bond (X14) (see Section 13.10.2);
- retention bond (X16) (see Section 13.10.3);
- an Incentive Schedule (X20) (see Section 20.8);
- Pricing Information (X22) (see Section 20.11.2);
- a Trust Deed (Y(UK)1) (see Section 13.10.4); and
- a Joining Deed (Y(UK)1) (see Section 13.10.4).

Whichever documentation is required from this list must be prepared at the appropriate time to ensure that information is properly passed between the parties at the required stage (be that during the tender stage, prior to the start date or during the currency of the works).

3.3 Contract Data

The Contract Data is a key element of the ECC and is divided into two parts. Part One is to be completed by the Client. It is necessary to carry out this exercise before the documents are sent to the tendering contractors for pricing. In completing Part One, the Client (or whoever completes the exercise for it) must carefully consider each entry and decide on the entry to be made, if any. The contents of the blank Contract Data can be used as a checklist by the compiler; it is judged good practice to consider every statement for each project rather than work from a version prepared for a previous project. Even for repeat streams of work, it is possible that certain statements will apply to some projects but not to others within the same work stream.

Part Two of the Contract Data is completed by the tendering contractor during the tender period. This part of the Contract Data consists of some basic information including details of the Contractor's key people and the Working Areas. More importantly, pricing information relating to the Prices, together with various rates and percentages for use in calculating payments and/or assessing compensation events (depending on the main option in use), is also included. The inclusion of the required information in Contract Data Part Two is vital if the Contractor's position in respect of payment is to be properly provided for.

The requirements for completing the Contract Data are considered in detail in Chapter 21.

3.4 The published documents

In respect of the ECC, the following volumes are published with the fourth edition (2017) of the NEC suite:

The Options: An Overview 21

- the complete Engineering and Construction Contract (NEC Panel 2017a)
- six merged versions of the ECC, one for each Main Option
- User Guide Volume 2 (NEC Panel 2017c)
- User Guide Volume 4 (NEC Panel 2017e)

The above nine documents are only part of the total of 43 documents that make up the suite at the time of writing. The author considers that anyone who understands and uses the ECC should be able to understand and use all of the other contracts in the family as they are so closely related and follow the same principles. Volumes 1 and 3 of the User Guides have not been mentioned above, as they cover all the family and not just the ECC.

3.5 Main Options: General outline

The first part of the flexibility within the ECC is generated by the user making the choice of one of the six[3] Main Options. In making this choice, the user selects the risk balance between the parties and sets the basis of the parties' obligations in relation to that risk. The titles of the six Main Options within the ECC are in Section 3.1 above.

The principle difference between these Main Options lies in the payment mechanism. By varying the payment mechanism, the allocation of risk between the parties is allocated differently between the Client and the Contractor. It is generally recognised that Main Option A carries the least risk for the Client while Main Option F carries the highest risk; the other Main Options following a sliding scale in between these two extremes.

Main Option A provides a priced contract where the total of the price tendered by the Contractor against each activity represents the amount it will be paid for that work, including all matters which are at the Contractor's risk. The Client only carries the risk of the matters identified as being compensation events and/or its insurable liabilities.

Main Option B is another priced contract but this time with a bill of quantities. Under this option, the Contractor is paid the actual quantity of work carried out at the rates in the bill of quantities. There is therefore some uncertainty over the final price as any inaccuracies in the bill of quantities will be corrected in the re-measurement process, being at the Client's risk. The Client also carries the risk for those matters identified as being compensation events and/or its insurable liabilities.

Main Option C is the first of the target cost options. Target cost contracts are used in varying circumstances, including, but not limited to:

- work that is not fully defined or detailed;
- where the risk is perceived to be high; or
- where the Client is seeking to encourage efficiency gains over a series of similar contracts.

Under Main Option C, the Contractor tenders the Prices (the target) backed up by an activity schedule together with relevant percentages and rates. The percentages and

---

[3] Five in the ECS.

	Main Option					
	A	B	C	D	E	F
Definition – Prices	11.2(32)	11.2(33)	11.2(32)	11.2(33)	11.2(34)	11.2(34)
Definition – Price for Work Done to Date	11.2(29)	11.2(30)	11.2(31)	11.2(31)	11.2(31)	11.2(31)
Definition – Defined Cost	11.2(23)	11.2(23)	11.2(24)	11.2(24)	11.2(24)	11.2(25)
Definition – Disallowed Cost			11.2(26)	11.2(26)	11.2(26)	11.2(27)
Pricing Document	Activity Schedule	Bill of Quantities	Activity Schedule	Bill of Quantities		
Additional Compensation Events		60.4, 60.5 & 60.6		60.4, 60.5 & 60.6		
Contractor's Share Mechanism			54.1 to 54.4	54.5 to 54.8		

Figure 3.1 Table comparing aspects of the six Main Options. The numbers in the table correspond to clause numbers in the Main Option. See Appendix 1, Table A1.1 for reference to the sections of this book where each clause is discussed.

rates are used for calculating the Defined Cost and Price of Work Done to Date (PWDD) (see Section 13.4.5), being what the Contractor is paid based on the resources employed to carry out the works. At the end, the PWDD is compared to the Prices (adjusted for compensation events). If the PWDD is less than the Prices, then the Contractor receives a share of the gain; if the PWDD is greater than the Prices, the Contractor pays its share of the overspend (see clause 54 and Section 13.7).

Main Option D follows the same approach as Main Option C, except that a bill of quantities is used against which the final Total of the Prices is re-measured. The re-measurement aspect provides less certainty of final prices than the use of the activity schedule; hence, the Client carries more risk under Main Option D than under Main Option C. It must also be said that, in practice, this option probably requires more administration than any other.

Main Option E is a cost reimbursable contract where the Client simply pays for all the resources utilised by the Contractor to carry out the works, following a formula employing various tendered rates and percentages, subject only to the disallowing of costs resulting from the Contractor's inefficient use of resources.

Main Option F provides a management contract option under which the Contractor is paid a fee for carrying out prescribed duties. Each subcontract is entered into between the Contractor and the subcontractor and the Client pays the actual cost of each such subcontract to the Contractor. The Contractor's fee will increase in line with any increases in the cost of the subcontracts.

Figure 3.1 gives a simple comparison of the key additional clauses of the six Main Options. It can be seen from this table that whilst there are common factors between some of the main options, no two are identical. It is this variance that creates the changing risk profile as the user considers one option against another. The primary change is how the definition of the term *Price for Work Done to Date* acts to determine the amount due for each payment. These differences can be seen in more detail by referring to Sections 13.4.3–13.4.5).

Chapter 4
Spirit of Mutual Trust and Cooperation

4.1 Introduction

This chapter examines all of core clause 10, not just the part of it referred to in the title of the chapter. When people come to the ECC for the first time and are shown this clause, most of them find that their attention is immediately focused on the phrase 'spirit of mutual trust and cooperation', which is the obligation in clause 10.2.

As this chapter demonstrates, there is far more in practice to the two clauses than just this phrase. Indeed, as readers will see, this phrase is not the most important part of the clauses, despite the fact that it creates more discussion than probably any other phrase within the ECC. This discussion is warranted as there is no doubt that the clauses seek to impose something that you will not find in a more traditional style of construction contract and which is akin to a principle that is not welcomed by the legal system in the United Kingdom. This same principle is, however, recognised and forms a major plank of the civil code within other legal systems around the world.

4.2 The clauses

Both clause 10.1 and 10.2 consist of a single sentence. However, the obligations that they create are substantial and underlie the whole of the ECC. Both begin by identifying the four major personalities (see Sections 5.2–5.5) that are named in Contract Data Parts One and Two (excepting the Adjudicator). By naming the Client, the Contractor (together the Parties), the Project Manager and the Supervisor together in this way, the ECC ensures that they are all bound by the same major underlying pair of obligations.

4.2.1 Shall act

The two most important words in both clauses come immediately after the names of the four personalities: 'shall act'. The word *shall* acts as a mandatory (see Section 1.4) command. Its application in tandem with the word *act* serves to make the carrying out of the actions required by the ECC a mandatory requirement; every personality is obliged to perform that act in the way that the ECC requires that act to be performed and in the

---

A Practical Guide to the NEC4 Engineering and Construction Contract, First Edition. Michael Rowlinson.
© 2019 John Wiley & Sons Ltd. Published 2019 by John Wiley & Sons Ltd.

time that the ECC requires that act to be performed. By entering into a contract with this underlying mandatory requirement, the Client and the Contractor are promising each other that they will act in the ways required. To fail to do so would constitute a breach of the contract.

Although the Project Manager and the Supervisor are not parties to the contract, they are appointed by the Client to act on its behalf and will carry out a major part of the administration of the procedures. By accepting an appointment from the Client requiring them to administer an ECC, the Project Manager and Supervisor are accepting that there is a mandatory obligation on them, acting for the Client, to act as stated in the contract. Should either of them fail to act in this way, they will cause a breach of the contract by their Client.

4.2.2 The first requirement

Clause 10.1 imposes a clear obligation on the personalities concerned to act as stated in the contract. This requirement is clearly separate from the second (see below). The clear obligation is to follow the project management procedures set out in the contract in the way required and to the timescales imposed. This obligation stands alone in clause 10.1 and carries equal importance with the second requirement as set out in clause 10.2.

4.2.3 The second requirement

The requirement in clause 10.2 is the one that usually attracts the attention of those reading the contract for the first time and often stays on the mind of experienced users for a considerable time thereafter. This requirement is that which gives this chapter its title, that is, to act in a spirit of mutual trust and cooperation. As with clause 10.1, this obligation applies to the same four named personalities.

The requirement was initially introduced in the second edition in response to a suggestion made by Sir Michael Latham in his report *Constructing the Team* (1994, paragraph 5.20.4), which had been jointly commissioned by the Government and the UK construction industry. Sir Michael's comments actually went further, as he also suggested, in the same sentence, that there should be a commitment by the parties to trade fairly with each other and with their suppliers and subcontractors. While the ECC requires subcontracts (see last bullet point of clause 26.3 and Section 8.3) to contain the requirement to work in the required spirit, this requirement is not imposed in any way when it comes to suppliers.[1]

This obligation stands alone in clause 10.2 and carries equal importance with the first requirement as set out in clause 10.1.

What the 'spirit of mutual trust and cooperation' actually means is considered in Section 4.3.

---

[1] By using the NEC4 Supply Contract (NEC Panel 2017f) and the NEC4 Supply Short Contract (NEC Panel 2017g) it becomes possible to easily pass on the spirit to suppliers, as it is an obligation in clause 10.2 of both these contracts.

4.2.4 Discretionary actions

Users must recognise that while the words *shall act* as discussed above clearly impose a mandatory obligation to follow the requirements of the contract, some of the actual procedures within the ECC are discretionary rather than mandatory (see Section 1.4). Such specific discretion will of course always take preference over the general mandatory requirement. Where such discretionary procedures exist, these will be identified and commented on in the appropriate places later in this guide.

4.3 What does it mean?

The question that will always arise when discussing clause 10.2 is what does the phrase 'a spirit of mutual trust and cooperation' actually mean? In answering this, all that can be said with certainty is that there is no clear meaning. What can be considered when interpreting the meaning of these words is the consideration of the term 'mutual cooperation and trust' by His Honour Judge Humphrey Lloyd QC in the case of *Birse Construction Ltd v St David Ltd* (1999) BLR 194.

The judge commented that although the words, which were included in a separate Partnering Charter, were clearly not legally binding, they were '… intended to provide the standards by which the parties were to conduct themselves and against which their conduct and attitudes were to be measured'. Later in the judgement, he also said that '… one would not expect, where parties had made mutual commitments such as those in the Charter, either to be concerned about compliance with contractual procedures if otherwise there had been true compliance with the letter or the spirit of the Charter' and that '… an arbitrator (or court) would undoubtedly take such adherence … into account in exercising wide discretion to open up, review and revise etc. …'.

These comments must be interpreted in light of the difference in circumstances between the facts in that case and the way in which the ECC is set up as a contract. In the case above, the Partnering Charter was not a contract document. In the ECC, the term forms part of the conditions; it is clearly part of clause 10.2. The ECC therefore creates a position whereby those named not only act as stated in the ECC (clause 10.1) but also do so in the required spirit. If the terms of the ECC are not complied with, then the contract and the law provide remedies for the injured party. The judgement above makes it clear that any third party tasked with making a decision about a dispute would take into account adherence to the spirit when deciding any grey areas in the application of the terms and conditions. In practice, when making a decision about a dispute that has been referred to them, adjudicators not only follow the terms and conditions of the ECC but often take significant account of the behaviour of the parties.

An example of a Court following this guidance occurs in the judgement of Deeny J in *Northern Ireland Housing Executive v Healthy Buildings (Ireland) Limited* (2017) NIQB 43, in relation to a contract based on the NEC3 Professional Services Contract, where the phrase in question is included in clause 10.1 of that contract and clause 15 concerns the early warning mechanism (see Section 6.4 for the NEC4ECC equivalent). The Judge made the following observation:

[43] First of all, it is a cardinal principle of contractual interpretation that one should look at the agreement overall. This particular contract begins with the agreement that the employer and the consultant shall act "in the spirit of mutual trust and co-operation" (10.1). It seems to me that a refusal by the consultant to hand over his actual time sheets and records for work he did during the contract is entirely antipathetic to a spirit of mutual trust and co-operation. Further clauses in the contract such as Clause 15 reinforce that spirit. I find that the overall sense of the contract with its emphasis also on the assessment of compensation events is strongly against the defendant here.

So in answer to 'what does it mean?' we can say with some certainty that the behaviour of a party or one of its appointed representatives will influence the outcome of any dispute. It therefore follows that continuing adherence to the spirit, together with compliance with the procedures contained throughout the contract, will benefit a party when the other party is acting in a different way.

4.4 Practical issues

4.4.1 Co-location

If you ask users of the ECC what the main secret behind establishing mutual trust and cooperation is, the most common reply that you will receive is 'co-location'. By this, users mean the co-locating of the Project Manager, the Supervisor and the Contractor in the same offices from which they jointly manage the project as a single team. This may seem to be a step too far for many who are maybe new to the ECC, and indeed, it is a sophisticated approach, but there is no doubt that it works.

Assuming that the arrangement above is a step too far for most, then it is still important for the Project Manager and the Supervisor to base themselves on the Site whenever the size of the project gives the opportunity to do so. Even if they have their own office set up separate from the Contractor, the fact that they are on the site means that face-to-face communication between them and the Contractor will occur on a regular basis; this will hopefully allow the depth of understanding that is necessary to build trust and cooperation to develop. If the size of the project does not warrant the Project Manager and Supervisor being based on the site, then they should make the effort to visit regularly in order to enhance the communications. It is suggested that turning up once a month for a progress meeting will not be enough. A minimum of one visit per week by the Project Manager is probably necessary if the working relationship is to blossom; more frequently would be preferable.

When parties do feel that they are sophisticated enough to share the actual office space, matters can develop to the point where the Project Manager and the Contractor's senior person within the site team often share the same room within the offices and work from a common set of files. This often extends to construction staff, planners, quantity surveyors and others in the sharing of office space, files and computer systems. Provided that each party respects the commercial confidentiality[2] required in connection with certain

---

[2] The level of commercial confidentiality relevant to a project will vary depending on the Main Option that has been chosen.

matters, then such sharing does not present any problems in practice. Indeed, the benefits are considerable in allowing all parties to achieve their goals.

The inclusion of subcontractor's staff, especially for major contributors to the construction, such as the mechanical and electrical subcontractor as a common example, has also been found to enhance the co-operation and the smooth running of the project.

The benefit to the development of mutual trust and cooperation of co-locating is of course dependent on the attitudes and behaviour of all the individuals involved. In practice, businesses sometimes need to recognise that some individuals are not suited to this way of working and need to be reassigned to other projects where their skills are better used.

4.4.2 Attitudes and behaviour

The second most important factor in establishing a spirit of mutual trust and cooperation is the attitude and behaviour adopted by the parties involved. If people enter into these arrangements with an attitude and belief that the others involved in the project are going to be self-serving, then they will adopt a defensive attitude. It only takes one party to have a defensive attitude in order for this approach to spread quickly, as everyone else takes up such a stance in order to equally protect their position. Once defensive attitudes are adopted, it is very difficult to change people's outlook. While these attitudes persist, the chances of establishing mutual trust and cooperation are lost.

On the other hand, when people adopt a behaviour that is open and indicates that their primary interest is achieving the successful completion of the project with everyone achieving their business goals, others (including any doubters in the group) follow that lead. The people most important in setting this behaviour within the supply chain are the Client, Contractor and Project Manager. The right behaviour is best set by those individuals who are most senior in the organisations involved; staff follow the lead of their bosses. It is very difficult to get middle and junior management to adopt a behaviour that is different from that set by senior management.

Without the right attitude and behaviour, both within and between the organisations involved, it will be difficult to engender a spirit of mutual trust and cooperation. In order to do so, many find that a good starting point is to understand each other's business goals from their involvement in the project.

4.4.3 Appreciating each other's goals

The importance of appreciating the business goals of the other organisations involved in the supply chain was referred to in Section 4.4.2. What is often found is that, where businesses do not appreciate what the goals of the other parties are, they make an assumption that puts the goals of those other businesses opposite to their own. Where the business goals of organisations in contract with each other are opposed, it acts to immediately create a position where both parties are pulling in the opposite direction. Such situations provide an ideal breeding ground for conflict.

By sitting down at the commencement of a project, all the key players can outline their goals from the project. When these are shared, it is often found that there is significant alignment between organisations, especially the Client and the Contractor. For example, a client organisation typically wants the project delivered on time, to budget and to the specified quality. If the client organisation has an end-user for the project, it also wants that end-user (the client organisation's customer) to be satisfied with the product. By achieving these goals, the client organisation achieves its financially related goals.

On the other side of the contractual relationship, the contracting organisation wants to build the project on time, on a right-first-time basis (i.e. no defects) and within its tender price. It also wants to satisfy its customer (i.e. the client organisation) and any end-users. By achieving these goals, the contracting organisation enhances its reputation, hence increasing its chances of repeat business with the same client organisation and with others who hear of its good performance. As a consequence of achieving these goals, the contracting organisation should also make a profit and improve its chances of making profits from other projects in the future.

Presented as above, it becomes clear that the business goals of both organisations are actually closely aligned and that the success of one is dependent on the other. By gaining this understanding, the two partners in the project will realise that working together will be of benefit to them both. By taking the lead with this newfound mutual appreciation, it is a much shorter journey to achieving real mutual trust and cooperation.

4.4.4 Communicating

All of the practical issues relating to mutual trust and cooperation discussed above can only happen if those involved communicate with each other. Co-location is a major factor in enhancing interparty communications. The adoption of the right attitudes and behaviour encourages good and positive communication. Appreciating the goals of those you are doing business with is the result of good focused communication.

The ECC is a construction contract that places a high level of emphasis on good communication. It is the foundation to achieving a spirit of mutual trust and cooperation, but also to the foresight required to make the project management principles that run throughout the contract work effectively and with benefit to the project. Communication under the ECC is discussed in detail in Sections 6.1–6.3 of this guide and referred to constantly throughout.

If participants in projects do not communicate clearly and effectively, they will not foster a spirit of mutual trust and cooperation and they will fail to effectively project manage the works and achieve their goals. In short, conflict will be ever-present and disputes will be bubbling away below the surface, waiting to erupt.

4.4.5 Good faith

At the end of Section 4.1, I made reference to a legal principle that does not form part of the UK legal system but is common in other jurisdictions around the world. The legal principle that I was referring to was that of 'good faith'.

The common law in the United Kingdom had been firm on this point for some years. However, in March 2012, a first instance judgement in the High Court, in the case of *Compass Group UK and Ireland Ltd (trading as Medirest) v Mid Essex Hospital Services NHS Trust* (2012) EWHC 781 (QB) appeared to open the door to such a principle either being interpreted from requirements to 'cooperate' in express terms of a contract or to be implied into contracts with similar obligations. Immediately following this judgement several articles appeared in various industry related journals excitedly reporting this development. The excitement was short-lived as a little over a year later, in April 2013, the case went before the Court of Appeal, who overturned the point and confirmed that 'good faith' does not apply.

Chapter 5
The Cast of Characters

5.1 Introduction

As for all construction contracts, the ECC names roles to be played by either the parties to the contract or agents appointed by either or both of the parties. The personalities so identified act in predetermined ways in carrying out duties under the contract to facilitate the administration of that contract.

The ECC names five such personalities and six classes of people or organisations, each of which are discussed in this chapter. In addition, other personalities and classes of people or organisations who may be involved in a project, whether anywhere in the world or within the United Kingdom, are also identified and discussed.

While two of the main personalities (Client and Contractor) will often be organisations, it is considered important that individuals should be named for the roles of Project Manager and Supervisor. This requirement is not clearly stated in the core clauses, but the spaces for the entries in Contract Data Part One clearly require both the name and the address of the each nominated individual to be identified. That it should be an individual named rather than an organisation is confirmed by the worked example at Appendix 1 to Volume 2 of the User Guide (NEC Panel 2017c). In practice, it is common to name a consultancy or a position in the Client's own organisation within one or both of these roles. Given the importance of these two roles to the successful administration of the ECC, it is suggested that the naming of individuals in this role, both of who have sufficient closeness to the project and the necessary skills and gravitas to undertake the role, is a major benefit to the project. It is also found in practice to be a major benefit in aiding the establishment and growth of the spirit of mutual trust and cooperation. It is much easier in practice to establish this trust and cooperation between individuals representing corporations rather than between the corporate entities themselves.

5.2 The Client

The Client is one of the two Parties (see definition at clause 11.2(13)) to the Contract. It is identified in Contract Data Part One. The Client's role under the ECC is fairly traditional in terms of the construction industry in that it hands over the day-to-day administration and project management tasks to its representatives who, under this contract, are the Project Manager and the Supervisor. This approach is based on the general assumption

A Practical Guide to the NEC4 Engineering and Construction Contract, First Edition. Michael Rowlinson.
© 2019 John Wiley & Sons Ltd. Published 2019 by John Wiley & Sons Ltd.

that the Client is not a construction professional. In order to cater for the complex interaction of the numerous facets of constructing a project of this scale, often as a one-off and outside the Client's usual sphere of business operations, it is advisable that the Client employs suitably qualified and experienced construction professionals to manage the project for it.

There are still a number of duties that the Client must carry out, however, together with several that it has the option to carry out should the circumstances so require. These are set out in Table A2.1 of Appendix 2, which also identifies whether each duty is mandatory or discretionary.

This table identifies a total of 126 rights, duties and obligations that are allocated to the Client under the ECC. The allocation of the Client's rights, duties and obligations across the core clauses and Options are tabulated at Figure 5.1. The reader will note that the majority of these relate to insurance (core clause 8), termination (core clause 9) and dispute resolution (options W1, W2 and W3). Of the matters relating to dispute resolution, only a portion of these could be included in a single contract as only one of options W1, W2 and W3 would be used. Depending on the combination of options chosen, the Client will have between 54 and 88 duties allocated to him.

This leaves a small balance of key duties that fall to the Client and which, in most contracts, are the only duties of any great practical importance. These are that the Client is required to allow access to and use of the site at the appropriate time or times (see clause 33.1 and Section 12.6.1) and to pay the amount certified by the Project Manager (subject to provisions for damages and withholding) at the time required by the contract (see clause 51 and/or Y(UK)2 and Sections 13.2 and 13.9). If a Project Bank

Character	Core Clauses	Main Option	Dispute Resolution	Secondary Option	Total
Client	40	0	W1: 24 W2: 31 W3: 14	17	126
Project Manager	108	46	W1: 4 W2: 2	29	189
Supervisor	15	0	0	0	15
Contractor	125	41	W1: 25 W2: 32 W3: 14	64	301
Senior Representatives	0	0	W1: 2 W2: 2	0	4
Adjudicator	0	0	W1: 14 W2: 18	0	32
Tribunal	0	0	W1: 4 W2: 4 W3: 4	0	12
Dispute Avoidance Board	0	0	W3: 10	0	10

Figure 5.1 Table of Obligations by Character and Source.

Account is being used (see secondary Option Y(UK)1 and Section 13.10.4) the actions in respect of payment are increased due to the operation of this facility.

The obligations regarding insurance are of course important and, in many instances, required by law. Clients who regularly procure construction schemes will be familiar with such requirements, and infrequent Clients will no doubt take professional advice on such matters. Once the necessary insurance policies are in place (with the exception of renewals) the only time these provisions would come into play is if an insurable event actually occurred.

In addition to the express requirements set out above, there is also an implied term that the Client would not prevent or hinder the Contractor in his performance. This is a well-established term in all types of construction contracts following the decision in *London Borough of Merton v Stanley Hugh Leach Limited* (1985) 32 BLR 51.

By reference to the list of compensation events (see Section 15.3) it is also possible to determine other implied terms, a breach of which would result in an express remedy for the Contractor. These compensation events will be described in more detail later in this guide. In summary, the Client will:

- provide something as shown on the Accepted Programme (clause 60.1(3));
- work within times shown on the Accepted Programme (clause 60.1(5)); and
- work within the conditions stated in the Scope (clause 60.1(5)).

The Client is, of course, responsible for a large number of other items under the ECC but the duties that take care of these are carried out on its behalf by either the Project Manager (see Section 5.3) or the Supervisor (see Section 5.4) as described below.

It could be said that the most important task for the Client is to appoint suitable people in the roles of Project Manager and Supervisor. The importance of these two roles is such that they may have an impact on the success of the project. In making these appointments, the Client has freedom of choice. The Client can select people from within its own organisation, which regular Clients often do (e.g. government agencies, utility companies, property developers, etc.), or select someone from one of the many professional consultancies which provide such services. There is no restriction on the profession of the person selected as Project Manager or Supervisor by the Client, thereby giving the Client flexibility in its choice.

Throughout the duration of the Contract the Client has the option under clause 14.4, at its sole discretion and choice, to replace the Project Manager or the Supervisor after it has notified the Contractor of the identity of the replacement person. Clients should note that the way this clause is worded means that they must notify the Contractor first and then make the replacement.

5.3 The Project Manager

The Project Manager is appointed by the Client. The Project Manager is identified in Contract Data Part One. The Project Manager's role is the key role within the Client's team, such that the Project Manager will manage all the main administration and duties under the Contract for the Client, except for those covered by the Supervisor.

To this end, the ECC places considerable authority in the hands of the Project Manager. It assumes, by implication, that the Project Manager has the Client's authority to make all the decisions and to carry out all the actions that are required within this role under the ECC.

The full list of all the duties and obligations for the Project Manager are set out in Table A2.2 of Appendix 2, which also identifies whether each duty is mandatory or discretionary. This table identifies a total of 189 matters that are required to be carried out by the Project Manager under the ECC. The allocation of the Project Manager's duties and obligations across the core clauses and Options are tabulated at Figure 5.1. The reader should note that over one-third of these matters relate to Options as opposed to the core clauses. The breakdown of activities shows that:

- 108 matters are contained within the core clauses;
- 46 matters are contained within the Main Option clauses (the actual breakdown being Option A: 3; Option B: 5; Option C: 11; Option D: 14; Option E: 6 and Option F: 7);
- 6 matters relate to the Dispute Resolution options (Option W1: 4 and Option W2: 2); and
- 29 matters are contained within the Secondary Options.

This allocation means that the Project Manager will be involved in managing between 111 activities (Core Clauses, Main Option A, Dispute Resolution Option W3 and no Secondary Options) and 155 activities (Core Clauses, Main Option D, Dispute Resolution Option W1 and Secondary Options X4, X7, X10, X13, X14, X16, X21, X22 and Y(UK)1) depending on the Main, Dispute Resolution and Secondary Options chosen for the project.

The skills base required to carry out all of these activities is wider than the specialities of most, if not all, construction professionals. The role of Project Manager involves (at least) activities and duties normally carried out by a project manager (contract administrator), designer (can be several different disciplines), planner, quantity surveyor, estimator, health and safety manager and insurance expert. In order to be able to carry out all of these activities on most projects, in terms of either expertise or volume, most Project Managers will need assistance.

The ECC provides for this by making provision at clause 14.2 for the Project Manager to delegate any of its actions. It is necessary for the Project Manager to first of all inform the Contractor what actions are being delegated and the identity of the delegate or delegates concerned. The simplicity and flexibility of this provision enables the Project Manager, in practice, to exercise complete freedom about how they organise their delegates. For instance, they could delegate different functions to specialists within that function (e.g. the payment functions (core clause 5) and financial side of the compensation event assessments (core clause 6) to a quantity surveyor). Alternatively, delegate by sections of the project (e.g. Building One to delegate A and Building Two to delegate B, etc.).

From the way in which clause 14.2 is worded, the Project Manager could even delegate the power to delegate to the delegates appointed, limited of course to the delegation of the delegated duties only.

By using the power to delegate in the ways described above (including mixing these approaches), flexibility is achieved to allow the development of a management team operating under the role of Project Manager that best suits the individual project for the benefit of the project and the advantage of all those who are involved in the project.

Project Managers are appointed by the Client, either under a contract of employment (where the Project Manager is an employee of the Client) or under a contractual arrangement (between the Client and the Project Manager's employer). The Project Manager must therefore take account of any constraints imposed by this employment when operating the ECC. For example, in the case of a limit on the amount that the Project Manager may authorise as a compensation event assessment, it is the responsibility of the Project Manager to ensure that all the acceptances required under the ECC are given in time. In order to facilitate this, the Project Manager and the Client will need to ensure that the system of approval in their arrangement will allow the Project Manager to comply with the timetables within the ECC; otherwise, the procedures will be hindered and relationships will suffer. If these are not complied with, then the Contractor has the ultimate right to raise the matter with the Adjudicator (see Section 5.7).

Having set out the above warning of potential clashes between the Project Manager's appointment and the ECC, it must be highlighted that the Project Manager is free to seek the Client's views as much or as little as their relationship and the Project Manager's contract with the Client requires.

Indeed, it is true to say that the Project Manager's role is somewhat different to that of the Certifiers under the more traditional forms of contract. Project Managers will on occasion find that they must consider matters and make decisions based on the Client's best interests, while on other occasions they must act fairly and reasonably between the parties as an independent certifier.[1] The way in which the Project Manager acts in each circumstance is determined, by implication only, by the terms and conditions of the ECC. The Project Manager and Client must always remember that, should the Contractor disagree with any action or decision of the Project Manager, it is always entitled to refer that matter to the Adjudicator.

To summarise, the Project Manager's role is to manage the contract for the Client with the intention of achieving the Client's objectives for the completed project. In order to carry out all of their duties, it is essential that Project Managers should have sufficient time and be close enough to the detail of the project to carry out their duties effectively. To this end, the Project Manager has considerable authority, which, of course, comes with considerable responsibility; not only to the Client but in the way that the Project Manager acts under the Contract and to the Contractor. The way in which Project Managers exercise their authority can be a major factor in fostering the spirit of mutual trust and cooperation and the success or failure of the individual project. In this respect, both the Client and the Project Manager should remember the requirement of the ECC that they work with the Contractor in a spirit of mutual trust and cooperation.

Clause 14.3 gives the Project Manager the power to issue an instruction to change either the Scope (see definition at clause 11.2(16) and Section 22.2) or a Key Date

---

[1] See judgement in *(1) Costain Ltd, (2) O'Rourke Civil Engineering Ltd, (3) Bachy Soletanche Ltd, (4) Emcor Drake & Scull Group plc v (1) Bechtel Ltd, (2) Mr. Fady Bassily* [2005] EWHC 1018 (TCC).

(see definition at clause 11.2(11) and Section 12.4.4). Providing that the instruction given by the Project Manager is in accordance with the contract, the Contractor has no choice but to obey (see clause 27.3 and Section 7.5.2). This power brings with it other connotations, which are considered in more detail in Section 7.5.

5.4 The Supervisor

The Supervisor is also appointed by the Client and is identified in Contract Data Part One. The role of the Supervisor is a job title that is not found in other major standard forms of contract used in the construction industry. A frequently asked question is, what does this role actually entail and who is this person?

As with the Project Manager, an individual should be named rather than an organisation. The following paragraphs concentrate on the scope and requirements of the role.

The main duties of the Supervisor are set out in core clause 4 (Quality management; see Chapter 9) and it is from this that the general nature of this role becomes apparent. The role of the Supervisor is, primarily, akin to that of a resident engineer or architect in checking that the works are in accordance with the contract. Depending on the size and nature of the works, it might be that a clerk of works or inspector or similar person could be appointed to this role.

The role can be divided into testing/inspecting and defects management. The testing part of the role includes inspections, both on and off site, and is covered by clauses 41 and 42. The tests and inspections to be carried out and who carries them out will be identified either in the Scope or required by the law applicable to the contract. The Supervisor must therefore be sufficiently experienced to understand and identify all these requirements.

The full list of all the duties and obligations for the Supervisor are set out in Table A2.3 of Appendix 2, which also identifies whether each duty is mandatory or discretionary. This table identifies a total of 15 matters that are required to be carried out by the Supervisor under the ECC, all within the core clauses. The allocation of the Supervisor's duties and obligations across the core clauses and Options are tabulated at Figure 5.1.

The comments made in the preceding section in respect of the appointment of the Project Manager, the ability to delegate actions, relationship with the Client and the requirement to work in a spirit of mutual trust and cooperation, all apply equally to the Supervisor. Similarly, the Client may replace the Supervisor under the provisions of clause 14.4 as also referred to in connection with the Project Manager.

It must be noted that the Supervisor's role to issue the Defects Certificate (see definition at clause 11.2(7), clause 44.3 and Section 9.6) is significant in that it is this action that brings to an end most of the obligations of parties under the ECC and triggers the process for making the last payment.

5.5 The Contractor

The Contractor is one of the two Parties (see definition at clause 11.2(13)) to the Contract. The Contractor's identity is recorded in Contract Data Part Two. The Contractor's key people (see Section 7.5.1) are also identified at this point.

The Contractor's main responsibilities and related matters are covered in detail in Chapter 7. Notwithstanding the contents of that chapter in common with the Client, the Project Manager and the Supervisor, a full list of all the duties and obligations for the Contractor are set out in Table A2.4 of Appendix 2, which also identifies whether each duty is mandatory or discretionary. This table identifies a total of 301 matters that are required to be carried out by the Contractor under the ECC. The allocation of the Contractor's duties and obligations across the core clauses and Options are tabulated as shown in Figure 5.1. The reader should note that over 58% of these matters relate to Options as opposed to the core clauses. The breakdown of activities shows that:

- 125 matters are contained within the core clauses;
- 41 matters are contained within the Main Option clauses (the actual breakdown being Option A: 3; Option B: 3; Option C: 8; Option D: 10; Option E: 7 and Option F: 10);
- 71 matters relate to the Dispute Resolution options (Option W1: 25, Option W2: 32 and Option W3: 14); and
- 64 matters fall within the Secondary Options (14 of which are in Option Y(UK)1).

This allocation means that the Contractor will be involved in managing between 142 activities (Core Clauses, Main Option A or B, Dispute Resolution Option W3 and no Secondary Options) and 231 activities (Core Clauses, Main Option D or F, Dispute Resolution Option W2 and Secondary Options X4, X7, X8, X9, X10, X13, X14, X15, X16, X17, X20, X21, X22, Y(UK)1 and Y(UK)2) depending on the Main, Dispute Resolution and Secondary Options chosen for the project.

As the Contractor is an organisation rather than an individual, then, unlike for the Project Manager and Supervisor, there is no need for a clause for delegation of these duties. Whatever skills necessary to enable the Contractor to fulfil his obligations that it does not have within its organisation is obtained by either employing additional people on a fulltime or contract basis or by subcontracting to others (e.g. designers, see Chapter 8). Whether such skills come from within its organisation or from an outside source, the Contractor will be responsible for everything done under its name in Providing the Works (see clauses 11.2(15) and 20.1 and Section 7.2).

5.6 *The Senior Representatives*

The people undertaking the role of Senior Representatives for both the Client and the Contractor are identified by completing the relevant entries in the blank Contract Data, in Part one for the Client (see Section 21.3.9) and Part two for the Contractor (see Section 21.4.6). Both entries allow for two people to be identified for each party but there is no compunction to stick with this number. As long as at least one person per party has been named, then the relevant provisions can be fulfilled.

The Senior Representatives have a role to play in attempting to resolve or narrow disputes that have occurred under Option W1 (see Section 19.2) and Option W2 (see Section 19.3).

Given that this is the role, then whoever is named by either party must have the authority of the party that person is representing to make a binding agreement with the other parties Senior Representative. Unless the Senior Representatives have such authority, then they will be incapable of fulfilling their role.

Whilst there is no express provision allowing the Senior Representatives to be replaced by the party that they are representing, there is no practical reason why they should not be so changed, as long as such a change is not made mid-process when a live dispute is under consideration. Should a party make such a change to replace one of its Senior Representatives during a period of time that the incumbent Senior Representative in question is acting in connection with a dispute, that could be seen as a breach of the spirit of mutual trust and cooperation required by clause 10.2 (see Section 4.2).

A full list of all the duties and obligations for the Senior Representatives are set out in Table A2.5 of Appendix 2, which also identifies whether each duty is mandatory or discretionary. This table identifies a total of four matters that are required to be carried out by the Senior Representatives under the ECC, split equally between Option W1 (see Section 19.2) and Option W2 (see Section 19.3).

5.7 *The Adjudicator*

The blank Contract Data provides entries to allow either the naming of an individual to act as Adjudicator or for a suitable nominating body to be identified, to which an application for a nomination could be made by whichever party elects to invoke the procedure. In practice, there is nothing to stop both entries being used, i.e. name a preferred Adjudicator and then provide for a nominating body to make the appointment should the named individual not be available for some reason. Indeed, clauses W1.2(3) (Section 19.2.5) and W2.2(5) (Section 19.3.5) both anticipate this position as a possibility. This combination has often been expanded by not simply naming a single adjudicator backed up by a nominating body but by naming a preselected panel containing suitably qualified individuals from several relevant construction disciplines. This was the method chosen for the contracts entered into for the construction of the Olympic facilities in London for the 2012 Games as well as other major projects, especially in the infrastructure sector, e.g. for the CrossRail projects.

In the United Kingdom, parties whose contract constitutes a 'construction contract' under the Housing Grants, Construction and Regeneration Act 1996 (or its amendments or successors[2]) should be wary of prenaming a single adjudicator. It has been suggested that, where an adjudicator is named in contracts (and especially a series of contracts) by a client, this could be a breach of the rules of natural justice: since the adjudicator is reliant on the client for a continued source of work, it may create the impression that the named adjudicator could favour the client. While such an adjudicator would in all likelihood take all steps to remain impartial, the mere impression of potential bias to an outside observer is all that is needed to render that adjudicator's decision unenforceable.

---

[2] The Housing Grants, Construction and Regeneration Act 1996 was amended by the Local Democracy, Economic Development and Construction Act 2009 for all new contracts entered into from 1 October 2011. At the time of writing no further amendments are foreseen but that does not prevent them from being made.

The Adjudicator is appointed jointly by the Client and the Contractor, although the Adjudicator's involvement is only called for by one of them. The Adjudicator only becomes involved when a dispute has arisen that the Parties have been unable to resolve between themselves under the provisions of clause 10.2. The Adjudicator will act under the provisions laid down by the dispute resolution option that forms part of the conditions (see Chapter 19) and reach a decision within the stated timescale. The Adjudicator will ideally enter into the NEC4 Dispute Resolution Service Contract (NEC Panel 2017h) with the Parties under which the Adjudicator's fee and expenses are determined.

A full list of the Adjudicator's actions, for both dispute resolution options, is included in Table A2.6 of Appendix 2 and is tabulated in Figure 5.1.

5.8 The Tribunal

The Tribunal is a body, rather than a person, albeit the body concerned will be made up of one or more people. The bodies available to fulfil this role will differ depending on the jurisdiction whose law is relevant to the contract.

To use the United Kingdom as the example, the choices are the Courts or Arbitration. Which is to apply is identified in Contract Data Part one (see Section 21.3.9), which also requires further entries to be completed if Arbitration is the chosen Tribunal. The further entries concerning Arbitration require:

- the arbitration procedure to be used to be identified;
- the place where the arbitration is to be held to be identified; and
- the body who will nomination the arbitrator(s) if the parties do not agree the choice.

A full list of the Tribunal's actions, for all three dispute resolution options, is included in Table A2.7 of Appendix 2 and is tabulated in Figure 5.1. The Tribunal will act under the provisions laid down by the dispute resolution option that forms part of the conditions (see Chapter 19). In each case, four actions are allocated to the Tribunal.

5.9 The Dispute Avoidance Board

The Dispute Avoidance Board (DAB) is another body who could play a role in the contract. It will only be active if Option W3 is selected (see Section 21.3.9). This body consists of either one or three members, as identified in the Contract Data. In the main it will be used outside the United Kingdom, with its use in the UK limited to any project to which the UK Housing Grants, Construction and Regeneration Act 1996 (as amended) did not apply.

Entries are required in the Contract Data, mainly in part one, but also in part two if the DAB is to consist of three members. A full list of the DAB's actions is included in Table A2.8 of Appendix 2 and is tabulated in Figure 5.1. The DAB will act under the provisions laid down by Option W3 if selected (see Section 19.4), under which it has a total of 10 actions.

5.10 Subcontractors

The first of the six classes of people or organisations to be considered are Subcontractors. The types of people or organisations who qualify as a Subcontractor are defined at clause 11.2(19). This definition is considered in detail in Section 8.2, to which the reader is referred.

5.11 'Others'

The second of the six classes of people or organisations in the ECC is referred to as 'Others', as defined at clause 11.2(12). In simple terms, any person or organisation that is not covered by Sections 5.2–5.7 and 5.9 above or is not an employee, subcontractor or supplier to the Contractor lies within the classification of 'Others'. In essence, the group of people known as 'Others' under the ECC is every other person or organisation in the world outside of that group directly linked to the Contract.

Before examining the practical relevance of this class, it is worth noting that in order to avoid confusion the word 'others' (lower case 'o') is not used in the ECC. Where the ECC needs to refer to other people outside of the classification 'Others', then the term 'other people' is used to distinguish between the two (e.g. see clause 15.2). This is an important point to note as, otherwise, users can be led to think that the ECC contains an error (which in this respect, it does not).

The practical implication of the definition of this term is that it refers to numerous people or organisations who will either be affected by the project, be able to influence the progress or success of the project or be directly involved in the project. Such people or organisations would include, for example:

- designers not directly employed by the Client or employed by or contracted to the Contractor;
- end-users who are not the Client;
- local authorities who are not the Client[3];
- utility companies bringing services into the Site not employed to do so through the Contractor (known in the UK as statutory undertakers);
- other contractors and suppliers working directly for the Client; or
- funding bodies and lenders.

It is clear from the compensation events listed in the ECC that it is the Client who will take the risk of the performance of those people or organisations who fall within the classification of Others and who have an involvement with the project, whether in relation to a statutory obligation or right or because of the way that the Client has procured the relevant service (see Section 15.3.5).

---

[3] If the Client is a named department of a local authority, then other departments within that same local authority would be classified as Others. For example, if the Client is the education department, then the highways department would fall within the classification of Others.

5.12 Named Suppliers

The third of the six classes of people or organisations was created by and only applies when secondary Option Y(UK)1 is being used (see Section 13.10.4). The term 'Named Suppliers' refers to all those Subcontractors (see Section 5.10) and suppliers of the Contractor who are signed up to the project bank account and as a result will receive all their payments for work carried out under their contract with the Contractor direct from that account.

Some of these named Suppliers will be named by the Contractor in Contract Data part two when it is submitting its tender. Other Named Suppliers may be proposed by the Contractor as the work proceeds and if accepted by the Project Manager are added to the category by signing the Joining Deed.

5.13 Designers

While designers play a vital and important role in every project, they are not referred to within the ECC either as a class or by reference to specific roles such as Architect or Engineer. They cannot however be ignored and the fact that design information for the project will have to come from somewhere must be considered. These are the fourth of the six classes or people or organisations to be considered.

The ECC is drafted in order to leave the option as to how the responsibility for design is allocated between the Client and the Contractor. How this allocation process works is addressed in Section 7.3. At this juncture, users need to recognise that designers will be involved. Whoever is responsible for them – whether it be the Client, the Contractor or both (in part) – needs to take appropriate steps to ensure that the designers are employed in such a way that the production of design information is carried out in a timely manner in order to ensure that the progress of the construction works is not delayed or otherwise impeded by the design process.

For any design that the Client is responsible for, in practical terms, it is advisable for the Client to make its principle agent, the Project Manager, responsible for managing and coordinating the production of design information for issue to the Contractor. The Client will need to ensure that the scope of the service provided by the Project Manager covers this responsibility in the contract between them. The Client should also ensure that the contracts between it and the designers, however many there happen to be, oblige the designers to submit information to the Project Manager for issue to the Contractor to the timetable required. Designers must also provide responses to queries forwarded to them within timescales that will enable the Project Manager to meet the period for reply (see clause 13.3 and Section 6.2) under the ECC between the Client and the Contractor. Where there is more than one designer, which is often the case, the Client should also appoint one of the designers as the 'lead designer' to coordinate that aspect of the project.

It should be recognised that since many organisations that provide design now also have departments that act as project managers on schemes, it is possible that the designers and Project Manager may come from the same organisation. This should not deter the Client, but it may find that matters are better organised and controlled if it enters separate contracts with the organisation concerned for the separate functions. Taking this approach will allow easier separation of the two different services should conflict occur between the two disciplines.

Should the Client intend that the project be constructed under what is often referred to as a 'design and build' arrangement, it may employ designers either to develop the performance specifications that the Contractor will need to satisfy or possibly the initial design. Before appointing such designers, the Client needs to decide whether they will stay in contract with the Client throughout the duration of the project (acting as his 'checkers') or whether such designers and the design that they have carried out while employed by the Client will be novated to the Contractor. The preferred option, especially where it involves novation, needs to be set out in the contract between the Client and the designer.

5.14 Principal Designer

In the United Kingdom, where the project is notifiable under the Construction Design and Management (CDM) Regulations 2015,[4] the Client is required to appoint a suitable person in the role of Principal Designer. This appointment should be made early in the process so that the Client's statutory obligations are fulfilled. In the event that the Client does not appoint someone in this role of Principal Designer, then the Client will be deemed by the Regulations to have assumed the role for itself.

The person filling the role of Principal Designer does not necessarily have to be a designer in the common sense of the meaning. The Regulations allow the Principal Designer to be someone who instructs someone else to carry out the design. This would allow the Project Manager (see Section 5.3) to take this role. If this is the option that the Client elects to take then the appointment of the person who is also being appointed as Project Manager should be separate from the appointment as Principal Designer. That is, the person concerned should be given two separate appointments, one as Project Manager and one as Principal Designer. By taking this approach, the Client will be able to demonstrate, as required by the Regulations, that it has appointed someone in that role rather than just making the duties part of a wider role.

It is not the purpose of this guide to set out the requirements to be satisfied by this role, but simply to set out how the Principal Designer is identified within the ECC.

To this end, there is no entry within the Contract Data under which to identify the person appointed to this role. Neither is the role referred to at any point within the terms and conditions. It is suggested that the identity of the Principal Designer, and the requirement for the Contractor to take account of the information issued by the Principal Designer and to provide information back to that person as required, is set out as a constraint

---

[4] These regulations came into force on 6 April 2015, replacing the existing 2007 Regulations of the same name. The role of CDM Coordinator under the 2007 Regulations has been replaced with the Principal Designer role under the 2015 Regulations.

within the Scope (see Section 22.2.7). By including the obligation in this manner, compliance with these requirements will become part of the Contractor's obligation to Provide the Works (see Section 7.2).

5.15 Principal Contractor

In the United Kingdom, whenever the Client is required by statute to appoint a Principal Designer, it is also necessary to appoint someone as Principal Contractor. As with the Principal Designer, there is no reference to this role within the terms and conditions of the ECC and no place for the entry to be made in the Contract Data. The solution is the same as above, i.e. by making an entry as a constraint within the Scope identifying who will undertake this role. It will normally be the Contractor but, in circumstances where the Client contracts with more than one contractor to undertake works on the same Site within the same time frame, the Client must identify which one of them will be the Principal Contractor. Where the Contractor under a Contract is not the Principal Contractor for the Site, it is still necessary to identify who will be undertaking the role of Principal Contractor for that Site within the Scope for that Contract.

5.16 Practical issues

This chapter has discussed the roles identified by the ECC as well as several that are not identified but that are necessary, depending on the jurisdiction being worked in, to facilitate the successful completion of a project. It must be recognised that they are roles identified and required by the ECC; these roles should not be confused with job titles given to individuals under their contracts of employment. This comment is made because the role of Project Manager under the ECC is often confused with the job title of project manager used by many organisations.

Moving on, it is true to say that the most important ingredient in a successful project is the people who are involved. It follows that for the Parties to a Contract (Client and Contractor) to give the project the best chance of success, the selection and appointment of the right people to the various roles required by the ECC is a key factor. As the ECC is a process-driven contract and requires those involved to work in a spirit of mutual trust and cooperation, it is suggested that all participants should have received some level of training in the procedures and that new teams would benefit from some form of team-building. It is becoming increasingly common for Clients to recognise this need and to set out the requirements for formal training within the Scope; it only takes one person not working in the way that the ECC requires, or under the spirit that it attempts to engender, to spoil a project.

Chapter 6
Communications, Early Warnings and other General Matters

6.1 Introduction

This chapter considers the contents of core clause 1, which is simply titled 'General'. Behind that title lies what many would consider to be the fundamental basics of a well-run project, without which those involved will not engender the required spirit of trust and mutual cooperation. These fundamentals give this chapter the main part of its heading, i.e. communications and early warnings. While they are not the key tool in managing a project (that honour goes to the Accepted Programme), they are mechanisms which bring the benefits of a good programme to the forefront of good project management.

Following a detailed examination of these fundamentals, this chapter then considers other matters that are included in core clause 1 (with the exception of clause 14, which was reviewed, as appropriate, in Chapter 5 in connection with the Client, Project Manager and Supervisor).

6.2 Communications: The clause

Clause 13.1 states that every communication required by the contract is made in a form that can be read, copied and recorded. In isolation, this should be encouraged as it provides the parties with the audit trail that comes from having everything written down in one form or another. The wording does not require everything to be in the form of a letter or other such traditional means of communication. The phrase used easily encapsulates all modern forms of transmitting information between parties, the most common these days being e-mail.

The last sentence in clause 13.1 acts as a navigation link to the Contract Data, where an entry will record what language is to be used in the administration of the contract. This is one of several pointers in the document to the suitability of the ECC to be used anywhere in the world.

It is important to ensure that communications, especially contractual type notices (instructions, certificates and anything to do with termination or dispute resolution),

---

A Practical Guide to the NEC4 Engineering and Construction Contract, First Edition. Michael Rowlinson.
© 2019 John Wiley & Sons Ltd. Published 2019 by John Wiley & Sons Ltd.

are sent to the correct address as notified by each person to the others. Clause 13.2 caters for this requirement in two ways. First, clause 13.2 recognises that a communication system may be used. If this is the case then the Scope will specify the system to be used and all communications will be sent via that system. The clause is constructed so that where such a system is in use then all communications must go through that system to be effective. Any communications sent outside that system will not be considered to be effective. Secondly clause 13.2 provides that when a communication system is not being used then a communication is not effective until it is received at the last address notified to the other party by the recipient for the receipt of communications. To accompany these requirements, clause 13.6 details to whom the Project Manager and Supervisor issue the certificates required by the contract.

The contract expressly requires the Project Manager, Supervisor and Contractor to reply to a communication that the contract requires them to reply to. This requirement, at clause 13.3, is required to be complied with within a time known as the period for reply, a period which has to be entered into the Contract Data. The exception to this period is where the contract states a period for certain types of reply.[1] The Project Manager and the Contractor have the right to agree an extension to the period for reply to individual communications, providing that the extension is agreed before the original period expires (see clause 13.5). The benefit of this proviso is that the Contractor can identify whether there will be any impact on the programme before such an extension is agreed. This, combined with the necessary agreement of the Contractor, should act to prevent the programme being affected by any unilateral decision to take longer over a reply.

Clause 13.4 deals with how the Project Manager replies to anything that has been submitted to him for acceptance. Unless the reply is acceptance, which must be communicated to the Contractor, then the Project Manager must reply by stating reasons; it is not enough to simply say 'no'. The reasons provided must be in sufficient detail to enable the Contractor to rectify the previous submission and re-submit it to the Project Manager. This clause introduces the term 'a reason for withholding acceptance'. This term indicates to the Project Manager the grounds that can be used for withholding acceptance, the first being that more information is needed in order to assess the Contractor's submission fully. Further reasons are given elsewhere in the contract in connection with specific submissions that are required from the Contractor.[2] Clause 13.8 then confirms that the use of one of the permitted reasons for withholding acceptance will not constitute a compensation event.[3]

What many consider to be good communication practice – to only include a single issue in a piece of correspondence – is covered by clause 13.7 (although it only refers to notifications required by the ECC rather than to all forms of communication) in order to prevent important notices not becoming apparent to the receiver. It also serves to prevent a delay in the reply to any issue which can be provided quickly; responses to more

---

[1] For example, clause 31.3 requires the Project Manager to reply to a programme submitted to him for acceptance within two weeks.
[2] See clauses 16.2, 16.3, 21.2, 23.1, 24.1, 26.2, 26.3, 31.3, 40.2, 55.4 (Option A), X4.2, X10.3(2), X13.1, X14.2, X16.3, X21.3, X22.2(2), X22.3(3), Y1.4 and Y1.6 for the other permitted reasons.
[3] See clause 60.1(9) which identifies that it is a compensation event for withholding acceptance for a reason not stated in the contract.

straightforward issues should not be delayed as a result of being raised in a multi-issue communication.

6.3 Communications: Practical issues

6.3.1 Good practice

Good communications are the key to any successful project and to any successful relationship. In the context of clauses 10.1 and 10.2, unless the personalities communicate well, they will not be able to act as stated in the contract or in the required spirit of mutual trust and cooperation. Anyone who doubts this fundamentally basic point should simply think about what happens if you don't communicate with your loved ones and friends.

That said, the requirements of clause 13.1 (if taken literally) can in the author's view make good communication harder and endanger any chance the parties had of operating in the desired way. The need for everything to be in a form that can be read, copied and recorded can lead to people sitting at their computers sending e-mails which are copied to all and sundry. Anyone who has been on a communications skills course will no doubt be familiar with the analysis that approximately 90% of the messages we communicate comes from nonverbal transmitters given through the voice (but not by the words) and by our body language. These elements don't come across in e-mails. Since most people are better at saying what they mean than writing it down, the use of e-mails (which are usually on the informal side) tends to mean that the full message is not properly conveyed. As a result of this, e-mail ping-pong starts, which is both time-consuming and wasteful.

It is suggested that, in practical terms, the best way to communicate is face-to-face. In the case of the common types of question-and-answer-type communications, all that is then required is for the people involved to record in some permanent form the question that was raised and the answer that was given or agreed. Where these points need to be communicated to others, this record can then be distributed in an appropriate way.

This approach to communication works very well for teams that are located in the same place while carrying out projects. It is beyond question that teams who are co-located and who communicate verbally and then record the pertinent parts of the discussions are those who establish the best working relationships and build the most trust in each other. On smaller projects where it is not economically viable or necessary to co-locate everyone involved in the project, it is suggested that the team needs to work at finding a way to communicate effectively such that trust and cooperation are built and enhanced.

6.3.2 Single-issue communications

If the requirement at clause 13.7 to include every notification in a separate communication is expanded to include all communications, so that no communication covers more than one subject, then this practice can assist the parties in communicating promptly and effectively. By adopting this practice, it allows the person responsible for replying to deal with each subject in a separate reply. This implies that straightforward queries can be replied to quickly and in isolation. If such a straightforward subject was part of a communication covering several subjects, it would not be uncommon for the reply to the straightforward matter to be delayed while the more involved and detailed subjects in the same communication were researched and also replied to. Such approaches cause inefficiency in the communication process.

It is also possible to see that this requirement at clause 13.7, when taken literally, can result in unnecessary paperwork being created, a situation that users should seek to avoid. For example, clauses 61.1 and 61.2 require the Project Manager, when issuing an instruction, to also notify the change as a compensation event and instruct the provision of a quotation by the Contractor. This can be interpreted as requiring three separate communications when clause 13.7 is applied. That said, the User Guide, Volume 4 suggests that except for notifications and certificates other communications could potentially be communicated together. In practice, for say a compensation event notification, it is far better to include a statement on the Project Manager's Instruction to say whether or not the matter is a compensation event and whether a quotation is required. This satisfies the ECC's goal in a concise and succinct manner. In the author's opinion what is important is the communication itself and the parties having clarity.

6.3.3 Electronic communication systems

Clause 13.2 recognises that it is becoming increasingly common for users of the ECC to employ one of the commercially available software packages that assist in the administration of projects. There are a number of such packages available which are written to cater for the requirements of the ECC. It is not the purpose of this guide to identify or discuss the merits of such packages, other than to record that users consider them to be of benefit and that they provide assistance with the timings required. One of the ways several such packages act is to prompt people that action is required of them and to issue reminders when the timescales for such actions are due to expire. If it is the intention of the Client that one of these packages is used then, as required by clause 13.2 the requirement is stated in the Scope (see definition at clause 11.2(16) and Section 22.2). This then becomes a constraint on how the participants should communicate.

6.3.4 The ECC communication forms

When the third edition was reprinted in April 2013, one of the new documents produced at that time was entitled 'How to … use the ECC communication forms' (NEC Panel 2013e). The book dealt with communication and provided example forms for most, but not all communications. That book is not part of the fourth edition.

Instead the information is incorporated into Volume 4 of the User Guide, being introduced in the guidance to core clause 13 (see Section 6.2). This identifies a number of standard forms, which have been drafted neutrally to be capable of use by both parties as appropriate. Illustrative examples are then provided at the end of the guidance to some of the core clauses.

To assist readers further Appendix 3 to this book lists the communication forms identified on page 6 of Volume 4 of the User Guide. In the first table in Appendix 3 the forms listed in the User Guide are identified. The second table in Appendix 3 then identifies every clause to which each of the forms in the first table would apply and the user for that form in respect of that clause. The third table in Appendix 3 identifies the forms that will be required to manage the communications but which are not listed in Volume 4 of the User Guide. As with the second table every clause to which each of these additional forms would apply and the use for that form in respect of that clause is also included.

Users of these tables should note that the requirements for all communications under Options W1, W2 and W3 have not been included in Appendix 3, albeit that some of the more vital notifications have been included. As these three Options cover dispute resolution, it is recommended that should the mechanisms in whichever of these three Options applies need to be instigated then all communications under them should be by letter. Dispute resolution is a serious matter and should not be managed by a series of pre-drafted forms. It should be managed by appropriate letters and notices written specifically for the matter under consideration at that time.

6.4 Early warnings: The clause

The opening paragraph of Volume 1 (Establishing a procurement and contract strategy) of the supporting documents (see Section 2.11) identifies the intent of the NEC family of contracts to facilitate sound project management principles using foresight as opposed to hindsight. The window on such foresight is created by the Accepted Programme (see Section 12.2). The mechanism introduced to encourage participants to apply foresight and to share that foresight with the rest of the team is called early warnings.

As it is a mechanism under the contract, it places an obligation on the Contractor and the Project Manager to comply with the requirements of clause 15. Failure to do so on the part of the Contractor will result in sanctions against him under the contract.[4] Failure to do so on the part of the Project Manager will result in the loss of any opportunity on behalf of the Client to consider mitigation that could be put in place to minimise the impact of an event. It would also constitute a breach of the contract by the Client in not ensuring that the Project Manager carried out the required duties, which in turn would be a compensation event as identified at clause 60.1(18) (see Section 15.3.18).

In order to comply with this requirement, the Contractor and the Project Manager are each required, by clause 15.1, to notify the other as soon as any matter which could impact on the Prices[5], Completion (see clause 11.2(2) and Section 12.5.2), a Key Date (see clause 11.2(11) and Section 12.4.4) or the quality of the works (see definition of

---

[4] See (1) clause 63.7 (following clause 61.5) and (2) the definition of Disallowed Cost (clause 11.2(26) in main options C, D and E and clause 11.2(27) in main option F).
[5] See clause 11.2(32) in main options A and C, clause 11.2(33) in main options B and D, and clause 11.2(34) in main options E and F and Section 13.4.3–13.4.5.

Scope at clause 11.2(16) and Section 22.2) comes to their attention. The Project Manager and the Contractor are also permitted to use this mechanism to notify the other of any matter which would increase the Contractor's cost, but which does not come within the four categories identified (see Section 6.5.4). In order to prevent unnecessary paperwork, clause 15.1 also clearly states that it is not necessary to notify of a matter which has already been notified as a compensation event as an early warning.

It is at clause 15.1 that the Early Warning Register is introduced. This term is defined at clause 11.2(8), where it identifies that the Early Warning Register requires only two pieces of information. The first is what the risk is or might be and the second states what actions are to be taken to avoid or reduce the risk. Development of the Early Warning Register begins at conception of the project by the Client and its consultants. Risks that could impact on the project, as opposed to business risks that affect the viability of the project (identified throughout the design and development phases), are entered into the Early Warning Register. Once the project is readied for tender, the Client/Project Manager references the Early Warning Register as it stands at that time in Contract Data Part One and provides the tendering contractors with a copy. As part of the tender submissions, the tendering contractors enter into Contract Data Part Two any further project risks which they consider adding to the Early Warning Register provided by the Client.

This starting point, of all the risks included at Parts One and Two of the Contract Data consolidated together in one document, is known as the Early Warning Register, which is prepared, in accordance with clause 15.2 by the Project Manager and issued by the Project Manager to the Contractor no later than one week after the starting date (see Section 12.5.1). The Project Manager records every early warning notified in the Early Warning Register. From the requirement that the Project Manager adds early warnings to the Early Warning Register and the requirement at clause 15.4 that the Project Manager revises the Early Warning Register, it is clear that this important document is maintained by the Project Manager.

Clause 15.2 requires the Project Manager to instruct the Contractor to attend a first early warning meeting within two weeks of the starting date. The same clause then prescribes that early warning meetings are then held if either the Project Manager or Contractor instructs the other to attend one or no later than the interval stated in the Contract Data. So in the absence of any requirement from either the Project Manager or the Contractor there will be a regular meeting at preset intervals. This should mean that those concerned are keeping the notified risks under regular review. Clause 15.3 then tells the attendees what they should cooperate with. In simple terms, this is to share all their ideas about how the effect of a risk notified in an early warning matter, whether it has occurred or not, could be reduced to the benefit of all those who would be affected should that risk arise, i.e. mitigation. Either party can bring other people to the risk reduction meeting if the other party agrees. These invited individuals, who would probably be specialists with knowledge of potential risks, contribute to the deliberations and assist in the process. The first two points for consideration at the risk reduction meeting are the core of what is required. The third point confirms that any actions decided upon should be in accordance with the obligations under the contract. The fourth point requires that risks that have been avoided or passed are deleted from the Early Warning Register.⁶ The fifth and final one emphasises the need to review previously recorded actions and thereby keep them up to date, with revisions as required.

The Project Manager records the decisions made at early warning meetings and issues a revised register as necessary. He also issues any necessary instructions for changes to the Scope which may be required as a result of such decisions.

6.5 Early warnings: Practical issues

6.5.1 Purpose

The purpose of early warnings is to give the Project Manager and Contractor foresight of any matter that may cause a problem in the future. Having identified a potential problem, they then monitor that problem and make contingency plans for what they will do should the problem materialise. These contingency plans could include making changes to the works or method of working that would allow the potential impact to be engineered out of the project, thereby mitigating the effects of a problem before it can become a major issue that acts to the detriment of the project. Once it is apparent that the matter will be a problem, the plans can be put into effect.

6.5.2 Risk management in practice

The process for dealing with early warnings at clause 15 is formal and linear. In practice, people working on projects are holding early warning meetings all the time. Every time two people discuss how they are going to do something, they are discussing the management of risk. This cooperative type of approach is especially evident on projects where the Project Manager, Supervisor and Contractor are co-located, usually in offices within the Working Areas. In such circumstances what the team need to do is identify a system whereby they can capture all the potential risks and introduce them into the monitoring and planning procedure.

It is not suggested that every minor risk is recorded. Most minor risks will be managed out of the process by someone taking an expedient step that does not need recording in the system. The system needs to be able to capture all those risks that may become a problem and that need monitoring and considering by those involved. To monitor these types of risks, project teams often arrange regular formal early warning meetings held at an appropriate frequency.⁷ These formal meetings review every item on the Early Warning Register and include a review of the current Accepted Programme. By combining these two key documents and adding their own professional skills, those involved apply foresight to the project to the benefit of all those who will be affected in respect of any

---

⁶ See comments in Section 6.5.7 regarding the practical implications of not deleting risks that have passed.
⁷ Weekly is not unusual, especially on larger projects or at critical times on any project.

matter that has been identified as a potential risk. This satisfies the general philosophy that the ECC requires and the specific requirements of clause 15.3.

Whilst clause 15.2 requires regular meetings at the intervals stated in the Contract Data the whole process is likely to work better if that frequency is varied depending on the requirements and quantity or importance of risks that have been notified and which could become actual issues in the immediate future. The author is aware of projects on which early warning meetings have been held at daily intervals in the weeks leading up to when critical activities are planned to be carried out.

6.5.3 Recognising the purpose of the early warning register

Anecdotal evidence and copies encountered by the author suggest that many users of Early Warning Registers do not grasp that the purpose of this document is to manage risks to time, cost and quality matters that will affect the outcome of the project. It is not a tool to manage business risk for either party. For example, the risk to the viability of a commercial development from potential reductions in rental income is not a matter for the Early Warning Register. This example is not a risk to time, cost or quality in respect of the building being constructed by the team. It is a risk for the Client in the longer term that its development might not provide the returns it hoped for. At the other end of the scale, the risk that the van bringing part of the workforce to site breaks down is not a risk that should be recorded[8]; this is part of the Contractor's normal working risk.

6.5.4 Increases to the Contractor's cost

With regard to either the Project Manager or the Contractor notifying a matter which would increase the Contractor's cost, this embodies the philosophy of the ECC that all concerned should work to achieve each other's business objectives. In this respect users should beware that, under the cost reimbursable methods of payment (Main Options C, D and E), any increase in the Contractor's cost will result in the Client paying more.[9] It is therefore in the Client's interest for such matters to be brought into the open and for everyone to work together to reduce such increases. An example might be that a specified material has increased in price. By identifying and approving a suitable alternative at the same price as the original, this increase will be negated. Another example could be that the Contractor has been carting surplus excavated material to a local tip but that tip is now full and not taking any further material. The only alternative is not only further away but also charges more per load, thereby increasing the Contractor's costs on two fronts. If the Project Manager can issue an instruction changing the earthworks profile in, say, the landscaping areas so that the amount of cart away is reduced, then the cost to the Contractor decreases. The change also constitutes a compensation event, so the Prices are reduced and both the Client and the Contractor should benefit.

---

[8] The author has seen this risk entered onto a risk register by a Contractor.
[9] Unless main options C or D have been set up as guaranteed maximum prices (GMP) (see Section 13.8.3) and that GMP is exceeded.

6.5.5 Volume of early warnings

A lot of users comment that, on occasions, a large number of early warnings are raised. Some project managers consider that some contractors issue far too many. Project managers should remember two things in such circumstances: firstly, it is better to have matters which could result in potential problems brought to their attention than not and, secondly, in the event that a contractor does not raise an early warning it could face the sanctions identified in Section 6.4 above.

As the failure of a contractor to issue early warnings can result in sanctions against him, users must appreciate that on a practical level the issuing of early warnings by contractors, often in large numbers, acts as a shield to protect them from these possible sanctions. Early warnings issued by contractors therefore have two uses: one to aid the parties in gaining foresight into potential problems and one to protect the contractor from the sanctions.

6.5.6 Contents of the Early Warning Register

The definition of the Early Warning Register (clause 11.2(8)) only requires two columns. In practice, many users feel that this is not enough and populate the register with other information such as who identified the risk and when, who the risk would be to and what the potential cost and time implications might be. The definition does not prevent the inclusion of information in addition to the two required matters; users should therefore develop the Early Warning Register in a format which suits their requirements.

6.5.7 Removing expired risks

Clause 15.3 states that risks which have passed should be removed from the register. To remove an item in any circumstances could lead to the information being lost and prevent the parties from benefiting from that foresight in relation to lessons learnt or for reference on similar future projects. Such information is recognised by most as being of value so, in practice, most users do not delete risks after they have passed; instead they use, say, shading tools available in common software to indicate those risks which have passed so that they do not keep referring to them. At the end of the project, they can then remove all shading to reveal a store of information for the future.

6.6 Other matters: The clauses

Core clause 1 contains several other matters as well as communications and early warnings that all assist in the general administration of the ECC or cover things that do not easily fit within one of the other core clauses. Matters concerning the Project Manager and Supervisor (clauses 14.2 to 14.4) were considered in Chapter 5 (see Sections 5.2–5.4); all other aspects of this core clause will be considered here.

6.6.1 Navigation tools

Two key tools for users to navigate their way around the ECC are set out at clause 11.1. The first of these is a useful link between the clause text and the Contract Data. This is achieved very simply by formatting words or phrases within the text in *italics* where there is detailed information within the Contract Data about that word or phrase. For example, the phrase 'boundaries of the site' appears in the text in *italics*; in Contract Data Part One, section 1, there is an entry which on completion states what the 'boundaries of the site' are. The downside of this navigation tool is that it does not give any clue as to whether the detail will be found in Part One or Part Two of the Contract Data. There is one general aid in that any word or phrase referred to in core clauses 1–9 (but not the additions created by incorporating a main option), in respect of which the Client is to provide the data, will be in the same numbered section within Contract Data Part One. Users will learn to find other information quickly by using the ECC regularly and learning where the various entries are located. This process can be assisted by good preparation of the Contract Data (see Section 21.2).

The second navigation tool relates to the use of Uppercase for any term that is defined in the ECC. All such definitions are located at clause 11.2. In this guide, the definitions are discussed with the clauses where relevant.

6.6.2 Contractor remains responsible

The common law position that the Contractor remains responsible for its work and liable for its design[10] (if any) despite any acceptance (whether of the work or by means of a communication) by the Project Manager or the Supervisor is confirmed at clause 14.1. This is a well-established principle but the industry still encounters people who, having done something wrong, try to pass liability on the grounds that someone in the contractual chain above them has accepted or approved the work previously. It is common, especially where lawyers have altered the contract and the Contractor has design liability, to see a provision that covers this point added to clause 21. In the author's view, such an addition is not necessary as clause 14.1 is wide enough in its coverage to negate the need for such duplication.

6.6.3 Contractor's proposals

The Contractor is given an option at clause 16.1 to propose that the Scope is changed so that the Employer pays less for the completed Works. The Contractor makes this proposal to the Project Manager, following which the Project Manager Consults with the Client and the Contractor. Whether separately or together is not prescribed.

Within four weeks of the Contractor's Proposal the Project Manager is required to reply to the Contractor by clause 16.2. Three potential replies are listed, these being an

---

[10] See clause 21.1 and Section 7.3 for details of how design (if any) by the Contractor is introduced.

acceptance of the proposal accompanied by an instruction changing the Scope, notification that the Client is considering the proposal and requires a quotation for the proposed instruction or confirmation that the proposal is not accepted. The Project Manager is allowed to give any reason for the proposal not being accepted.

What is missing from these two sub-clauses is any incentive for the Contractor to make such proposals, but that does not mean that they are missing from the ECC as a whole. How the Contractor's incentive comes about depends on the main Option in use. For Options A and B clause 63.12 provides for the Contractor to receive the value engineering percentage, which in turn is stated in Contract Data Part One as being 50% by default or a different percentage if inserted in the entry. Should either of Option C or D be in use, then clause 63.13 would apply resulting in the Prices not being reduced. As a result, the Price for Work Done to Date (see Section 13.4.5) would stay the same, creating a more advantageous result for the Contractor when the Contractor's share mechanism at clause 54 (see Section 13.7) is applied. There is no equivalent mechanism in either Option E or F.

6.6.4 The Site and the Working Areas

A provision is included at clause 16.3 for adding to the Working Areas, a term which is defined at clause 11.2(20). In order to appreciate the need for this provision it is necessary to understand the relationship between the Site and the Working Areas together with the use of the latter term within the ECC. While there are several clauses which refer to the Site,[11] this term does not have a clause all to itself except for its definition. It is stated at clause 11.2(17) as being the area within the 'boundaries of the site' (a term for which detail is included in Contract Data Part One) and the volumes above and below the ground which are affected by the works. The definition of Working Areas refers to the same phrase, but in italics, leading users on this occasion to Contract Data Part Two. Here the Contractor can complete the entry which says that the working areas are the Site and whatever the Contractor enters in the blank space.[12] Following the definition, these areas should only be areas required to construct the works and which will be used solely for work related to the project concerned. The provision for adding to the Working Areas is to provide the flexibility to the parties so that this area can be extended should such an extension be required in order to provide the works. The Project Manager is given a reason for not accepting such a proposal from the Contractor, where either the extension is not necessary in order to provide the works or will be used for work not under the contract.

6.6.5 Requirements for instructions

Clauses 17.1 and 17.2 set out the method for resolving ambiguities and inconsistencies in the first case and illegal and impossible requirements in the second case. In both cases,

---

[11] The other clauses which refer to the Site are 11.2(18), 20.2 (Option F), 30.1, 31.2, 33.1, 60.1(2), 60.1(5), 60.1(7), 60.1(12), 60.2, 72.1, 73.1, 80.1, 92.2, W3.1(5), W3.1(6), W3.2(4), W3.2(5) and X2.1.

[12] See Section 6.7.2 for comment on what areas should and should not be included in this entry.

the Contractor or the Project Manager informs the other should it find such a matter. In the first case the Project Manager states how the matter notified should be resolved. This gives the Project Manager a wider range of options than just issuing an instruction and may result in the ambiguity of inconsistency being resolved without the need for any change to the Scope. In second case, once any such matter has been identified, the Project Manager is obliged to give an instruction to resolve the issue by changing the Scope.

6.6.6 Corrupt acts

What constitutes a Corrupt Act is defined at clause 11.2(5). The definition itself is straightforward and does not require any explanation. Many jurisdictions have legislated for such acts; for example, the Bribery Act 2010 in the UK.

Clause 18.1 simply requires that the Contractor does not do any corrupt act. This is followed by clause 18.2, which requires that the Contractor takes immediate action to stop any corrupt act being committed by a Subcontractor or supplier if the Contractor is either aware of or should have been aware of that act. The reference to acts that it should have been aware of puts the Contractor into a potentially difficult situation, but in most cases no different than legislation demands.

As the ECC treats all a Subcontractor's employees as being the Contractor's employees, this prohibition then extends to Subcontractors (see Section 8.2). As a result, as stated in clause 18.3, this same requirement must be included in any Subcontract and any contract for the supply of Plant and Materials (see definition at clause 11.2(14)) and Equipment (see definition at clause 11.2(9)) placed by the Contractor.

There is no sanction in this clause, but a breach of the mandatory obligation will trigger the reason for termination at clause 91.8 (see Section 18.2.3).

6.6.7 Prevention

A principle known as 'Prevention', which was introduced into the third edition family, is set out at clause 19.1. Volume 4 of the User Guides says that this principle is, in effect, the equivalent of the French term *force majeure* used in other contracts. The French term is very difficult to define. The ECC attempts to define this by spelling out that the event that gives rise to this clause being applied is one which: had such a small chance of occurring that it would have been unreasonable for an experienced Contractor to have allowed for it; could not have been prevented by either party; and will stop the Contractor from completing the works either at all or by planned Completion, regardless of what steps could have been taken. Should such an occurrence arise, the Project Manager must issue an instruction about what to do. Such events will in practice be very rare as can be seen by the last line of the corresponding compensation event at clause 60.1(19) (see Section 15.3.19), which states that this event can only arise in connection with something that is not one of the other compensation events.

6.6.8 Law and interpretation

The four subclauses of clause 12 set out the following basic matters of interpretation and law:

- that singular means plural and vice versa (clause 12.1);
- that the law of the contract, as detailed in Contract Data Part One, governs the contract (clause 12.2);
- that unless provided for by the contract its terms and conditions can only be changed by an agreement in writing signed by the parties (clause 12.3); and
- that the contract is the entire agreement between the parties (clause 12.4).

These points are all easy to understand. Users must remember that the provision at clause 12.4 does mean that any document not forming part of the contract cannot be relied on by either party. It is therefore important to ensure that everything that either party intends to form part of the contract must be included by one means or another. The most common place to reference such items is the Scope.

6.7 Other matters: Practical issues

6.7.1 Completing the Contract Data

The use of terms in *italics* as a primary navigation tool to the Contract Data provides the person putting together a contract using the ECC with a cross-check that every entry required has been included. It can be laborious (but effective) to do but, by going through the clauses with a copy of the Contract Data to hand, the person doing the preparation can check that they have included a suitable entry against each relevant statement in the Contract Data. To this end, users must recognise that the flexibility created by the use of Options means that not every entry within the blank Contract Data will be used; those that are not used should be discarded (see Chapter 21).

6.7.2 Extent of Working Areas

One issue that has caused problems in practice is the extent of what should be included in the Working Areas. The extent of the Working Areas is determined initially by the Contractor making in an entry in Contract Data Part Two, which is then submitted with its tender. The practical application of using the working areas to extend the Site beyond its boundaries is to allow the Schedules of Cost Components to be operated, as generally items are only included in the cost if they are in the Working Areas.[13]

---

[13] See Chapter 14 for details of the Schedules of Cost Components.

This application is important where there is no room within the Site for the Contractor to locate its offices, welfare facilities, stores, etc. This type of situation is commonly encountered on road and other infrastructure projects which are linear in nature (e.g. pipelines or power cables). It is also encountered on cramped urban sites where the footprint of the building fills most if not all of the site footprint. In such circumstances, the Contractor often acquires the use of an adjacent or nearby piece of land to use for their temporary areas. The intent of the term Working Areas is that such areas will be included which will be used solely in connection with the project.

What the term Working Areas does not envisage is that it should include the Contractor's regional or head offices, workshops, yards, stores, etc. or any such similar areas belonging to any of the Contractor's subcontractors. The author has seen all these areas and more, included against the entry for this term. Clients, advised by their project managers, should not enter into contracts containing such entries. Such entries should be negotiated away before the contract is entered into if the term is to be given its true meaning and application.

That said, the author has also encountered circumstances where it has been to the Client's advantage or to the advantage of the project for some or all of a facility belonging to the Contractor, which is also being used for work on other contracts and therefore strictly not capable of being classified as Working Areas, to be classified as Working Areas. When such circumstances have arisen, the Parties have made a clear endorsement to the Contract to the effect that the additional area concerned is agreed to be part of the Working Areas, notwithstanding that it contravenes the definition of Working Areas at clause 11.2(18) (see Section 6.6.4).

6.7.3 Prevention

The remaining matter to consider practically is what type of event may give rise to a prevention event under clause 19.1. First, it will have to be something that does not qualify as one of the other compensation events (see clause 60.1(19) and Section 15.3.19). It will also have to be something that prevents the Contractor from completing the Works either at all or by planned Completion, no matter what steps could be taken. The only recent event that I can think of which occurred in the United Kingdom and which (in parts of the country only) would constitute such an event was the last major outbreak of foot and mouth disease in 2001. This stopped construction projects in many rural areas, as neither the contractor nor the employer were allowed access to the exclusion zones for fear of spreading the infection to other areas. The nature of this clause is that it would have to be something major of this type to qualify.

Chapter 7
The Contractor's Main Responsibilities

7.1 Introduction

The Contractor's obligations are set out throughout the ECC. Although many clauses contain requirements which impose an obligation on the Contractor, it is within core clause 2 (entitled as for this chapter) that the Contractor's primary obligation is set out. Core clause 2 also contains several important other obligations which will be covered in this chapter. The provisions concerning subcontracting, which also form part of core clause 2, are considered separately in Chapter 8.

7.2 Providing the Works

The Contractor's main responsibility, stated at clause 20.1, is to Provide the Works. This term is defined at clause 11.2(15) as doing everything necessary to complete the works and to do it in accordance with the contract. This is said to include doing everything that is incidental to that obligation, including, but not limited to, all the actions required and all the services to be provided by the Contractor. Users must also remember that the Contractor is also under an obligation to complete the works by the Completion Date, as stated at clause 30.1 (see Chapter 12). Indeed, these two obligations taken together encapsulate what the Contractor has to do.

Clause 20.1 requires this providing of the works to be done in accordance with the Scope. This simple requirement is key to what the Contractor is obliged to provide under the ECC and to the controlling of the works by the Project Manager and the Supervisor. In the author's opinion, the Scope is the most important document under the ECC. The need for and content of the Scope is considered at Chapter 22 so all that needs to be done here is to note the definition, which is found at clause 11.2(16). This definition identifies two things: what the information is and where it is found. The first of these determines that the information describes and specifies the works on the one hand and, on the other, sets out any constraints on how the Contractor fulfils this obligation. The second identifies that this information is either where the Contract Data states it is or is in an instruction given under the contract.[1]

---

[1] In most instances this will give rise to a compensation event as clause 60.1(1); see Section 15.3.1.

A Practical Guide to the NEC4 Engineering and Construction Contract, First Edition. Michael Rowlinson.
© 2019 John Wiley & Sons Ltd. Published 2019 by John Wiley & Sons Ltd.

In Providing the Works, the Contractor remains responsible for everything it is to provide regardless of any acceptance, approval, confirmation or otherwise that may be given by the Supervisor or the Project Manager (see clause 14.1 and Section 6.6.2). For complying with this obligation, the Contractor will be compensated in accordance with the Contract.

Depending on the Main Option in use, there are up to three other subclauses of clause 20 that may form part of the conditions. The only Main Option to include all these subclauses is F (Management contract). Clauses 20.2 is exclusive to Main Option F. Clause 20.2 requires that the Contractor manages all aspects of the design and works, both temporary and permanent. The Contractor is required by this clause to subcontract all of these aspects, except for the parts identified by the Contract Data to be done by the Contractor.

Clauses 20.3 and 20.4 form part of Main Options C, D, E and F.[2] The former requires the Contractor to inform the Project Manager of all practical implications of the design and the Contractor's subcontracting arrangements. This brief requirement brings out the openness of these four types of contracts. The Client is in effect paying for what happens, subject to certain rules over risk allocation and the sharing of over- or underspend against a target, all depending on exactly which main option has been selected.

There is therefore no reason why there should not be complete transparency of all aspects of the service being provided by the Contractor for the Project Manager, as the Client's agent. Those involved just need to keep in mind where contractual responsibility lies for each aspect, how overcomplicated levels of information or protracted discussions can increase costs for the Client and that Project Managers may, inadvertently in most cases, create compensation events if they impose their preferences on such arrangements.

Clause 20.4 requires the Contractor to forecast the total Defined Cost[3] for the works at intervals stated in the Contract Data. He is required to do this in consultation with the Project Manager. As it is done in consultation, there is no requirement for the Project Manager to accept the forecast when submitted (although each forecast should be accompanied by an explanation of the changes since the last forecast). The purpose of this requirement is to give the Project Manager the ability to keep the Client informed of the likely outturn cost of the project. Providing this forecast is kept up to date and the compensation event assessments are being done promptly then, even for Main Options C and D, the calculation of the outturn cost is easy, thereby allowing the Client to plan its finances.

7.3 Contractor's design

Unlike other popular families of standard forms of construction contracts, the NEC family of contracts does not contain a purpose-written design and build or design and construct form. Neither does it contain any special supplements for contractor-designed portions or anything else of that nature.

---

[2] Main Options A and B do not include any of the four additional subclauses to clause 20; only clause 20.1 is operative when one of these main options is selected.

[3] See clauses 11.2(24) (main options C, D and E) and 11.2(25) (main option F).

Instead, the ECC contains clause 21.1 which simply states that the Contractor designs such parts of the works as stated in the Scope. While amazingly simple and very elegant, this provision is also extremely effective. It provides users with total versatility in respect of how little or much of the project, if any, the Contractor is to design. By this simple clause the Contractor can have no design responsibility, total design responsibility or any point in between. It is worth repeating that the key is the Scope; as stated above, this is covered in more detail at Section 22.2.10.

Clauses 21.2 and 21.3 concern the submission of the Contractor's design particulars to the Project Manager for acceptance. Again, reference to the Scope is made where a submission/acceptance procedure could be set out as required to suit the particular circumstances. Reasons for not accepting the design are given for the Project Manager to apply if necessary. The submission of the design in parts is allowed, providing that the parts can be assessed independently. This is, of course, necessary to allow works to progress in parallel with the design process as is increasingly common in construction. Users should be very aware of the last sentence within clause 21.2, which in the document is set separately from the rest of that clause and which clearly acts to prevent the Contractor from proceeding with any work that it has designed until its design has been accepted. It is common, certainly in the United Kingdom, for contractors with design responsibility, to proceed with works for which they have not received design approval; this is often to be said to be at the Contractor's risk. Under the ECC such a step is expressly forbidden; taking such a step would be a breach of the contract.

How the Client may use and copy the Contractor's design is set out at clause 22.1. Again, reference to the Scope is made in the context of either limiting or expanding the basic provision provided by the clause. In essence, this is a very simple copyright provision and should suffice; it is very common to see this provision completely re-written by lawyers, however, who insert the more usual long-winded type of provision common in other contracts and especially prevalent in the world of collateral warranties. Whether such a change is required is not something that is worth debating here. One aspect of this provision that the Contractor must pay attention to is creating an obligation for any Subcontractor with design responsibility to provide the same rights to the Client as the Contractor must give.

Clause 23.1 also covers submission of the Contractor's design but this time in connection with Equipment.[4] In such cases, the Contractor only submits its design when instructed to do so by the Project Manager. Unlike clause 21.2, there is nothing to prevent the Contractor from proceeding with the works if its design for any item of Equipment has not been accepted. As stated at clause 14.1, the Contractor remains responsible for the works in any event.

7.3.1 Reasonable skill and care

For any Contractor or subcontractor who has design responsibility for permanent works, it is essential in practice that the contract includes Secondary Option X15. This secondary

---

[4] See definition at clause 11.2(9) which identifies that Equipment is anything which is temporary, including temporary works and items such as scaffolding and formwork (unless permanent).

option acts to set reasonable skill and care as the test by which the Contractor's liability for defects due to its design is measured. The wording also makes it clear that the test will be benchmarked against the skill and care normally used by professionals dealing with works similar to the works under the particular project. This is the normal position that would be expected from a professional designer.

This clause is important where the contracting party has design responsibility; otherwise it would be responsible for satisfying a stricter test known as 'fitness for purpose'. Such a term would be implied by section 14(3) of the Sale of Goods Act 1979. Most contractors find that it is not possible to insure themselves against a fitness for purpose obligation; most will therefore shy away from a contract that seeks to impose such a condition. Others, maybe through ignorance or maybe because they decide to take the risk, will accept a fitness for purpose obligation; employers may be well advised to consider whether such a contractor is the type of organisation that they wish to do business with.

The second sub-clause in Secondary Option X15 acts to create a compensation event should the Contractor correct a defect for which it is not responsible under the contract. This clause is anticipating the position that the Contractor may correct a defect which arises from its own design, but where the Contractor used reasonable skill and care and is therefore not liable for that repair. In such circumstances, although the Contractor has been required to correct the defect under core clause 4 (see Chapter 9) the Contractor is not responsible for the cost of correcting such a defect.

Clause X15.3 concerns the use by the Contractor of what is referred to as material and which the Contractor provided under the contract. Essentially, this material is the design documents which the Contractor will have produced in fulfilling its design obligation. It will consist of not just drawings but also specifications, schedules, calculations, reports and the like. The clause allows the Contractor to use this material on other projects, not only for the same Client, unless ownership of that material has been transferred to the Client under Option X9 (see Section 7.3.3) or as stated otherwise in the Scope. In many ways this is the reverse of clause 22.1 (see Section 7.3).

The Contractor is also required, by clause X15.4, to retain copies of the material for what is called the period of retention, the duration of which is stated in Contract Data Part One. No default is provided for this period; it is suggested that it should be no longer than the limitation period (see Section 11.3). How the copies are to be retained, i.e. hard copies or electronic files or both or some other way, is to be specified in the Scope.

Clause X15.5 concerns an insurance requirement, which is covered at Section 11.3.

7.3.2 Undertakings to the client or others

The undertakings that are being referred to in the heading are what are commonly called 'collateral warranties' in the United Kingdom. The heading is also the title of secondary Option X8. The purpose of a collateral warranty is to create a contractual relationship between two parties who would not otherwise be in contract with each other. The word 'collateral' is referring to the agreement being subsidiary to another contract, while the word 'warranty' indicates that a legally binding and enforceable promise has been created.

There is an alternative to the use of that provision in UK law, being The Contracts (Rights of Third Parties) Act 1999, which can be used by the inclusion of Option Y(UK)3 (see Section 20.11).

The Option refers to three classes of undertakings, these being:

- Contractor undertakings to Others (see clause 11.2(12) and Section 5.11) (as clause X8.1);
- Subcontractor undertakings to Others (as clause X8.2); and
- Subcontractor undertakings to the Client (as clause X8.3).

Clause X8.4 requires that the form of these documents is to be set out in the Scope, which is normally done by providing a draft of the undertaking required. Clause X8.5 then provides that the Client prepares the required undertakings and sends them to the Contractor. The Contractor then signs and returns those that it is to provide and makes arrangements for the Subcontractors to sign any that they are to provide. Signature and return is required to be completed within three weeks of the documents being received by the Contractor.

7.3.3 Transfer of rights

If Option X9 is operative the effect is that the Contractor's rights over the design materials that the Contractor prepares are owned by the Client. In such instances, the Contractor will also have to make sure that all its Subcontractors (see Section 8.2) provide the same transfer of rights. The Contractor will have to obtain the necessary documents that transfer those rights. Such a requirement from Subcontractors could go several tiers down the supply chain.

When Option X9 is in use the only exception to the requirement is as may be stated otherwise in the Scope (see clause 11.2(16) and Section 22.2).

Many Contractors and more so Subcontractors will not agree to transfer ownership of their copyright and intellectual property rights to the Client whatever the circumstances. This is because many of these organisations regularly design the work that they carry out. The work is specialist in nature and were they to give away the ownership of the copyright and/or intellectual property then they would not be able to use it on future projects. This could have the potential effect of preventing them from trading effectively in the future.

It is possible to get around this problem by creating different classes of materials. In simple terms, only the rights to any unique design that has been especially created for the project will be handed over. This is often facilitated by classing all existing copyright and intellectual property that the designer has produced in the past and repeatedly uses on projects is called 'background property' (or something similar), which is then excluded from the rights transferred to the Client under this mechanism. The documents transferring the rights would have to be drafted by lawyers with specialist knowledge to protect all those concerned.

7.4 Information modelling

7.4.1 Introduction

The 2013 reprint of the third edition included, as part of the 'How to' series of books to help users one entitled 'How to … use BIM with NEC3 Contracts' (NEC Panel 2013h).[5] The operative part of this book provided a Z Clause for incorporation into the ECC. The effect of that model Z Clause was to incorporate, by reference, the CIC Building Information Protocol, first edition 2013 (CIC 2013). This Protocol provides for the necessary protection and limits of liability for users employing Building Information Modelling (BIM) on their projects.

The fourth edition removes the need for any Z clause by making this Secondary Option available to users. By using this Secondary Option users are provided with all the tools needed to employ BIM on their project. That the title of the Option excludes the word 'Building' from the generally used term for this type of technology is probably to reflect the fact that the NEC4 suite of contracts is used for many projects which construct something which would not be called a 'building'. That approach is fine as the technology is as suitable for a civil engineering project as it is for a building project.

7.4.2 Definitions and documents

The Option includes five definitions that are unique to the Option and only apply within the Option. Taking these in number order the first one is the Information Execution Plan (clause X10.1(1)). Reading this definition immediately brings to mind the Accepted Programme (see clause 11.2(1) and Section 12.1) as in the way that this plan is either identified in the Contract Data or produced within a set time from the Contract Date if not so identified (see clause X10.4(1)). Further the Plan is also subject to revision and each accepted revision supersedes the previous accepted Plan.

The second is Project Information (clause X10.1(2)), which is said to be the information provided by the Contractor which is used to either create or change the Information Model, a term that is itself defined at clause X10.1(3). This third definition tells us that the Information Model is the combined electronic model of all the Project Information (clause X10.1(2)) and other similar information that has been provided by both the Client and the other Information Providers (Clause X10.1(5)). The term *Project Information* is limited to that information provided by the Contractor, which would include the Contractor's Subcontractors, but not information provided by the Client or by Others (see clause 11.2(12) and Section 5.11). This distinction within the term Project Information is important when considering liabilities (see Section 7.4.4). The form of the Information Model is as stated in the Information Model Requirements (clause X10.1(4)). As can be seen from the preceding sentences the third definition takes us to the fourth and fifth definitions. These three definitions are linked and interdependent.

---

[5] The 'How to' book also provided a similar Z Clause for the Professional Services Contract (NEC Panel 2013i) and the Engineering and Construction Subcontract (NEC Panel 2013j).

The Information Model Requirements (clause X10.1(4)) are set out in the Scope and identify the requirements for the creating the Information Model and how any changes thereto are to be made. There are several documents available from other BIM related systems that can be used to identify the type and nature of the requirements that will need to be included. Some guidance is available on page 72 of Volume 2 of the User Guide.

The fifth definition, Information Providers (clause X10.1(5)), identifies these as the people or organisations that are identified in the Information Model Requirements (clause X10.1(4)) and who contribute to the Information Model (clause X10.1(3)). When the Information Model Requirements are written for inclusion in the Scope, it may only be possible to identify the Providers by class, especially when some or all of the design will come from the Contractor (see Section 7.3). It must also be expected that the Information Model Requirements may require amending during the duration of a Project. As they are part of the Scope then an instruction from the Project Manager under clause 14.3 (see Section 5.3) will facilitate such a change.

7.4.3 Operating the process

The starting line for this process is the Information Execution Plan. As with the Accepted Programme, there is either a Plan identified in Contract Data Part Two, or a Plan is provided within the period stated in Contract Data Part One (see clause X10.4(1). The acceptance procedure and reasons for not accepting the submitted Plan are given in clause X10.4(2). The period for acceptance or nonacceptance is two weeks. Two reasons are given for non-acceptance, these being that the Plan does not comply to the Information Model Requirements or that it will not allow the Contractor to provide the Works (clause 20.1 and Section 7.2). Clause X10.4(3) allows for revised Plans within the period for reply after instructed by the Project Manager or when the Contractor elects to provide a revision.

The Contractor and the other Information Providers collaborate with each other in the manner stated in the Information Model Requirements (clause X10.2). Part of this collaboration is to provide early warnings (see clause 15 and Section 6.4), an obligation that is stated at clause X10.3 and that extends to the Project Manager as well as the Contractor. The early warning should consider anything which would affect the creation or use of the Information Model. Throughout this process, the Contractor is required to provide the Project Information set out in the form identified in the Information Model Requirements and in accordance with the Information Execution Plan (clause X10.4(4)).

Should any compensation event which occurs causes an alteration to the Information Execution Plan then, in accordance with clause X10.5, the Contractor includes the effects of the change in the quotation submitted pursuant to clause 62.3 (see Section 16.6.2).

7.4.4 Copyright and insurances

An important aspect of a procedure such as this one is to ensure that the liabilities, whether insurable or not, and the ownership of the copyright within the Project information and the Information Model are all fully set out.

As far as ownership of the copyright is concerned, clause X10.6 sets out that it is the Client who owns the Information Model (i.e. the integrated collection of all the Project Information and other similar information) and the Contractor's rights over the Project Information. This ownership by the Client can be varied by a statement setting out the alternative position in the Information Model Requirements. The same clause requires the Contractor to secure such ownership from its subcontractors and to provide the necessary documentation transferring that ownership.

The Client is responsible for any fault or error in the Information Model that has not been caused by a Defect in the Project Information, and for any fault in any information provided by Information providers other than the Contractor (and its Subcontractors) (clause X10.7(1)).

The Contractor is only liable for faults or errors in the Project Information if the Contractor failed to exercise the skill and care expected of a professionals providing such information (clause X10.7(2)). This is akin to the same duty of care as set out in clause X15.1 (see Section 11.3). Clause X10.7(3) requires that the Contractor maintains insurance against such a failure (professional indemnity insurance); again this is akin to the requirement for such insurances at clause X15.5. Should both Option X10 and X15 be in use, which would be common when the Contractor has any design liability, then it is suggested that a single insurance policy should cover both requirements.

7.5 Other matters

This section considers clauses 24, 25, 27, 28 and 29 (with the exception for clause 25.3, which, being an issue related to Time, is covered in Chapter 12).

7.5.1 People

People are the subject of clauses 24.1 and 24.2, each of which has its own specific purpose and requires a common-sense approach to its interpretation. To this end, Contract Data Part Two requires the Contractor when tendering to insert the name, job, responsibilities, relevant qualifications and experience of its key people. The ECC does not identify which job roles such key people should be identified for. If there are specific job holders to be identified, then the Client and/or Project Manager are free to specify such requirements either in the Scope or instructions to the tenderers. Indeed, many tenders contain such express requirements. Such information is then provided by the tendering Contractor.

Clause 24.1 requires the Contractor to either employ the key person identified in the Contract Data to do the job stated or a replacement person who has been accepted by the Project Manager. The Project Manager is given one reason for not accepting the nominated person, that being that the replacement person's relevant qualifications and experience are not as good as the person who is to be replaced.

It is possible that, on a strict reading of the clause, some replacements who are proposed by the Contractor could be rejected on the grounds that their relevant qualifications and experience are not as good as the named person. This rejection could be made notwithstanding that the replacement person's relevant qualifications and experience makes him or her a suitable person for the job in question. The Guidance Notes to the contract, although not part of the contract, make it clear that the reason given in the clause in question does not act to preclude the acceptance of a person who is suitable for the job.

When coupled with clause 10.1, it is suggested that the reason given enabling a proposed replacement person to be rejected is there to enable the Project Manager to reject people who are not suitably qualified or experienced for the job in question. The reason is not meant to be interpreted literally and applied in a manner that would lead to an unjust result.

Clause 24.2 permits the Project Manager to instruct the Contractor to remove an employee from the project. When giving this instruction, the Project Manager must give reasons. However, the permissible reasons are not specified in the contract. Turning again to the Guidance Notes users will find that three possible reasons are given relating to (i) security, (ii) health and safety (communicable diseases) and (iii) disorderly behaviour prejudicing the employing party's operations. From a Project Manager's perspective, one could easily add matters such as substance abuse, alcohol-related matters and lack of suitable certification to the list, all of which could be considered reasonable.

The problem with this provision is that, the way it is drafted (and this is confirmed to be the case by the Guidance Notes), the Project Manager could invoke the provision for any reason whatsoever. Something stupid and petty, such as being a supporter of the wrong football team in the eyes of the Project Manager, could be cited as the reason and the Contractor would have no option but to comply. It is suggested that the exercise of this power for such a petty reason as the example given, while permissible under the clause in isolation, would be contrary to the requirement to work in a spirit of mutual trust and cooperation as required by clause 10.1.

No sensible person could object to this provision being used to remove someone who is a danger to others and the works or who is detrimental to the satisfactory completion of the project. In practice, most project managers find that they do not need to use this power; a quiet word with the contractor results in the contractor taking the necessary action, if such action hasn't already put in motion, to remedy the position without the need for the formal instruction. That, surely, is how in practice it should be.

What would no doubt be detrimental to the project as a whole would be if either of the provisions discussed above were used in a way that would be contrary to the requirements of clause 10.1. The conclusion must therefore be that these provisions are only used sparingly and sensibly when the circumstances dictate, remembering always the overriding principle.

7.5.2 Co-ordination

The content of clauses 25 and 27 are in many ways similar in terms of the type of issues they address. They are about co-ordination with others and the basic responsibilities that would be expected in any contract. Taking these clauses in the order they appear in the ECC (excluding 25.3 as stated at the start of this section), we start with clause 25.1. This requires the Contractor to cooperate with Others (see Section 5.11) in both providing and obtaining information which they need to share in relation to the works. It also requires the Contractor to share the Working Areas with the Others as set out in the Scope. While the clause is nothing more than common sense, it emphasises again the importance of the Scope.

A further reference is made to the Scope in clause 25.2, this time in relation to the setting out of services and other items that the Contractor and Client are required to provide to each other. The accompanying provision that the Contractor pays whatever sum the Project Manager assesses should the Contractor fail to provide something may seem one-sided. The ECC is drafted this way so that, should the Client fail to provide something, then this is covered by a compensation event (see clause 60.1(3) and Section 15.3.3) providing that the requirement was shown on the Accepted Programme (see clause 31.2 and Section 12.2). Given the reference to the Scope in this clause, the compiler of that document should give careful thought to the information to be given. In particular, where the Client is to provide something, it is worth considering when the Client will be able to make such provision and how much notice may be required before the item can be delivered to the Working Areas. Careful thought to this area can minimise the risk of the Client being put under obligations that it cannot comply with, by dates it cannot achieve, being included in the programmes submitted by the Contractor.

Clause 27.1 is a short clause requiring the Contractor to obtain approval of its design from Others as may be necessary. This requirement is so close to the provisions of clause 25.1 that it must be questioned why they were not set out together. An example of the requirement in clause 27.1 is where the Contractor is responsible for designing the lift shaft and lift beam and where the lift will be installed by Others under a direct contract with the Client. Clients and Project Managers must ensure that the particulars they include in the Scope about such work by Others includes sufficient details to indicate to the Contractor that, in this example, it needs to obtain the approval of the lift contractor for the design of the shaft and beam. As it is the Client who will be responsible for the actions, inactions or defaults of Others, it is in the Client's interests to ensure that such information is provided.

The requirements of clause 27.2 are self-explanatory. Users should note that the requirement is not limited to items in the Working Areas and will therefore apply equally to things in stores, yards, workshops, etc. belonging to the Contractor, its subcontractors and its suppliers, regardless of the location of those facilities.

The clear requirement at clause 27.3, that the Contractor obeys any instruction which is in accordance with the contract and given by the Project Manager or the Supervisor, is extremely important. The Contractor is not given any right of objection in this clause. In the event that the Contractor considers that an instruction it has received is not in accordance with the contract, then its only options are to either use clause 10.1 (and discuss the issue with the Project Manager or Supervisor and the Client if necessary) or to refer to the Adjudicator. This clause is not linked in any way to the compensation event procedure at core clause 6. The Contractor obeys the instruction and gets on with the work as is necessary for the satisfactory completion of the project. The compensation event procedure is carried out in parallel with the implementation of any change; this procedure does not act to delay the works, contrary to a totally incorrect myth put forward by some people. To allow such a situation to arise would create so much delay it is inconceivable that the drafters would ever have contemplated such a provision.

The subject of clause 27.4 is health and safety and, again, reference is made to the Scope. Users should not ignore the requirements of any statute in respect of health and safety. Certainly in the United Kingdom, a party cannot escape the requirements of the law in respect of such matters. Even if they are not mentioned in the Scope, then the statutory requirements will apply by means of the law. That creates a position in the United Kingdom where the health and safety provisions that need to be considered and included in the Scope are those which are additional to the law. This is important, as the industry sector in which the Client is engaged often properly requires anyone working within that sector to comply with industry-specific health and safety requirements. It is these requirements that need to be set out in the Scope.

7.5.3 Assignment

Assignment is where a right or benefit, but not an obligation, is transferred to a third party by the party to the contract who has that right under the contract. As an example one of the more common rights that is transferred, especially by subcontractors and some suppliers, is the right to be paid, which is assigned to an invoice factoring company.

Under clause 28.1 neither the Client nor the Contractor can assign any right or benefit without notifying the other of their intentions. The assignment does not require the acceptance of the other party. The only restriction applies to the Client, who cannot assign to any party who does not intend to act in a spirit of mutual trust and cooperation (see clause 10.2 and Section 4.2.3).

7.5.4 Disclosure

The title of clause 29 could easily be replaced with 'Confidentiality' as that is the prime aim of the clause. There is an equal obligation in clause 29.1 on both the Client and the Contractor to not disclose any information they have obtained in relation to the works unless it is necessary for that Party to carry out their duties under the contract. This will allow either Party to disclose information when applying for consents, licenses, permits and other such things or when procuring subcontractors, suppliers or work from statutory undertakers as examples. Anything not required to be disclosed in order to facilitate such matters must be kept confidential.

Related to the above, clause 29.2 restricts the Contractor to only publicising the works with the Client's agreement. This will require a communication from the Contractor to the Client detailing what publicity it intends to issue and then a reply agreeing to that publicity. Such permission would have to be sought for each publicity event or release. There is nothing unusual in this type of clause.

7.6 Practical issues

7.6.1 Importance of the Scope

What employers and project managers should learn from core clause 2 is the importance of the Scope. This document is referred to in the context of setting out and establishing the Contractor's obligations in so many respects; it should quickly become clear that accurate and complete preparation of the Scope is essential if the price is going to reflect what the Client wants to have built. If the Client is open to the idea (most are not) that the price they accept as the tender will have no correlation to that which they pay (save that the latter will be far higher than the former), then the Scope is not so important as all the vital information that is missing can be added throughout the duration of the project. This, of course, will create numerous compensation events that will give rise to an increase in price and an extension of the time to complete the project.

The only way to control the price and the time reasonably close to the tendered price and time required is to ensure that the Scope is accurate before the tender is issued. I say 'reasonably close' as I have not seen a project that did not have any change or met anyone who has encountered such an occurrence and I doubt that I ever will. That is not a criticism but a practical observation that originates from the unique nature of most projects; construction is an industry that rarely builds the same thing twice (with the exception of the mass housing market) and therefore, most projects are prototypes. This inevitably leads to some amount of change, of which participants should not be afraid.

7.6.2 Forecasting the Defined Cost

It appears that in practice, the preparation and submission of the forecast of total Defined Cost required by clause 20.4 (Main Options C, D, E and F only) is something that is not done as required. Providing that the other provisions of the contract are being complied with and, in particular, the inclusion of resources in the Accepted Programme as required by the penultimate bullet point of clause 31.2 (see Section 12.2), then the preparation of this forecast should be straightforward. This forecast in simple terms is nothing different to the cost to complete exercise regularly prepared by contractors for their own internal use. All this clause requires in practice is that the forecast of the outturn cost is submitted to the Project Manager, so that the Client can be kept appraised of the forecast cost of the project.

7.6.3 Identifying the design responsibility

The implementation of clause 21.1 by reference to design within the Scope is, in practice, carried out in two ways: one good and one bad. The good way is where a single clear section or heading is included in the Scope, which clearly identifies and sets out all the matters which the Contractor is to design. By detailing the requirement in this way, there can be no doubt or question about the extent of the Contractor's liability. The bad way is where the specification includes, as part of the general text, individual statements about portions of the works that the Contractor is required to design. This approach relies on the Contractor finding every such entry for himself, thereby creating an amount of doubt and potentially mistrust. Mistrust is not helpful, as it is contrary to the requirements of clause 10.2.

7.6.4 Common amendments by lawyers

Core clause 2 contains two of the provisions that appear in practice to attract the most attention from lawyers when it comes to making amendments to the ECC. It has become rare to see a contract based on the ECC in which clauses 21 and 22 have not been amended by the Client's lawyers. Clause 21 is routinely expanded to include matters such as the level of care to be taken, that approval or acceptance of the design does not relieve the Contractor of its responsibility[6] and expansions of other matters already covered. The very simple copyright-type provision at clause 22 is regularly replaced with considerably longer and more involved provisions of that type and/or intellectual property rights provisions, both written in the legalistic-type language that the drafters of the ECC so studiously avoided. While most users accept such changes and they are legally correct and safe, it doesn't make them right for a contract that is trying to engender a different attitude to contracting.

7.6.5 Free issue plant and materials

Having criticised the common changes made by lawyers to clauses 21 and 22, I am now going to suggest a minor addition to clause 25.2, which employers who provide free issue materials and plant may find to be of considerable benefit and which may assist in removing an area of doubt from users who operate in such sectors. Clause 25.2, as covered in Section 7.5.2, provides for the Project Manager to assess the cost to the Client of the Contractor not providing any of the services or other things that the Contractor is to provide. A compensation event (Section 15.3.3) acts to compensate the Contractor should the Client fail to comply with any similar obligation that the Client may have. The other things that the Client often provides under this clause, in some industry sectors, include Materials and/or Plant. These are passed into the Contractor's care for incorporation into the works. Despite the best efforts of the Contractor to securely store such items, it is sadly not unknown for them to be stolen or damaged before they are incorporated into the works. Replacements are then provided by the Client. This clause does not allow the Project Manager to assess the cost of such replacements in the assessment under this clause. By adding the words 'or by not taking care of anything provided to it' in the third line between 'provide' and 'is', the Project Manager can address the replacement cost through the payment mechanism.[7]

---

[6] This is despite this point being adequately covered at clause 14.1.
[7] See clause 50.3 and Section 13.4.2 for details of how such amounts are covered in the amount due.

Purists will say that such a position is provided by the ECC through clause 81.1, which requires the Contractor to insure all the liabilities identified therein. As the risks carried by the Client (as listed at clause 80.1) include Plant and Materials supplied by the Client to the Contractor until such items have been received by the Contractor, then the analysis that the liability arises through clause 81.1 is correct.[8] However, employers may find that my suggested amendment in practice is neater and simpler from their perspective.

---

[8] See Chapter 11 for more details on the application of the insurance provisions.

Chapter 8
Subcontracting

8.1 Introduction

Subcontracting is a way of life in the construction industry today. Many main contractors do not employ any direct labour, and those that do only employ the trades that they have a constant demand for and can therefore keep busy without the risk of standing time. As a result, having some or all of the work carried out by another organisation is a common and accepted practice and is referred to as subcontracting. Different contractual arrangements manage this process in different ways, so it is important to know the specific requirements under the contract in use.

It must be recognised that there are no provisions in the ECC for nominated or named subcontractors. If subcontracting is to take place, which in all likelihood it will, then it will be at the Contractor's prerogative and not the Client's or the Project Manager's. Accordingly, if the Client wants to influence which subcontractor completes an element of the works or even select a subcontractor in this regard, it will need to either be very restrictive in the way the Scope is written or amend the Contract in such a way that it maintains the overall intent.

8.2 Definition of a Subcontractor

The definition of a Subcontractor at clause 11.2(19) identifies three qualifying classes of person or organisations who constitute a subcontractor under the ECC. The first of these is someone who constructs or installs part of the works, a definition that does not pose any problems as this class is easily recognisable as being the common definition of a subcontractor. It should be noted that by restricting this class to someone involved in only a part of the works, the ECC prevents the Contractor from subletting the whole of the works to a single person or organisation.

The second class covers designers (see Section 5.13) who design part or all of the Works, with the exception of any design done by a supplier of Plant or Materials (as defined at clause 11.2(14); see Section 14.2.3).

The third class is someone who provides a service necessary to Provide the Works (being the phrase used to describe the Contractor's principle obligation (see clause 11.2(15) and Section 7.2)). The key word in the definition of this third class is *service*, which acts to include in this class anyone not covered by the first class but who provides some type of service. That said, the definition of this third class includes two exemptions that serve to remove anyone providing the hire of Equipment or providing the supply of people who will be paid for on a time basis. The first exemption clearly refers to hire companies and the second one to labour agencies.

8.3 The core clauses

Firstly, clause 26.1 makes it clear that the Contractor remains responsible for Providing the Works when it subcontracts any part of it. The clause also states that the contract applies as if any subcontractor's employees and equipment were that of the Contractor's. This is the situation you would expect; the Contractor cannot sublet its responsibilities or obligations to the Client under the Contract between the two of them to a third party.

The remainder of clause 26 deals with the identification of potential subcontractors to the Project Manager, and the acceptance of proposed subcontract documents, including terms and conditions. Clause 26.2 requires the Contractor to submit the name of each proposed subcontractor to the Project Manager for acceptance. In reply to such a submission, the Project Manager is given one reason for not accepting the proposed subcontractor: that the appointment of the proposed subcontractor will not allow the Contractor to Provide the Works. It is difficult to see what factors might exist that would allow the Project Manager to use this reason. The use of any other reason would potentially trigger a compensation event (see clause 60.1(9) and Section 15.3.9), so Project Managers must tread carefully when considering such matters.

By the final sentence of clause 26.2 and its two attendant bullet points, the Contractor is prohibited from appointing a proposed subcontractor until the Project Manager has accepted that subcontractor and the proposed subcontract documents to the extent that such acceptance is required by the contract. Contractors should be aware of this prohibition as, at clause 91.2 (see Section 18.2.3), the appointment of a subcontractor for substantial work before the Project Manager has accepted the subcontractor is Reason 13 of the reasons allowing the Client to terminate the Contract. The only point for debate in respect of this reason is: what is substantial work as opposed to unsubstantial work?

Clause 26.3 follows a similar theme and, in practice, runs parallel to clause 26.2 in that the Contractor is required to submit the proposed subcontract documents (less any pricing information) for each subcontractor unless either an unamended NEC contract is proposed or the Project Manager has agreed that no such submission is necessary. The Project Manager is given two reasons for withholding acceptance of the subcontract documents. The first of these reasons is the same as that allowed under clause 26.2, and the second is that the conditions do not include a statement requiring the parties to work in a spirit of mutual trust and cooperation.

8.4 Provisions in the Main Options

Main Options C, D, E and F all include clause 26.4, which, following clauses 26.2 and 26.3, requires the Contractor to submit the pricing information as part of the submission of the proposed subcontract documents required by clause 26.3 (see Section 8.3). The only exception to this requirement is where the Project Manager has agreed that no submission is necessary.

8.5 Practical issues

8.5.1 Shortcuts to the process

While necessary, the requirements of clauses 26.2 and 26.3, and clause 26.4 where one of the Main Options C, D, E or F is in use, might be considered to be bureaucratic. Users of the contract should not forget that they are required to work in a spirit of mutual trust and cooperation. By using this principle, the Contractor and the Project Manager can manage these requirements between them. They must discuss how the Contractor intends to sublet portions of the works and agree what submissions shall be made to the Project Manager in advance in respect of the requirements of the two or three subclauses that form part of their particular contract.

In practice, the Contractor will have established some form of procurement schedule for the packages of work it intends to sublet. In addition, the requirements at clause 31.2 (see Section 12.2.2), which set out the contents of the Accepted Programme include the procedures set out in the Contract. By linking these two together, it is arguable that the Contractor's procurement schedule for subcontractors should form part of the family of documents that make up the Accepted Programme in order to comply with this requirement. I say arguable because it is likely that the Accepted Programme requires less information than most Contractors would probably put on their procurement schedule.

However, following the spirit of mutual trust and cooperation, there is no reason for the Contractor not to share these details (apart from any commercially sensitive information) with the Project Manager. The Main Option in use will of course determine how sensitive information relating to, say, tender allowances and the subcontract tender prices actually is. For Main Options A and B financial information would be sensitive; for Main Options C–F the amount to be paid to each subcontractor would have to be reported to the Project Manager in any event as clause 26.4 (see Section 8.4) requires the pricing information to be provided.

By sharing not only the procurement schedule but also the names of the proposed subcontract tender lists with the Project Manager as early as possible in the process, the Contractor can create an environment of trust and allow any potential problems to be identified early; hence, facilitating the management of such matters without delay to the progress of the works on site.

8.5.2 Accepting the supply chain

A common complaint from employers and project managers is that, although they have to be asked to accept each subcontractor appointed by the contractor, the process stops there and does not provide for situations where the subcontractors sublet work to other contracting bodies (sub-subcontractors). It is not unknown for there to be three or even four layers of providers below the accepted subcontractors, each of which would constitute a subcontractor in their own right had they been appointed directly by the contractor. This leads to positions where the employer finds that a sub-subcontractor to whom the project manager would have withheld acceptance is actually working on the project providing part of the works. In the event that employers or project managers are concerned about the prospect of this happening, the only solution available in practice is to use Option Z to add provisions to clause 26 requiring that all sub-subcontractors must be subject to the same acceptance procedure as for subcontractors.

8.6 Options for forms of subcontract in the NEC4 family

In respect of subcontractors who will construct or install part of the works rather than designers, the family includes two options.

The first choice is the NEC4 Engineering and Construction Subcontract (ECS) (NEC Panel 2017i). Anyone who is familiar with the ECC will recognise the contents of the ECS instantly. The ECS is, on the face of it, identical to the ECC once the names of the relevant personalities have been changed (in the ECC substitute Client, Project Manager and Supervisor with Contractor, and Contractor with Subcontractor) and main Option F has been removed.

In the finer detail there are in fact some changes in periods, for example, in clause 62.3 regarding the submission of and reply to quotations and clause 51 regarding payment. Note, however, that the periods for payment in Secondary Option Y(UK)2 in the ECS are identical to those in the ECC version of the same secondary option, resulting in identical payment timetables. Main Contractors will no doubt wish to use the facility in the Subcontract Data Part One to amend these periods in order to allow them a few days to process the payment from the Client and clear funds in the bank.

The other option provided by the NEC4 family is the Engineering and Construction Short Subcontract (ECSS) (NEC Panel 2017j). The ECSS is said to be for subcontracts 'which do not require sophisticated management techniques, comprise straightforward work and impose only low risks on both the contractor and the subcontractor'. The ECSS simplifies most, but not all, of the processes contained within the ECS and so becomes more appropriate for subcontracting organisations that do not possess sufficient management resources or skills to carry out many of these requirements. In particular, the ECSS simplifies programming to the requirements set out in the Scope and simplifies the management of defects (among other matters).

It does, however, in the Scope (item 1), refer to design by the Subcontractor, but does not include clauses equivalent to secondary Option X15 (see Section 11.3) to reduce the Subcontractor's liability for reasonable skill and care. Neither does it have any entry prompting the need for professional indemnity insurance from the Subcontractor. In the event that a Subcontractor appointed using the ECSS is going to be required to carry out design, then both the Contractor and that Subcontractor need to consider an amendment to provide both parties with the necessary protection associated with design liability. There is a space at the bottom of page 5 of the Subcontract Data for such additional conditions to be inserted.

What the ECSS does not do to any great extent is simplify the procedures related to compensation events, although the content is different. The Subcontractor is required to notify compensation events within three weeks of becoming aware of the event and is subject to the same condition precedent as the Contractor under the ECC should it fail to so notify. What is not included is the deemed acceptance procedure present in the ECS should the Contractor fail to reply. The combination of these two would appear to make the compensation event procedures more onerous on the Subcontractor under the ECSS than it is under the ECS.

The other major difference between the ECSS and the ECS is the presentation of the Subcontract Data. In the ECSS this is found at the front and contains clear sections enabling the parties to fill in all the relevant details including a Price List, Scope and Site Information, together with all the other details necessary to make the ECSS work. Guidance is given where needed in each section of the Subcontract Data to assist the parties in completing the relevant information, which, if followed, should not present any difficulties.

Regarding designers, clause 26.3 of the ECC refers to the proposed subcontract conditions as being an NEC contract. This wide definition includes reference to the NEC4 Professional Services Contract (PSC) (NEC Panel 2017k) or the NEC4 Professional Services Short Contract (PSSC) (NEC Panel 2017l), which, as part of the NEC4 family, provide a perfect choice of vehicle for both the Contractor and the Client under which to sublet the design.

Users of the ECC should have no problem understanding the PSC or the PSSC, although the terminology and names are different; for example, the designer is called the Consultant (as opposed to Contractor or Subcontractor). Further the payment provisions flowing from the Main Options are different but recognisable.

Chapter 9
Quality Management

9.1 Introduction

This chapter covers core clause 4 where the contractual procedures for quality management, testing, inspections and the management of defects are set out. The procedure in core clause 4 is a skeleton that establishes responsibility for the various actions required, liability in the circumstances that can be encountered and the procedures that are required to be followed. It is within this core clause that the majority of the Supervisor's actions are included.[1] The procedures set out in this clause require that relevant information is included in the Scope (see Chapter 22), without which aspects of the procedure will not work, as there are no default provisions.

The core clause also introduces two terms that are not used in other popular standard forms of contract within the construction industry. These both relate to defects and will be explained. The contents of the clause can be considered in three distinct parts.

9.2 Quality management system

The Contractor is required by clause 40.1 to operate a quality management system. In keeping with the skeletal nature of the clauses in core clause 4, the requirements for what is required both in and from the quality management system are required to be set out in the Scope (see Section 22.2.19). Nonetheless many contractors already operate such a system for which they hold accreditation to one or more of the recognised standards for such systems. So whilst the Scope could be very prescriptive as to the requirements for such a system, it could also be very simple by just requiring a system that was accredited to a recognised standard.

Two separate submissions are to be made by the Contractor as stated in clause 40.2, including the requirement that they are submitted within the period stated in Contract Data Part One after the Contract Date (see clause 11.2(4) and Section 12.3). The first requirement is for a quality policy statement. What this might consist of is not stated but could easily be seen as the equivalent of the general statement committing to quality that many contractors have and which can be found on many a website. This statement

---

[1] The Supervisor's only other actions are all contained in core clause 7; see Chapter 10.

is subject to acceptance by the Project Manager, who is given a single reason for not accepting the statement, being that it will not allow the Contractor to Provide the Works (see definition at clause 11.2(15) and Section 7.2).

The second statement is referred to as being a quality plan. Whilst, as for the quality policy statement, there is no statement as to what might be required, the requirements within the clauses suggest that the plan will be more comprehensive and project specific. The plan is subject to the Project Manager's acceptance in the same way as for the statement. In addition, the last part of clause 40.2 allows the Contractor to change the quality plan, subject to submission to the Project Manager for acceptance. Further, clause 40.3 allows the Project Manager to instruct the Contractor to correct a failure to comply with the quality plan. Such an instruction is confirmed by clause 40.3 as not being a compensation event. These three specific provisions in relation to the quality plan provide enough confirmation of the importance attached to this document.

9.3 Tests and inspections

Clauses 41 and 42 concern tests and inspections, both on and off the Site, but only as required by either the Scope or the applicable law (see clause 41.1). The Scope should also clearly define whether any materials, samples or facilities are required in connection with any of the required tests and which of the Parties provides those items (see clause 41.2). The practical effect of these two short subclauses is to move the emphasis away from the terms and conditions and into the project-specific information set out in the Scope.

In practice, it is necessary to include a detailed test and inspection plan within the Scope. This could either be as a separate and clearly identified collection of all such tests and inspections in one plan or by including these requirements throughout the specification for the various elements of the project. It is suggested that the former contributes to establishing the spirit of mutual trust and cooperation as it brings early clarity to the requirements. This plan should also clearly set out which party is responsible for providing anything consumable or any other facility for use in a test or inspection. For example, in order to test a water-retaining structure, the requirement is usually to fill it with water and measure any losses as well as carrying out visual inspections for obvious leaks. The ECC requires that the Scope identifies which party is responsible for providing the water for such a test.[2] Where the Contractor is providing Scope in respect of its design (see Section 21.4.3), the required tests or inspections should form part of that Scope if not previously stated in the Scope provided by the Client.

In the event that a test or inspection has not been included in the Scope, then the Project Manager can always issue an instruction (as allowed by clause 14.3; see Section 5.3) adding the required test or inspection. This would constitute a compensation event with a resulting change to the Prices and possibly the Completion Date or a Key Date (clause 60.1(1) and Section 15.3.1). Users should note that it is only the Project Manager

---

[2] The author once worked on a project that required eight million gallons of water for the test to each cell of a concrete reservoir. An important factor in the pricing is which party is going to pay for that amount of water.

who can issue such an instruction, but it is the Supervisor who carries out and monitors tests and inspections. Such circumstances require them to liaise with each other. In the event that the Scope does not specify which party is to provide something required for a test or inspection, then the Project Manager will need to issue an instruction resolving this ambiguity within the documents (as required by clause 17.1; see Section 6.6.5). The assessment of any resulting compensation event would be determined by reference to clause 63.10 (see Section 17.5.2).

The carrying out of tests and inspections by either the Supervisor or the Contractor should be notified to the other in reasonable time before the test or inspection starts. There is no requirement for the party carrying out a test or inspection to wait for the other to attend, providing sufficient notice has been given. The tester carries out the necessary actions at the prescribed time, records the results and gives a copy to the other. If this is carried out in the spirit of mutual trust and cooperation required by clause 10.2, then that should be the end of the matter unless the test or inspection reveals a defect. One quirk of this procedure is that clause 41.3 states that the Supervisor may watch any test carried out by the Contractor (see final sentence of the clause). What I find strange is that this provision does not also refer to inspections, and why isn't the Contractor given the equal right to watch any test, and inspection for that matter, of the Supervisor? Surely the principle at clause 10.2 would suggest that both the Supervisor and the Contractor should be able to watch all tests and inspections carried out by each other or any outside agency if so specified by the Scope.

There is a requirement for the Supervisor to do tests and inspections without causing unnecessary delay to the works or a payment. One can see that there could be a difference of opinion over what is meant by 'unnecessary delay'. However, use of the Accepted Programme, the proper giving of the notices required as set out in the Scope and the spirit of mutual trust and cooperation should go a long way towards avoiding such differences.

When any test or inspection is required of Plant or Materials before the goods are brought to the Working Areas, the Supervisor has a vital role to play. Regardless of whether the Supervisor personally carries out the test or inspection or whether it is done by the Contractor or one of the Contractor's suppliers or subcontractors, clause 42.1 is quite clear that such Plant or Materials cannot be brought to the Working Areas until the Supervisor has notified the Contractor that the test or inspection has been passed. If the Supervisor does not issue the necessary notice to the Contractor that the off-site test or inspection has been passed, the Plant or Materials cannot be delivered and a delay could result. The Supervisor must therefore be aware of such requirements and make sure that the required notices are issued in a timely manner. It must be hoped that mutual trust and cooperation will mean that the Contractor reminds the Supervisor of this requirement before any delay is allowed to occur. (Recall, however, that the Contractor is not obliged by clause 42.1 to provide this reminder and cannot be held at fault under that clause if it does not.) On the other hand, clause 15.1 requires the Contractor to issue an early warning of the potential delay as soon as it becomes aware of the possibility that a delay may occur. There is a potential conflict of provisions here, which hopefully the spirit of clause 10.2 should avoid.

9.4 What is a Defect?

The third part of core clause 4 relates to Defects, a term that is defined at clause 11.2(6). The definition uses two bullet points to set out what constitutes a Defect. The first one simply refers, as you would expect, to something that does not comply with the Scope. The second bullet point concerns works designed by the Contractor; here the benchmark is the applicable law[3] or the design by the Contractor that the Project Manager has accepted (pursuant to clause 21.2; see Section 7.3).

It is suggested that this definition could give rise to conflicts within its interpretation. The first of these potential conflicts is within the first bullet point. When the Contractor has design responsibility, the Scope is within the documents referred to at Contract Data Part One (provided by the Client) and Contract Data Part Two (provided by the Contractor). Should there be an undetected conflict between these two sources of Scope, an item of work could be defective under one source but not defective under the other. In such circumstances, the Project Manager would have to issue an instruction (as required by clause 17.1; see Section 6.6.5) and the assessment of any resulting compensation event would be determined by reference to clause 63.10 (see Section 17.5.2).

The second and third of these potential conflicts are between the first and second bullet points. Here the Scope could conflict with either the applicable law or the Contractor's design which the Project Manager has accepted. It is suggested that the remedy is the same as for the first source of conflict in the preceding paragraph.

9.5 The Defect procedure

Under clause 43.1, the Supervisor has the power to issue an instruction to the Contractor to search for a Defect. This is the only situation in which the Supervisor has power to issue an instruction. The actions that the Supervisor can instruct the Contractor to undertake are set out in the clause. This type of provision is common in construction contracts and users should not find it difficult to follow. What are not set out in core clause 4 are the financial consequences of such an instruction. It is, of course, necessary for givers and receivers of such instructions to understand what the implications may be. As users would probably expect if no defect is found, a compensation event (clause 60.1(10); see Section 15.3.10) would arise, subject to an exception regarding timing. Alternatively, if a defect is found, then the compensation event would not arise.

With regard to any Defect that either the Supervisor or the Contractor finds, then they are both obliged to immediately notify the other. A quirk of the ECC that should be remembered here is that clause 13.7 requires every notification given under the contract to be given separately from other communications. Technically, this can be interpreted such that every Defect found should have its own separate notification, i.e. one Defect per notice (see Section 9.8.1).

The Contractor is required by clause 44.1 to correct every Defect whether it is notified or not. In practice, the Contractor would not notify most of the defects it found

---

[3] It is suggested that this primarily picks up the building codes in the relevant jurisdiction.

while constructing the Works; the Contractor would simply rectify them as part of the normal working procedure that would enable it to satisfy its obligation to Provide the Works (see Section 7.2) by the Completion Date.[4] It is suggested that the Contractor and the Supervisor may start to each identify and notify the other of defects as the state of Completion becomes closer, in order to allow a smooth handover of the project to the Client. The notification procedure itself will definitely happen immediately before and after Completion.

The requirement to notify defects is set out at clause 43.2. The relevance of this notification becomes apparent at clause 44.2 where an obligation arises for the Contractor to correct a notified Defect within the 'defect correction period'. This period commences either at Completion for any Defect notified before that time or at the time of notification for defects after Completion, subject to the proviso in clause 44.4 (see below). Defects are notified under this procedure until the 'defects date'.

Before continuing with the procedure, it is necessary to consider the two terms that users who are new to the ECC will not have encountered before. These terms are 'defects date' and 'defects correction period'. The *defects date* is a period of time stated in Contract Data Part One that establishes the period after Completion for which the defects procedure will apply. In other forms of contract and the industry in general, this is often referred to by terms such as 'maintenance period', 'defects liability period' or 'rectification period'. The use of the word *date* when *period* would be better can cause confusion until the intent is understood. The ECC does not contain any default period for this mechanism should the Client not complete the relevant entry in the Contract Data. It is therefore essential that this entry is completed; without it there is no way of making this requirement work, which would be to the disadvantage of both Parties. The 'defect correction period' is a separate period of time during which the Contractor is obliged to correct each defect after it has been notified, with the period commencing as stated in the previous paragraph.[5]

Under clause 44.4, the Project Manager is required to arrange for access to any part of the Works required by the Contractor to correct a Defect once the area in question has been taken over by the Client (see clause 35 and Section 12.6.3). The start of the defect correction period is delayed until such time that access has been arranged and provided but cannot be delayed beyond the defects date.

Figure 9.1 is a simple depiction of how the various terms that have been referred to in this section, both defined and undefined, fit together. Three defects are shown in this figure. The first exists at Completion, so its defect correction period begins at Completion. The second is notified some time after Completion but a long time before the defects date. The third is notified only a short time before the defects date, resulting in the defect correction period expiring after the defects date. This results in the timing of the issue of the Defects Certificate (see Section 9.6) being delayed until the end of this last defect correction period. It has not been shown in the figure, but should the Project Manager not be able to arrange access for the Contractor as required by clause 44.4 then the start of any defect correction period to which such a condition applies would be delayed until

---
[4] See clause 11.2(3) and Section 12.5.2. This could be a Sectional Completion Date if Option X5 was in use or a Key Date (see Section 12.5.3).
[5] See Section 9.8 for practical issues connected with these two terms.

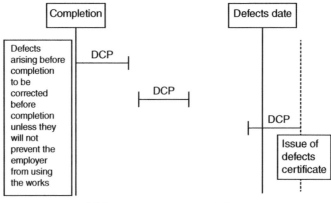

Figure 9.1 Illustration of the terms in the Defects procedure.

access is provided by the Client (but no later than the defects date itself). If access to correct a defect is not provided by the defects date, then the Contractor does not correct that defect (see clauses 46.1 and 46.2 and Section 9.7).

This defects procedure only applies to defects notified by the defects date. Any other defects which have not been found or notified are expressly excluded from the procedure by clause 44.3. The common law position regarding the liability for latent defects would apply to such defects.

9.6 *The Defects Certificate*

The defects regime makes the Supervisor responsible for the issue of one of the most important certificates required by the ECC, i.e. the Defects Certificate. It is the Defects Certificate that brings most of the Contractor's obligations under the ECC to an end and determines the timing of the last payment to the Contractor. That it is the Supervisor rather than the Project Manager who issues this important certificate distinguishes the ECC from other forms of contract, where the norm is that such an important certificate would be issued by the lead administrator for the Client. Supervisors should familiarise themselves with the requirements of this certificate and any liabilities that its issue may bring under the terms of their appointment.

The contents of the Defects Certificate, in principle, are set out in the definition of the term at clause 11.2(7) with which the Supervisor should be acquainted. In simple terms, the Supervisor must issue the Defects Certificate at the later of the defects date and the last defect correction period (see clause 44.3 and Section 9.5). The certificate should state either that the Contractor has corrected all notified defects or list all those that the Contractor has not corrected by that date. Although not stated in the definition, it is suggested that it would be helpful if, when listing any defects not corrected, the Supervisor could record whether or not the Contractor was afforded the necessary access to correct each defect listed by the defects date. This piece of information is part of the determination of how the adjustment for each uncorrected defect is assessed (see clauses 46.1 and 46.2 and Section 9.7).

9.7 Uncorrected Defects

Where the Defects Certificate lists any defects which have not been corrected at the time of issuing that certificate, clause 46 sets out that there shall be a financial adjustment to compensate the Client. How the adjustment is assessed depends on whether the Contractor was given access to correct the defect or not. If the Contractor was given access but did not correct the defect, then the Project Manager assesses the cost of other people correcting the defect (see clause 46.1). If the Client did not give the Contractor access to correct the defect, then the Project Manager assesses the cost to the Contractor of correcting the defect (see clause 46.2). The assessment in the second of these situations is likely to be lower than in the first. The clause states that, in whichever case applies, the Contractor pays the amount assessed.[6] There is no requirement that the Client has incurred the cost so paid or that the Client corrects the defect; the Client could choose to live with the defect and keep the payment received.

The ECC also provides a mechanism at clause 45.1 where the parties can agree that a defect should not be corrected in return for the Client receiving a reduction in the Prices and possibly (although it is difficult to see how this could apply) an earlier Completion Date. The clause actually refers to either the Contractor or the Project Manager proposing to the other that such a condition could be implemented upon successful agreement of the savings to be given. A project manager would be well advised to consult with the client and ensure that such a solution is acceptable before making or accepting such a proposal, unless the Project Manager's appointment gives *carte blanche* over such matters. If the proposal is acceptable, then the Contractor submits a quotation (see clause 45.2). If the quotation is accepted, the Prices are revised and the Project Manager gives an instruction changing the Scope such that the defect is no longer a defect. This provision is similar to provisions contained in other contracts and one that makes commercial sense in certain circumstances.

9.8 Practical issues

9.8.1 Communications policy

Section 9.5 above refers to the requirement to notify each Defect separately from every other Defect in order to satisfy the requirements of clause 13.7. In practical terms, this would become very cumbersome and is frankly unnecessary. Common sense and mutual trust and cooperation suggest that an appropriate way of notifying and recording Defects

---

[6] The amount to be paid is deducted from the amount due under the provisions of clause 50.3; see Section 13.4.2.

should be established for each project, using mechanisms that are administratively workable by those responsible. This is what the vast majority of users do in practice. The shared use of one of the computer-based defects management packages is becoming an increasingly common way of satisfying this requirement. This usually has the side benefit of improving the spirit of mutual trust and cooperation between the Supervisor, the Contractor and the Contractor's Subcontractors, who should also be invited to share the software. Supervisors may consider recommending to the Client and/or Project Manager that the system to be used should be set out in the Scope.

9.8.2 Defect correction period

In practice, the length of the defect correction period often causes parties a problem. There is no default length of time for this period. The worked example of a completed Contract Data in Appendix 1 of Volume 2 of the User Guides inserts '3 weeks' as the period; this is only a suggested period. It is suggested that there is little point in having a period which is too short. However, is it also impractical for the Client to have a period that is too long, as the purpose of the provision is to return the Works to a defect-free state so that the Client can enjoy the project that has been paid for. The Client therefore has a difficult balance to obtain between giving the Contractor a reasonable time to correct defects and the delay in achieving such a condition. Another practical issue which causes problems is where the time taken to obtain parts is longer than the defect correction period. Clients can manage this in one of two ways: they can have a pool of spares provided through the contract for their own use and allow the Contractor access to this pool to correct defects with (and then replace the parts used later), or they can set the defect correction period for plant requiring spares at a sufficiently long period to allow them to be obtained.

The problem with defects corrections periods being too short can of course always be overcome by the parties agreeing to a longer time than set out in the Contract Data for the correction of defects affected by such circumstances. The parties to a contract are free to make whatever agreements they like (subject to the applicable law), so there is nothing to prevent the spirit of mutual trust and cooperation acting to relax the requirements when it suits those affected to do so.

9.8.3 Priority system for defects correction

Another way of providing flexibility in the correction of defects in practice is to introduce a priority system where the length of the defect correction period varies depending on the urgency that is required to repair the defect. For example, some projects in the UK use a system where defects are given a priority rating between, say, 1 and 4. If a defect occurs between Completion and the defects date, it is allocated a priority as part of the notice issued notifying that defect. If the defect is such that it needs to be corrected urgently as it is seriously impacting on the day to day use of the facility, then it is given a priority 1 status. This could require that the defect is either temporarily or permanently repaired within, say, one day in order to avoid unnecessary disruption. Any temporary repair can then be permanently repaired at a time suitable and convenient for the facility. Defects of less impact on the facility could be given priorities of between 2 and 4, each of which would have a different defect correction period.

Such a system can be adopted for any project regardless of its use in order to allow a variable timetable for the repairing of defects which is dependent on the criticality of each defect to the operation of the facility. The only proviso is that the prioritisation of defects needs to either have some pre-determined guidelines or needs to be applied carefully with an eye on the principle of mutual trust and cooperation.

9.8.4 Multiple defects dates

A variance on the different ways of using the defects correction system is to have more than one defects date. It has been common in practice under other contracts to see two or three different periods during which the contractors remains responsible for correcting defects under the contractual system. This is usually done by prescribing different periods for different elements of a project. For example, the defects date for the building may be 6 months while the mechanical and electrical installations are separated with a defects date of 12 months; soft landscaping may be allocated a defects date of 24 months. The application of the procedures within core clause 4 remains the same, but they are then applied to (in this example) each of the three different periods. The only real difference created is that the Supervisor would need to issue a Defects Certificate (see clause 44.3 and Section 9.5) for each of the three separate elements of the construction.

9.8.5 Searching for similar defects after one has been uncovered

In the event that the Supervisor issued an instruction under clause 43.1 (see Section 9.5) directing the Contractor to search for a defect and that search found a defect it is not unusual in practice for further searches to be instructed in order to search for the same or similar defects elsewhere in the Works.

Should such a second instruction be given and the search finds that there is no defect then in accordance with the provisions of the Contract the Contractor would be entitled to a compensation event under clause 60.1(10) (see Section 15.3.10). As the second instruction was given as a direct result of defects having been found by the search carried out under the first instruction many in the industry would think that the Contractor should not be paid in such circumstances. Indeed some other commonly used construction contracts expressly state that the Contractor would not be entitled to payment under such circumstances.

Users must recognise that is not the case under the ECC. If the Client wants it to be that way then an amendment would be required.

9.8.6 The quality-related documents

In practice it must surely be the case that the purpose of the quality-related documents (i.e. policy statement and plan; see clause 40 and Section 9.2) together with the management system have the sole purpose of ensuring that the required quality is secured. As such their purpose is not to set the quality required but to determine how that quality is monitored, checked and recorded. The quality required is a major part of the Scope (see Chapter 22) that will be determined by the drawings and specifications.

Given the extent of the systems operated by contractors and subcontractors alike it is suggested that clients should take advantage of such systems by allowing the contractors to use their own, providing it is accredited to a recognised statement. That the contractor's staff will be familiar with the system operated will be of benefit. To impose a bespoke system on the contractors will, in all likelihood, only serve to introduce extra cost in both training people on the system to be used and in inefficiencies caused by the learning process.

Of the three documents, it is clear that it is the quality plan that will require the most work, both from the Client's team and from the Contractor. For example, the design team can assist by preparing comprehensive lists of the quality check stages that are required for each part of the construction and which are within the specifications. The provision of such information as part of the Scope then enables the Contractor to include all those requirements within the specific quality plan drawn up for the project.

Chapter 10
Title

10.1 Introduction

The shortest core clause in the ECC is number 7, which considers title (as in legal title or ownership) of Plant, Materials, and Equipment (whether in the Working Areas or not) and also of any existing materials and objects within the site, whether known about or not when the works commence, together with any material provided by the Client. Although it is only a short clause, the issues that it tackles are important (especially to the Client) and stray into areas that, legally, can be both complicated and difficult.

That said the position on ownership of any materials that have been brought onto a site and incorporated into a structure is straightforward. As soon as any materials or other goods are incorporated into a structure, even though that structure is not complete, the materials or goods legally become part of the land. As such, they are now owned by the landowner, whether the landowner has paid for them or not. The title to those materials or goods passes from the supplier to the landowner, no matter how many intermediate contracts with other parties may exist between the supplier and the landowner.

As a result, as far as Plant, Materials, and Equipment brought to the site are concerned, core clause 7 only covers title to such things up to the time that they are incorporated into the works and therefore become part of the land.

10.2 The core clauses

Under clause 70.2 title to Plant and Materials passes from the Contractor to the Client the instant that such items are brought into the Working Areas; the title reverts to the Contractor should any Plant or Materials be removed from the Working Areas. However, Plant and Materials may only be removed from the Working Areas with the Project Manager's permission. This rule stands even if the Plant or Materials are defective or surplus to requirements. In respect of the same items outside the Working Areas, in accordance with clause 70.1, title passes to the Client upon completion of an action by the Supervisor in marking such items as being for the contract. Unlike many other forms of contract, these methods of creating a change in ownership do not rely on any payment being made by the Client to the Contractor in order for title to be gained.

A Practical Guide to the NEC4 Engineering and Construction Contract, First Edition. Michael Rowlinson.
© 2019 John Wiley & Sons Ltd. Published 2019 by John Wiley & Sons Ltd.

Contractors should be aware of these positions and ensure that their subcontracts reflect these provisions.

The Supervisor's duty to mark Equipment, Plant and Materials that are outside the Working Areas is set out at clause 71.1 and arises when specified circumstances arise and have been satisfied. These circumstances arise when the contract has identified such items for payment before they are brought to the Working Areas. The process for marking should be set out in the Scope and must be completed by the Contractor before the Supervisor marks the items.

The Contractor is made responsible by clause 72.1 for removing Equipment from the Working Areas that is no longer needed, unless the Project Manager allows it to be left in the works. This exception can only apply to items that are left in the works permanently, despite originally only having been needed for temporary purposes.

Clause 73 deals with title to items that are already on the Site when the works commence. By clause 73.1, the Contractor does not have title to any object of historical or other interest or of value that is found. Instead, if the Contractor finds any such object, it must inform the Project Manager who then issues an instruction about what to do.[1] The Contractor is expressly barred from moving the object unless it has been instructed to do so. Clause 73.2 deals with arisings from excavations and demolition and sets out that the Contractor has title to such items unless the Scope states something different. This provision means that the author of the Scope will need to consider and identify any materials that may arise from excavations and demolitions that the Client wants to retain the ownership of. If such materials are identified then the Scope will need to say what is to be done with them.

Finally, clause 74.1 creates a right for the Contractor to use any material that the Client provides for use in Providing the Works (see clause 11.2(15) and Section 7.2). That the word *material* does not have a capital initial means that it is not a defined word (see clause 11.1 and Section 6.6.1). This leaves us in a position that what could be covered by this clause is very wide. In practice, it covers not only free issue items that could be permanently incorporated into the works but also the use of intellectual information such as design which the Client has had produced and made available to the Contractor for use in the Contractor's own designs that the Contract requires to be completed by the Contractor. The Contractor's right to use the material can be passed to a Subcontractor. That such material can only be used for Providing the Works acts to impose a restriction that stops the Contractor from using the material for any other project or purpose.

10.3 Practical issues

10.3.1 Removing waste and defective materials

As the title to all materials passes to the Client when they are brought into the Working Areas, the parties will need to develop a system or include a statement in the Scope that enables title to pass back to the Contractor in respect of the waste, off-cuts, genuine

---

[1] This may be a compensation event as described at clause 60.1(7); see Section 15.3.7.

surpluses, and defective items that need to be removed by the Contractor. If the ECC is followed literally, the Project Manager would have to give permission for the removal of every such item from the Working Areas. In practical terms, this provision creates an administration burden that the parties will not see any tangible benefit from. Catering for it with a simple statement that removes the administrative tasks will therefore allow those involved to concentrate on managing the project.

10.3.2 Materials off site

In the event that materials are to be paid for while stored outside the Working Areas, in order to be legally sure that title will pass to the Client when such items are marked by the Supervisor then (in practical terms) the Supervisor will have to do a lot more than simply 'mark' them. It is suggest that in practice the Supervisor, as a minimum, will have to:

- check that the Contractor has title to the materials in order to pass it to the Client;
- ensure that the materials comply with the Scope;
- ensure that the materials are stored as required by the Scope;
- ensure that the necessary vesting certificates have been executed;
- ensure that the storage facility is secure;
- ensure that the materials are stored separately from goods belonging to the supplier or other third parties;
- ensure that the method of marking is secure and cannot easily be removed; and
- take photographs which clearly record the materials being so stored and marked.

The procedure to be followed by the Contractor and the Supervisor in respect of any such items should be set out in the Scope and include copies of any paperwork that is required to have been completed. Such procedures are not uncommon and should identify which Plant and/or Materials will be considered for treatment in this way.

10.3.3 The wording of the antiquities clause

In practical terms, clause 73.1 deals with antiquities. That such items are encountered during construction projects is not uncommon and, in sensitive areas, archaeological surveys, and pre-construction digs are often carried out. Users should, however, be aware that the wording used is slightly loose by the inclusion of the phrase 'or other interest'. In practice, this does not appear to cause any problems but other commentators on the ECC have speculated that someone could seek to exploit this phrase.

10.3.4 Ownership of arisings from excavations and demolition

Clause 73.2 is very simple in its construction but, as with the marking of materials, it requires far more detail to be included in the Scope. This is particularly important where a project is likely to create excavated or demolition materials of value. In these instances, where the Client wishes to secure the benefit of the value of such items, what is effectively required is a scrap policy. This should cover matters such as who is responsible for disposing of the scrap, accounting for the material in order to satisfy environmental, waste management and accounting requirements, and provisions for how any monetary receipts are to be managed, including which party takes receipt of the money and how it is catered for within the payment mechanism. This latter aspect is important and needs to be set out as to how it reconciles with the payment mechanism (see clauses 50 and 51 and Sections 13.2 and 13.4) and the Contractor's share (see clause 54 and Section 13.7) when either Main Option C or D is in use. Even though the Client may retain ownership, the Contractor can still be charged with the task of managing and selling such scrap and then such receipts can be accounted for through the payments.

10.3.5 Material to be provided by the client

The way clause 74.1 is written creates the potential for it to cover a wide range of things. The material could be something physical to be incorporated into the works. Such things are often referred to as 'free issue materials'. In some sectors, such items are common. If such materials are to be provided by the client, then the Scope will need to set out several related matters such as, for example:

- what is to be provided;
- at what physical place will it be made available to the Contractor, i.e. will it be delivered to the Working Areas or will the Contractor have to collect it and if so where from;
- when will it be provided;
- details of any special handling requirements;
- how much is it worth for insurance purposes;
- any requirements in connection with the fixing of it; and
- any post installation protection which may be required.

Conversely, the material to be provided could be something that is not physical, such as the intellectual property in some design. Just as with the physical free issue materials discussed above, the Scope will need to spell out relevant detail as to the nature of the material and how it will be provided, what it can be used for, and the like. It is also usually necessary to ensure that the copyright holder of such material has given a license for its use to the Client and that the license can be passed to the Contractor and/or its Subcontractors.

Chapter 11
Liabilities and Insurance

11.1 Introduction

The two key words in the title of this core clause identify exactly what the clause is about. It concerns the allocation of certain *liabilities* to one party or the other and then establishes what *insurance* each party is obliged to maintain to provide cover against the occurrence of the liabilities. There is no duplication with the financial and contractual risks[1] that are determined by other core clauses.

This chapter reviews these provisions and offers practical advice on what needs to be done by clients, project managers and contractors alike. In relation to insurance matters, the best practical advice that any business involved in the construction process can employ is that, where they have any doubt or queries in connection with such matters, these should be referred to your insurance broker or company. The advice you receive is usually well informed and often free as part of the overall service being provided to you. This is especially true in that vital period just before you pay your annual premium.

11.2 The core clauses

11.2.1 Allocation of the liabilities

On the face of it, the liabilities are allocated in a clear and concise way. Those risks carried by the Client are set out at clause 80.1, which, by means of bullet points, identifies nine main categories of liability allocated to the Client (some of which include a number of subcategories). A description of each of these nine main areas is given in Volume 4 of the User Guide (NEC Panel 2017e) and does not need to be expanded on here, with the exception of a comment on the insurance of the works and how the extent of the liabilities can be increased.

The fifth main bullet point allocates the liability for damage by identified outside influences to the Client. This liability is limited to the works, Plant and Materials (see Section 10.2); it does not cover anything not covered by these terms and owned by the Contractor, the major element of which will be all things that would fall under the definition of

---

[1] That is, those risks primarily determined by the payment and compensation event mechanisms.

A Practical Guide to the NEC4 Engineering and Construction Contract, First Edition. Michael Rowlinson.
© 2019 John Wiley & Sons Ltd. Published 2019 by John Wiley & Sons Ltd.

Equipment (see clause 11.2(9) and Section 14.2.2). The Contractor will therefore need to be certain that it is adequately insured for any loss to things in its ownership damaged by such outside influences. Users should also note that the outside influences listed do not include terrorism. It is therefore advisable for users to check with their insurance companies whether a change is required in practice to the core clause or their policies, should the parties wish to insure themselves against the consequences of such an event.

The sixth main bullet point contains the words 'taken over', indicating that the liability for the risk of damage to the works is assumed by the Client once it has begun to use it as set out in the procedures at clauses 35.1–35.3 (see Section 12.6.3). In most circumstances, this act of take over will occur on or just after completion of the works or sections of the works. The Client then needs to insure its asset in the normal business way. Responsibility for a small number of exceptions remains with the Contractor, but these will diminish with time and disappear once the Defects Certificate (see clause 11.2(7) and Section 9.6) is issued.

The liabilities that the Client is responsible for can be expanded by detailing additional liabilities in Contract Data Part One. These additional liabilities are incorporated into clause 80.1 by the simple statement that is the ninth main bullet point. Should there be no such statement in Contract Data Part One, then there would be no additional liabilities. However, many lawyers seem to consider it prudent to make a statement in Contract Data Part One to the effect that there are no such additional liabilities. Others amend clause 80.1 by deleting the ninth main bullet point. From a practical point of view, the first of these two options is the neater.

Clause 81.1 identifies the Contractor's liabilities in four bullet points. The first bullet point concerns third-party claims that arise as a direct consequence of the Contractor Providing the Works (see clause 11.2(15) and Section 7.2). The second refers to what is often called Contractor's all risk. This covers the works, any Plant or Materials (see clause 11.2(14) and Section 14.2.3) or any Equipment (see clause 11.2(9) and Section 14.2.2). The third is for any damage to the Client's property, whether owned or occupied, and which is not the works, where the damage is a consequence of the Contractor Providing the Works (see clause 11.2(15) and Section 7.2). These are the normal types of liability that contractors in the United Kingdom are used to insuring and for which most, if not all of them, carry as routine as they are required on every project, whether under an NEC4 contract or other standard form.

11.2.2 Recovery of costs

Clauses 82.1 and 82.2 provide equal and opposite provisions establishing that should one of the parties pay any monies to someone who is not a party as a result of a liability for which the other party is responsible, then the cost concerned is paid by the party with the liability to the other party. There is nothing odd or surprising about these two clauses.

The amount to be repaid by one party to the other is reduced if an event for which the party receiving the payment is responsible contributed to those costs (see clause 82.3). This is akin to the principle of contributory negligence in UK jurisdictions,[2] which

---

[2] That is, England, Wales, Scotland and Northern Ireland, where the exact provisions can vary although they follow the same general path.

no doubt exist in other parts of the world, especially those based on the UK's common law principles.

11.2.3 Insurance cover and policies

The insurances to be provided by the Contractor are set out in the Insurance Table that follows clause 83.3 and is first referred to in clause 83.2. The requirements set out in this Insurance Table can be reduced or increased by statements in Contract Data Part One, which state either that the Client will provide some or all of these insurances or that the Contractor will provide other insurances in addition to those in the Insurance Table. These two methods of altering the Insurance Table when used together can both remove and add requirements to the basic provision. In practice, users should pay careful attention when any such transfers of responsibility for insurances are made or where the Contractor is required to insure other matters not otherwise referred to. Clause 83.1 makes it clear that the Client must provide whatever insurances that are identified in Contract Data Part One to be provided by it.

The insurances are required by clause 83.3 to be taken out in the joint names of the Parties (see clause 11.2(13) and Sections 5.2 and 5.5), except for the fourth insurance in the table, which will be in the Contractor's name only. The reason for taking out the policies in joint names is that only someone named on a policy can benefit from that policy by receiving a payment from the insurer. The Client is therefore put in a position whereby it can benefit from a policy taken out and paid for by the Contractor. The required insurance policies are to cover events that are for the Contractor's risk. This clause also sets out the time the insurances are to be provided for as being from the starting date (see Section 12.5.1), which is when the Contractor starts to perform its obligations (and could be many months before it actually starts doing physical work on the Site), until the Defect Certificate (see clause 11.2(7) and Section 9.6) has been issued or when a termination certificate (under clause 90.1, see Section 18.4) is issued.

The Insurance Table, as a default position and therefore notwithstanding any alterations to the insurances to be provided by the Contractor following the application of the provisions of clause 83.2 (see above), requires four insurance policies to be provided. In simple terms and with reference to the UK market, these are the four standard insurance policies carried by construction companies in order to comply with both UK statute and accepted practice. The minimum amount of cover or indemnity to be provided is stated in the right-hand column of the Insurance Table. For the first two types of insurance this is stated as being the replacement cost, which for the first type of insurance must also include the cost of any Plant and Materials provided by the Client (i.e. free issued to the Contractor) to the value stated in Contract Data Part One. The third and fourth types of insurance require cover to the level stated in Contract Data Part One. In order to complete this entry, the Client or the Project Manager on the Client's behalf, needs to estimate the value of such items at a level that is sufficient to protect the Client (but not overstated, as this may have the effect of pushing up the Prices without providing any additional benefit).

Clauses 84.1 and 86.1 require the Contractor and the Client to demonstrate to each other that the insurances they are both to provide have been taken out before the starting date. These demonstrations are repeated at renewal dates (clause 84.1) or when the

Contractor instructs (clause 86.1). The purpose of these two clauses is common sense and consistent with the principle of mutual trust and cooperation. One disparity is that the Contractor must supply certificates signed by its insurer or broker while the Client is required to supply policies and certificates, but without a requirement for confirming signatures. Clause 86.2 further states that the acceptance of a Client's insurance policy by the Contractor does not act to relieve the Client of the responsibility to provide the insurances, which Contract Data Part One states it will provide.

The Project Manager accepts the Contractor's insurance certificates if two tests are met, these being (i) that the insurance complies with the contract and (ii) that the commercial position of the insurer is strong enough to carry the liabilities. The first of these tests is easy to consider but the second one is not so easy. It is suggested that the Project Manager can only take reasonable precautions by checking the insurer's current rating within the market.

The Parties are also required by clause 84.3 to comply with all the terms and conditions of the insurance policies. It is therefore important, in practice, to ensure that all those involved are aware of such terms and conditions such that neither party looses any of its rights as a result of one or other of the parties omitting to follow some procedure that constitutes a condition precedent (see Section 1.5).

All insurance policies, provided by either party, are required by clause 84.2 to include a waiver of the insurers' subrogation rights in order to protect the Parties, the employees and directors of the insured parties (i.e. the Contractor and the Client). However, this protection does not extend to the Project Manager or the Supervisor unless they are an employee or director of one of the insured. Subrogation rights are a legal right open to insurers to put themselves in the insured party's position and to then use that position to recover the damages they have paid to the insured from the person or persons who caused the damage. This waiver does not apply when an employee or director of an insured party has been involved in fraud.

In the event that a party does not provide evidence that it has taken out an insurance policy required by the ECC, then the other party can take out that insurance and recover the cost from the other party. These rights are set out in clause 85.1 where the Contractor has failed and clause 86.3 where it is the Client who is at fault.

11.3 Secondary options

The four classes of loss that the Contractor is required to insure against in the Insurance Table (after clause 83.3) do not include Professional Indemnity Insurance (PII). This type of insurance is required by clause X10.7(3) and/or clause X15.5, should either or both of Secondary Options X10 and X15 apply. The primary purpose of PII is to provide coverage where the insured carries risk in respect of the defective performance of a professional obligation, the most common one being in respect of design. Without such cover, the party who carries that risk (which could be the Contractor by means of a statement in the Scope and clause 21.1 (see Section 7.3)) would have to stand the cost of rectifying any such defect at its own cost.

The clauses themselves are very simple. They both set the time the insurance is required as being from the starting date (see Section 12.5.1) to the end of the period stated in Contract Data Part One. The common position in this respect in the United Kingdom is that, when the Contractor is required to provide PII, it is also required to maintain that cover for a period equal to the limitation period determined by The Limitation Act 1980. This is 6 years for contracts executed under hand and 12 years for contracts executed under seal or as a deed. Contract Data Part One must also be visited to insert the cover level required. This cover is said to be required for the stated sum on an each claim basis. Users should be aware that not all professional indemnity policies are written on this basis; some provide cover 'in the aggregate for any period of insurance'. If this is the basis of the Contractor's policy then a change will be required to the wording in Contract Data Part One. (This is more likely to be an issue with Sub-Contractors using the ECS.)

11.4 Practical issues

11.4.1 Insurance of 'free issue' Plant and Materials

In some sectors of the construction industry it is common for the Client to supply items which, under the ECC, fall within the definition of Plant and Materials (see 11.2(14) and Section 14.2.3). The provision of such free issue items is dealt with at clause 25.2 by reference to details set out within the Scope and commented on from a practical viewpoint at Section 7.6.5. In practice, many issues arise about who provides replacement material in the event of loss or damage and who pays for such replacements. The risk on the Plant and Materials is stated, at the fourth main bullet point of clause 80.1, to be the Client's until such time as the Contractor has received and accepted them. Upon receipt and acceptance, the risk transfers to the Contractor under clause 81.1 (third bullet point). By stating an amount for the replacement of such items in Contract Data Part One, they should be protected by one of the joint name insurance policies taken out by the Contractor.

In practice, the Scope should not only set out what items are to be provided by the Client but also the procedure for delivery and acceptance, any requirements for storage and the procedure and timescale for replacement. Unless robust procedures for recording the delivery and acceptance are in place, then the point of transfer of risk from the Client to the Contractor can become blurred and create problems for the parties in determining who was carrying the risk should any loss or damage occur. Further, when it is clear which party carries the risk, should there be any loss, then responsibility for any delays that may occur because of the time taken to obtain replacements also becomes clear.

In order for the items to be covered by the joint names policy taken out by the Contractor, the replacement cost needs to be stated in Contract Data Part One. Rather than stating a single lump sum, it is better in practice (especially where there is more than one item or type of material being provided by the Client) to identify types, quantities and cost of the items to give transparency. The list of items in Contract Data Part One to be insured should match the list of items to be supplied in the Scope.

11.4.2 Self-insuring by Clients

It is common practice (certainly in the UK) for large employers, especially those in the public and utility sectors, not to take out insurance in the traditional sense of paying a premium to an insurance company and then receiving payments to compensate them for any insured loss. These bodies are said to 'self insure'. Where a Client entering a contract under the ECC is such a body, clauses 86.1, 86.2 and 86.3 act to impose an obligation on the Client and provide a remedy for the Contractor which is not consistent with this approach. In practice, it is therefore necessary for Clients who fall into this category to be aware of the potential consequences of either not amending the contract, not making appropriate entries in Contract Data Part One or not (through discussion with the Contractor) making sure that the Contractor will not exercise its option under clause 86.3 (see Section 11.2.3).

11.4.3 Project insurance

Core clause 8 is drafted on the premise that the requirement to provide insurance should be allocated between the parties as most appropriate for each project. Given that the ECC is drafted for use anywhere in the world, this approach is understandable and provides the necessary flexibility to suit global markets. That said, in some jurisdictions (including the United Kingdom) the use of project insurance is on the increase.

Project insurance is a way of providing a single insurance policy to cover all the risks for all those involved in the project, including the Client, the Client's agents, the Contractor and the Contractor's supply chain.[3] It operates on a 'no blame culture', which is consistent with the operating principles upon which the ECC is founded. Where such an insurance policy is employed, a single premium is paid by either the Client or the Contractor. The cost of this policy should, in theory, be more than covered by a reduction in the amount every organisation, whose operations are covered by the policy, would otherwise have had to include in its pricing. The saving should be made, as the situation where every organisation provides its own cover usually results in individual risks being covered several times over by many organisations within the overall supply chain; this unnecessary duplication is ultimately paid for by the Client.

From a practical and commercial point of view, especially on large projects, it is becoming increasingly recognised that project insurance offers an advantage to a project. If this approach is selected for a project under the ECC, then revisions will need to be made, which may be possible through the entries in Contract Data Part One. Users taking this step for the first time are well advised to consult both their insurance brokers and their advisors to ensure that the contract and the policy are consistent with each other.

---

[3] The extent or parts of the Contractor's supply chain that are covered are usually defined and limited by the policy.

Chapter 12
Time

12.1 Introduction

As all construction professionals are aware, the management of time is a major issue in the industry for both clients and contractors alike. Ours is an industry where the old adage 'time is money' applies; it is guaranteed that if the parties don't manage time within expectations, then the cost will also escalate for one party or the other (or more likely both). Issues with time affect all types and sizes of projects.

Complexity is not a determining factor as to whether the project will run to time or not. In the UK, we only have to compare the time relative to the original programme that it took to build the new Wembley football stadium with the same factors for the new Emirates stadium built for Arsenal FC. Viewed from outside the project teams, the Emirates stadium was built on time and without any noticeable issues. Wembley stadium, on the other hand, was an unmitigated disaster which brought long delays together with a massive cost overrun. While Wembley is the larger of the two stadiums, that fact alone is insufficient to explain the differences in the times taken for the construction of these facilities.

Taking these facts into account, and as the ECC is based around good project management, it follows that the management of time is at the heart of the contract. The general obligations in respect of time, the required contents of the programme, the process for managing that programme and other matters related to time are all set out in core clause 3. If these obligations and processes are not complied with and followed, then those involved in the project will find that the administration of the other process required by the ECC becomes considerably more difficult. The Accepted Programme (see clause 11.2(1) and Section 12.3) created under this core clause is the centre of the management process; without it many of the other processes designed to facilitate the good management of the works are frustrated and will either be significantly reduced in their effectiveness or fail altogether.

Users will do well to remember the saying that 'to fail to plan is to plan to fail' or, to put it another way, the 6Ps ('prior planning prevents p***-poor performance'). Where users fail to follow the requirements of core clause 3, these maxims will probably come back to haunt them.

A Practical Guide to the NEC4 Engineering and Construction Contract, First Edition. Michael Rowlinson.
© 2019 John Wiley & Sons Ltd. Published 2019 by John Wiley & Sons Ltd.

12.2 The programme: Contents

Clause 31.2 of the ECC lists nine main bullet points of information which the Contractor is required to include on every programme that is submitted to the Project Manager for acceptance. Two of these main bullet points are each supported by four secondary bullet points. Collectively, this makes clause 31.2 the most involved clause in the ECC.[1] In this section, the required contents are considered and cross-referenced where applicable to other clauses and processes in the ECC.

12.2.1 Fixed and variable dates

The first, second and fifth main bullet points require a number of fixed and variable dates to be included on the programme. These will all be indicated by milestones as opposed to being activities with a duration. The fixed dates (of which there are four types) are all covered by the first bullet point, the first three of which will be found in Contract Data Part One with the fourth being in either Part One or Part Two of the Contract Data.

The first date required is the starting date, a term which is not defined in the ECC but is referred to by several other clauses.[2] This is not the term that is used to signify when the Contractor first goes onto the Site to commence physical work. The term *starting date* is used to indicate the date upon which the Contractor is to start to perform its obligations under the contract, which often include matters such as design and procurement before it enters onto the Site. What is required to be done after the starting date but before the Contractor goes onto the Site will vary from project to project; in some instances there will be no interval between the two. The starting date will be a fixed point in time that will not vary.

The second date is called the access date; again, this is not a defined term. In the ECC this term is used to establish the dates upon which the Contractor is allowed to access either all or parts of the Site. Accordingly, the term is often used in its plural form as there can be several such dates. These dates are principally referred to in clause 33.1 (see Section 12.6.1) and can be found set out in Contract Data Part One accompanied, where there is more than one access date, with a description of what areas of the Site the Contractor will be given access to on each such date. If there is only one access date then this signifies that access will be given to the whole of the Site on that date. The access date (or first access date) will, on occasion, be the same as the starting date. The first sub-bullet point to the seventh main bullet point prompts the Contractor to show a later date as an alternative to any access date where the Contractor does not require access to that part of the Site on the access date.[3]

The third date is a term that was introduced by the third edition, i.e. Key Date. This term is defined at clause 11.2(11) as being the date by which a related condition

---

[1] Technically, clause 80.1 is physically longer but in practice does not involve the Contractor and Project Manager in as much work or have as much influence on the success or otherwise of the project.
[2] See clauses 15.2, 20.4 (Options C, D, E and F only), 32.2, 50.1, 83.2, 84.1, 86.1, W1.2(1), W2.2(1), W2.2(3), W3.1(2), W3.1(5), X10.7(3), X15.5, X20.2, X22.2(1) and X22.2(5).
[3] This is potentially problematic; see Section 12.8.1.

is required to be achieved. Any key date and its associated condition are set out in Contract Data Part One. It goes without saying that there can be more than one key date but these do not appear to be widely used in practice. Should the Client or Project Manager have elected to use one or more of these dates, then they must be shown on the programme.[4]

While fixed at any one time, the fourth and final date can be changed by the mechanisms within the contract.[5] This is of course the Completion Date, which is a term defined at clause 11.2(3). This date can be found in Contract Data Part One where the Client has decided the date in advance and included it as a date the Contractor had to plan and price to achieve when preparing its tender. In the alternative, the Client can decide to allow the Contractor to offer the best fit of price and time as part of its tender (in which case the Contractor inserts the Completion Date into Contract Data Part Two before submitting its bid). Whichever method is used to establish the Completion Date at the outset, that date is shown on the programme.

The variable dates, described at the second and fifth main bullet points, are variable in the sense that they are not set out in the Contract Data but will be established by the planned durations and sequencing of the Contractor's work activities. The second main bullet point requires the Contractor to show the date by which it plans to complete the works (referred to as 'planned Completion'). This will be the date that the last activity of the critical path through all the activities that the Contractor will need to undertake to Provide the Works (see clause 11.2(15) and Section 7.2) is finished. At any time it may be before, on or after the Completion Date. It will represent at that time the earliest date that the Contractor forecasts that it will complete. The critical path should not include any float (see 'The Contractor's activities' below). This milestone for 'planned Completion' could move forwards or backwards in time at each revision of the programme. This date is important, as it is the marker in time that delay is measured against (see clause 63.5 and Section 17.3).

The date or dates required by the fifth main bullet point are, in part, the equivalent of planned Completion but in relation to any or each Key Date[6] that the Contractor is required to meet. The other part requires the Contractor to indicate when it will complete any activities it is required to carry out in order to allow the Client or someone who falls under the category of 'Others' (see definition at clause 11.2(12) and Section 5.11), although there is no contractual obligation to finish any such work in a specified time or by a specified date other than at clause 25.1 (see Section 7.5.2) requiring cooperation.

12.2.2 The Contractor's activities

The third main bullet point requires the showing of all of the operations (activities) which the Contractor intends to carry out in order to fulfil its obligation to Provide the Works (see definition at clause 11.2(15) and Section 7.2). When a programme is prepared in the form of a Gantt chart, it is these activities that are represented by all the bars on that

---

[4] See Section 12.4.4 for comment of the practical implications of Key Dates and Conditions.
[5] That is, by compensation events and in particular clause 63.5; see Section 17.3.
[6] See Section 12.4.4 for comment on the practical implications of Key Dates and Conditions.

programme, with the duration being indicated by the length of each bar. These activities are the normal substance of a programme and should not be unfamiliar to any construction professional. What needs more attention when preparing a programme is the meaning of the term 'Provide the Works' as defined by clause 11.2(15). This not only encompasses the activities for the carrying out of the physical construction work, but also, for example, all design by the Contractor, any actions in procuring the goods and services needed, the preparation for and completion of tests and inspections, the preparation and submission of all documents and records and anything else required by a particular project as set out in the Scope for that scheme.

When showing these activities, the sixth main bullet point requires the inclusion of four subcategories of information. The first of these refers to float. As established above in relation to 'planned Completion', there should be no float on the critical path leading to that date; otherwise it would not be 'critical'. This reference here to float can only be to float on activities that are not on the critical path. By showing this float, the programme makes it abundantly clear to those using it how long an activity which is not on the critical path can be delayed before it impacts on the critical path (and from there on causes a delay to the 'planned Completion' date).

The second of these sub-items requires the programme to show 'time risk allowances'. This is a term that many construction professionals are not familiar with. In effect, what is required is that the programme identifies what allowances have been made within the programme durations for matters which are at the Contractor's risk. A simple example would be the allowance the Contractor would need to include in its planning to allow for lost or downtime caused by weather conditions which would not constitute a compensation event (see clause 60.1(13) and Section 15.3.13). By showing such allowances separately, should anything happen which results in the programme being revised, due account can be made for such allowances including adjusting the length of these factors should the particular circumstances dictate that such an adjustment would be fair and reasonable in the circumstances (see clause 63.8 and Section 17.3.1).

The third sub-item concerns the inclusion of health and safety requirements on the programme. Such items would include any statutory requirements and the specific requirements set out in the Scope (see clause 27.4 and Section 7.5.2) that were of sufficient impact to warrant the inclusion of a separate activity on the programme. This will very much depend in practice on the nature of the project. For example, a simple building on a green-field site would probably not require any such activities to be included, whereas a project for the decommissioning of a nuclear reactor would probably require numerous health and safety activities on its programme. In other words, the programme should include whatever is suitable for the project.

The fourth and final sub-item refers to the procedures set out in the ECC being shown. Again, this will depend on the nature of the project and the extent of the actions the Contractor is required to carry out. If the Contractor is required to carry out design, then the procedure for submitting the design through to being able to commence the related construction activities (see clauses 21.1–21.3 and Section 7.3) would be shown. In many ways this is linked to the second sub-bullet to the seventh main point. This sub-bullet requires the programme to show when acceptances are required. In the main, such acceptances will come at the end of one of the procedures stated in the contract, for example, the acceptance of the Contractor's design without which it would be unable to start the work covered by that design (see clause 21.2 and Section 7.3).

12.2.3 The activities of the Client and Others

Just as the Contractor is required to show all its own activities on the programme, the fourth main bullet point requires it to show all the activities of the Client and Others. The timing and durations that are put into the programme are either as the Contractor last agreed with these parties, or as stated in the Scope. This wording makes it clear that any agreement as to timing or duration would override anything stated in the Scope. The Contractor therefore needs to keep a record of all such agreements and preferably obtain conformation of that agreement from the party concerned. Provision of a copy of that agreement to the Project Manager would help to remove any queries that might otherwise be raised and demonstrate the attention to detail that the Contractor is bringing to the programme; this in turn should assist the Project Manager in deciding whether the submitted programme should be accepted or not (see Section 12.3).

In the event that agreement is not reached between the Contractor and any other party carrying out activities on the Site, then the Scope acts as a fallback position. It is therefore vital that the compiler of the Scope ensures that timings and durations for all such activities are included in the Scope. If they are not, then the preparation of the programme could be frustrated by something outside of the Contractor's control.

One of the activities of the Client which it is important to show on a programme is when the Client should deliver Plant and Materials or other items that are required by the Contractor in order to allow it to carry out the works. That these requirements are required is covered by the third sub-bullet to the seventh main point. The requirement to show when Plant and Materials are required is straightforward. Users should remember that the Scope can be used to set out constraints in connection with the provision of such items such as notice periods, delivery times, volumes, etc. Unless these are included in the Scope, then it would appear that it would be open to the Contractor to dictate to the Client, via the programme, when it wanted such items to be delivered to the Site. The inclusion of the words 'other things'[7] in this sub-bullet point opens up what the Contractor can show on the programme. It is suggested that this term would encompass the provision of outstanding design information, etc. that the Client need to provide in order to facilitate the construction of the works. From the outset of a project, many contractors produce what are often referred to as 'Information Required Schedules' or by some similar term. The openness of this requirement allows the Contractor to incorporate that schedule into its programme and to link the requirements to the relevant design, procurement and construction activities as may be applicable. That the wording does not refer to the Scope means that the items that the Contractor can identify is not limited to whatever other things the Scope may state that the Client will provide.

In a similar vein, the fourth sub-bullet of this seventh main point refers to information required from Others, thereby allowing the Contractor to show the timing and

---

[7] See clause 25.2 (Section 7.5.2) for the origin of the Client's obligation to provide 'other things'.

interaction of not only work by Others but also any necessary information which they are to supply.

12.2.4 Supporting statements

The eighth main bullet point is the one that in practice is complied with less than any other. This requires that the Contractor includes with a programme a statement for every operation that sets out how it plans to do that operation and identifies the resources that it plans to use. Showing a statement on a programme is not something that is practical. What is wanted here is a supporting statement which sets out the detail requested. These statements should then be submitted with the programme and form part of the family of documents that make up the submission.

The clause requires such a statement for every operation. If a programme has, say, 100 activities, there would not necessarily be 100 statements. Where several 'activities' relate to the same 'operation', then a single statement should cover those several activities. Each statement should describe how the Contractor plans to do the work identified by that operation. It should also identify the principal Equipment (see definition at clause 11.2(9)) and other resources that are intended to be used on that operation. The statement does not need to be extensive. In the second edition of the ECC, the equivalent provision used the term 'method statement'. This was interpreted by many to be a detailed statement containing all the associated health and safety provisions and documentation, risk assessments, etc. The word *method* was removed when the third edition was published, thereby reducing people's view of the required contents. As it is only the principle Equipment that is required, then it is not necessary to identify all the small tools and similar items. It is, however, necessary to identify other resources to be used. The main items to be considered here must be people and materials, meaning that the supporting statements collectively result in a resourced programme being provided by the Contractor (see Section 12.4.8).

12.2.5 Other information

If the Project Manager considers that the preceding eight main bullet points do not cover all the information that the Client or the Project Manager wants to see on the programme, then the ninth main bullet point refers to any other information which is set out in the Scope as being required as part of all such submissions. It was suggested by the Guidance Notes to the Third Edition (NEC Panel, 2013k) that this could cover detailed method statements, as opposed to the general statements required by the eighth main bullet point. No such statement is included in the User Guide, Volume 4 (NEC Panel, 2017e) in relation to the last bullet point. Instead, the guidance related to the penultimate bullet point makes reference to method statements. Whichever of these last two bullet points is used to include this information is academic at the end of the day; the management of the project needs the information regardless of which requirement is used to provide it. This appears to be a reasonable place within the ECC procedures to include for such matters or any other similar types of requirement. However, beyond the expansion of the detail of the statements to be provided in support, there seems to be little other practical application for this last bullet point. That said, the author knows of one employer who has a continuous high value construction programme in a particular sector, who uses this entry to require 'resource levelling'[8] to be shown on programmes.

Where Main Option A is in use, clause 31.4 also requires the Contractor to provide information to the Project Manager that describes how activities on the programme and in the Activity Schedule relate to each other. This does not mean that there needs to be an activity on the programme for every activity in the Activity Schedule, or vice versa. The two are required to correlate with each other, which can be achieved by identifying which programme operation includes each activity on the Activity Schedule.

12.2.6 Summary

All in all, the requirements of clause 31.2 are extensive and comprehensive. They require attention to detail and the Contractor to consider not only the physical construction but also matters such as the involvement of the Client and Others, the management and timing of procurement and design and other influences that (if not considered) could have a negative impact on the completion of the project. These requirements are most certainly best practice in terms of planning and programming.

12.3 *The programme: Submitting, accepting and revising*

One of the underlying principles of the ECC is that the programme becomes an active part of the management process. In order that the programme is maintained as a worth-while tool within the overall project management procedure embodied within the contract, the ECC provides a rigid regime for the submission, acceptance and revision of this key document. This means that the document stays fresh and will work in a dynamic way as part of the management procedure. By considering the effect of progress and of other changes on a regular basis, the changes in the interactions between the many activities and work fronts can be seen in advance and managed effectively. On the other hand, programmes that are not updated and revised soon become stale. As such, they do not provide any useable foresight into the forthcoming activities and future coordination required.

For many projects, a programme is prepared and submitted by the Contractor as part of the tender. In these cases, that programme is identified in Contract Data Part Two. When the contract is formed, any programme identified in Contract Data Part Two

---

[8] 'Resource levelling' is where the resources required by the operations at a particular point are compared with the need for the same resources both before and after that point to ensure that there is a steady flow of work for specialist resources thereby avoiding peaks and troughs in demand. Peaks and troughs in demand for specialist resources usually results in not enough being available.

becomes the Accepted Programme. The term 'Accepted Programme' is defined at clause 11.2(1) as either the programme identified in the Contract Data (i.e. in Part Two) or the latest programme accepted by the Project Manager. The definition confirms that when any subsequent programme is accepted by the Project Manager, then that programme replaces the previous Accepted Programme and becomes, for the purposes of the contract from then on, the Accepted Programme. This replacement of one Accepted Programme by another creates the dynamics that are required to actively manage a project with foresight.

In the event that a programme is not identified in Contract Data Part Two, then clause 31.1 requires the Contractor to submit a first programme within whatever time is stated in Contract Data Part One for such a first submission to be made. When this period of time starts is not stated in the clause; instead, it is only by reference to the statement in Contract Data Part One that a starting point is given (this being the Contract Date; see definition at clause 11.2(4)). The Client or Project Manager needs to insert the actual period for the first submission into this statement and must note that no default period is provided. In practice, this period is often as little as two weeks. The practical reason behind such a short initial period is that a programme needs to be established early in order to start monitoring progress and to allow foresight to be applied to the ongoing development of the management process.

Whenever a programme is submitted for acceptance, clause 31.3 allows the Project Manager two weeks within which to notify the Contractor that the programme is either accepted, at which time it becomes the Accepted Programme, or give a reason or reasons for not accepting this submission (see also clause 13.4 and Section 6.2). Four reasons are given, in addition to the general reason at clause 13.4, which the Project Manager can use for not accepting a submitted programme. Of these four reasons, the first and third are subjective and potentially, in practice, problematic. The second and fourth reasons are objective in so much that they both refer to the programme not showing information, which is stated either in the contract (second reason) or the Scope (fourth reason). Although these two reasons are objective, for most programmes under consideration they provide any project manager who wants, for whatever reason, not to accept a programme with a way of taking this stance. Sadly, any project manager who adopts this approach will frustrate the operation of many mechanisms in the ECC to the detriment of the project. Practical comment on how this can be avoided and why is given in Sections 12.4.1 and 12.4.2.

The subjective (first and third) reasons for not accepting a programme are open to interpretation. From a practical angle, the only way that the Contractor and Project Manager can avoid either of these reasons being used is by discussion. Such discussion should remove any misunderstanding over the views that might result in the Project Manager citing either of the reasons and allow the programme to be prepared to avoid such a position or allow it to be amended in a way that is acceptable to all.

Clause 31.3 also provides for what can happen if the Project Manager does not reply to the submission of a programme within the two-week period allowed by the clause. In such circumstances the Contractor may notify the Project Manager of this failure. Having received that notification, the Project Manager then has a further week to reply in the manner required. If the Project Manager does not reply in one of the required ways by the end of this one-week period, then the programme in question is treated as being accepted and becomes the Accepted Programme.

Once a first Accepted Programme has been established, clause 32.2 requires that a revised programme is submitted to the Project Manager when any one of three trigger events occurs. The first two of these trigger events are elective, the first being an instruction from the Project Manager after which the Contractor must submit a revised programme within the period for reply (see clause 13.3 and Section 6.2). The second trigger event allows the Contractor to submit a revised programme when it elects to, the circumstances behind such a decision being entirely without restriction and at the Contractor's discretion. The third trigger event is on the expiry of a fixed period of time stated in Contract Data Part One. Regardless of how many programmes may have been submitted under the first two trigger events, the third trigger event will always require the submission of a revised programme. The most common interval stated in Contract Data Part One is monthly, which is usually used to align the programme submission with the monthly progress meeting.

The requirements for what is shown on each revised programme that is submitted for acceptance is set out in three bullet points at clause 32.1, in addition to all the requirements of clause 31.2 (see above). The first requires the actual progress achieved to be shown against each activity and that this progress is reflected to show its effect on the remaining work to be done. On its own, this requirement means that the remaining work is re-programmed based on the work still to be done. This means that the Contractor needs to consider what progress has been achieved on each operation and, as a result, may need to adjust the duration of the activity for the work remaining in that activity where the production rate is slower or faster than planned previously.

The Contractor also shows how it plans to deal with any delays and correct notified defects (second bullet). Whether the correction of defects is shown on the programme in practice will depend on the nature of the defect and their impact on the programming of other activities. The reference to delays is open and therefore covers delays caused by the Contractor or the Client or any other party. The Contractor is obliged to mitigate the effect of all such delays as best as it can but without implementing measures that would be considered as acceleration. This requirement does not negate the Contractor's entitlement to a proper evaluation of a compensation event and is in fact consistent with the way in which a compensation event is assessed (see clauses 63.5, 63.8 and 63.9 and Sections 17.3.1 and 17.3.2).

The third bullet gives the Contractor a free hand to make any other changes which it elects to make. In practice, this allows the Contractor to re-sequence the works, change the durations of activities, change the level of planned resources, divide an existing activity into two or more smaller activities or make any other kind of change. What the Contractor must always consider when it uses this right to amend the programme is that the Project Manager has to accept the revision once it has been submitted. In order to ensure that this happens, the Contractor therefore needs to explain to the Project Manager the reasoning behind such amendments in order to facilitate the smooth and effective operation of the acceptance procedure. It is in the interest of the Contractor, the Client, the Project Manager and the continuing good management of the project that revised programmes are accepted as soon as possible.

12.4 The programme: Practical issues

12.4.1 Accept the inaccuracies

From a practical perspective, it is essential that users of programmes always remember that, at the time it is produced, a programme is nothing more than its author's forecast of how long each activity will take to carry out and the inter action between all the activities. A forecast is an estimate; an estimate is the forecaster's best guess. It therefore goes without saying that a programme produced at the start of a project is unlikely to be accurate, hence the reason for revising it as time goes by.

Given that there are inbuilt inaccuracies from the outset, users must accept that, as the time elapses between revisions, the programme will become less accurate. This does not mean that the programme should be ignored. As long as the best available information was used to prepare the programme it will always, despite its inaccuracies, provide a window on the future which the team can use to manage the project. Remember that applying foresight gives more chance of predicting problems and solving them, or at least reducing the impact, before the problem becomes an issue that delays completion, increases cost or impacts on quality. These three potential outcomes of issues are the matters referred to in clause 15.1 as being the triggers that require an early warning to be issued (see Section 6.4). If you don't know what you are supposed to be doing next or don't appreciate the impact of one activity on others, then it becomes difficult to foresee problems. What tends to happen then is that any foresight is lost and the management of the project becomes reactive instead of proactive. When this happens, the advantages of using the ECC have been lost.

12.4.2 Proportionality

When Lord Woolf reviewed the way the court system worked in England and Wales towards the end of the last century, one of the principles that he suggested for inclusion in the Civil Procedure Rules that were put in place as a result of that review was one of proportionality. This principle underlies the concept in that procedure which makes the amount of cost incurred in taking a matter through the courts proportional to the amount in dispute. It is suggested that in practice this principle of proportionality needs to be applied to some of the procedures in the ECC and especially to the programme.

The major problem on many projects, especially those of lower value and complexity, is that the cost of providing the amount of information required to be included on the programme by clause 31.2 is out of proportion with the benefit of providing it all. Further, providing the depth of the information required and keeping it revised can cost far more than any benefit that may be generated for the project. For example, if a programme suitable for a small project can be prepared and maintained for say £5,000, as opposed to an expenditure of say £25 000 to fully comply with the ECC, there is no commercial benefit unless the additional expenditure creates efficiencies and savings worth at least

£20 000.[9] Measuring whether such benefit is created is difficult, so in most cases the level of expenditure committed to the programming will be based purely on experience.

In order to prevent unnecessary expenditure on management time producing a higher level of planning information than is worthwhile, users need to understand the uses of the programme in the ECC. By doing this they can restrict the amount of information shown on programmes prepared for projects that will not benefit from the full and detailed application of clause 31.2. By taking an approach where the level of information provided is proportional to the project, then good use of the programme can be made at a reasonable cost.

The problem for contractors with this approach is that if they base their pricing on a level of programming that will not fully comply with the ECC, but the project manager insists that a fully compliant programme is produced, then the contractor would make a loss on this exercise with potentially no gain. In practice, unless amendments to the contract or the Scope set out that a programme of lower complexity more proportional to the project is all that is required, contractors pricing projects have little option but to allow for compliance with the ECC. Not to make such a provision is to take a commercial risk.

From this it follows that if an employer or project manager preparing a tender considers that the project will not benefit from the full programming requirements then, in order to receive the benefit of a lower level of pricing from the tendering contractors, it will be necessary for amendments to be inserted in order that the contractors can reflect any saving with security. The inclusion of such an arrangement is rare. In practice, such requirements tend to be reflected in behaviour borne from regular trading rather than in contract amendments. Such situations are fine until a dispute arises.

For practical purposes, when considering whether the programming can be simplified but still provide the benefit that the procedures bring, examples of the types of compromises that can be employed are as follows.

Where Main Option A is being used, amend the definition of Price of Work Done to Date (PWDD) (see clause 11.2(29) and Section 13.4.3) so that a proportion of the value of part completed activities is included in the calculation of the PWDD. By taking this step, the number of activities on the programme and in the Activity Schedule can be reduced.

Where a number of activities will be carried out by the same gang of people and equipment as they move through the project (e.g. a pipe-laying gang working along a sewer run), instead of having a statement for every operation (eighth main bullet point) use a common statement for that gang. By taking this step the number of supporting statements will be reduced.

Along similar lines to the previous example, where a project will be carried out by, say, four teams, plot the path of each team along the relevant activities and then provide just one statement for each team. Again, this can be used to reduce the number of supporting statements.

Where a team or gang size may vary slightly depending on the task being carried out, accept statements which give the minimum, maximum and average number of people. Again, this can be used to reduce the number of supporting statements.

---

[9] The amounts quoted are for illustration only; they are not accurate representations of what it might cost to comply with the programming provisions or to provide a simple programme.

Where design and or procurement are issues to be included on the programme set them up as sub nets that can be monitored and updated separately, thereby making these processes easier.

In conclusion, it is better in practice to have a programme that can benefit the project as opposed to a programme that slavishly follows the requirements of the contract but is so complicated that it cannot be used as a simple but effective tool to assist in the management of the project or to provide foresight to the team. The programme should be capable of being used and managed without becoming a burden to the project and those tasked with its delivery.

12.4.3 Developing the programme

The way that clause 31.3 is worded, any programme which is not fully compliant with clause 31.2 can be rejected by the Project Manager. In practice, fully complying with clause 31.2 at the start of a project can be extremely difficult. For example, it is unlikely that when preparing the first programme or the first revision after the Contract Date that the Contractor will have completed its procurement of subcontract packages for work to be carried out in the later stages of the project. Further, if design is still ongoing by either the Client or the Contractor, it may not be possible to identify the exact method to be used for construction activities. In such circumstances, the Contractor will only be able to include approximate durations and timings for the activities and will not be able to provide supporting statements or resource levels for these operations. It is easy to see how a project manager could reject a programme under these conditions.

From a practical point of view, this aspect of the way the ECC is drafted is self-defeating. The ECC wants the parties to have a programme setting out the planned intent with which to monitor progress and which provides foresight to the users. To overcome this situation and satisfy the higher level intent, it is necessary to apply some simple common sense and cooperation to the development of the programme. By recognising the functions that the programme performs, the Contractor and the Project Manager can agree between them (using the principles in clause 10.1) that different levels of detail will be included on the programme for different parts of the contract duration.

The agreement might be along the lines that for, say, the three or four months immediately after the time the programme is produced, the Contractor will show all the information which clause 31.2 requires. For the balance of the contract period remaining, the Contractor will show its best intent and information available at that time and note where matters are still in development. At each revision thereafter, the Contractor will expand the level of detail as more information becomes available. By keeping the Project Manager fully informed of how the detail is developing and when additional information is likely to become available, the transparency gives the Project Manager the comfort that the programme being accepted is the best that is reasonably available and that the Contractor is actively developing the document. This approach can satisfy the goals of the contract, allows sensible development of the programme without it detracting from the real job of managing the construction and help to build the spirit of mutual trust and cooperation.

This method is recognised as being the practical approach by the guidance given in Volume 4 of the User Guide.

12.4.4 Key Dates and Conditions

The provision first introduced in the third edition of the ECC and called Key Dates gave Clients and Project Managers a third method of controlling when they require parts of a project to be completed ready for something to happen.

To recap, the first method is the imposition of a Completion Date (see clause 11.2(3) and Section 12.5.2) by which the Contractor must complete all the work required to be completed by that date. The second method is the use of one or more Sectional Completion Dates (see Secondary Option X5 and Section 12.7.1) where separate obligations can be created, requiring the Contractor to complete defined sections of the works by specified dates, in addition to the overall requirement to complete all the works by the Completion Date.

In order of priority, Key Dates sit below the other two methods. The term Key Dates in defined at clause 11.2(11) where the user is referred to one or more entries in Contract Data Part One that state the key date or dates. Associated with each key date is a condition that sets out what must be achieved by the key date. It is this condition which determines the extent of the obligation to be completed by the stated key date.

The guidance in Volume 4 of the User Guide helps users to understand that this concept is for use where the work to be completed does not constitute a section of the works on its own merits. The situations that the concept is intended for are those where the Contractor is required to achieve completion of elements of the works by specified dates, to allow someone else (the Client or Others) to then install something prior to the Contractor returning to that area to carry out further works. In other words, the Key Date concept is a contractually binding sequencing tool.

To give a practical example, the Client may require that a space (say a room or chamber) must be completed to a specified state but without its roof or cover by a stated (key) date. This requirement is to allow something that the Client requires to be installed by Others to be craned into the space and fixed to the floor or a wall at a predetermined time. After the Others have finished their work the Contractor must then return to the space, put the roof or cover on it and carry out other finishing-type work in the space around the item that has been fixed. The Contractor may even need to connect pipes or cables to whatever the Others installed. In this example, it is not possible to define the partly completed space as a Section and apply a Sectional Completion Date. Instead, the part completion of the space can be set out as a condition and the related date attached to it and called a key date.

12.4.5 Float

Clause 31.2 requires that the Contractor shows 'float' (see first sub-bullet to the sixth main bullet point and Section 12.2.2) on the programme. As discussed in Section 12.2.2, this cannot be a reference to float on the critical path as critical paths do not by definition include float. In practice, that leaves us with two other types of float. The first of these is any float between the end of the critical path and the Completion Date. This type of float is referred to in several ways such as 'total float', 'project float' or 'terminal float', among others. The way the programming requirements of the ECC work, any such 'float' is represented by the difference between 'planned Completion' (see second main bullet point and Section 12.2.1) and the Completion Date (see clause 11.2(3), the first main bullet point at clause 31.2 and Section 12.2.1). The way many construction contracts are drafted, contractors would be reluctant to show float of this nature; should the employer under such an arrangement cause a delay, the contractor would not be entitled to an extension of time.[10] In the ECC however, the Contractor is not only required to show such 'float' but can do so in the knowledge that such 'float' belongs to the Contractor (see clause 63.5 and Section 17.3).

The second type of float to consider here is float on noncritical activities. If an activity is not on the critical path then it must have float; if an activity has no float, then it must be on the critical path as its completion within the planned duration and at the planned time is critical to the overall programme. The ECC requires such float to be shown in order that the amount of delay that may be incurred on a non-critical activity is clear to users of the programme before that activity impacts on the critical path. So if a non-critical activity has two weeks float, the users know that if that activity were to suffer a delay of up to two weeks then there would be no impact on 'planned Completion'. However, should the same non-critical activity suffer three weeks delay, then there would be an impact on the critical path (probably of one week in duration, but possibly more or less depending on the interactions between other activities). This means that if the Client causes delay to a non-critical activity then the Contractor's float on that non-critical activity would be used without any compensation for the Contractor. Should the Contractor subsequently encounter a delay event on that same activity that was at its risk, then it would not have the comfort of any float with which to cushion himself against delay damages.

In practice, the first of these two types of float is shown as a function of clause 31.2. The second type needs to be actively shown, however, and is far more important if the programme is going to provide the best information available to the team managing the project. If programming software (see Section 12.4.9) is used, then showing float on non-critical activities is easy to do providing that all the necessary links are included on the programme.

12.4.6 Time risk allowances

It appears that the showing of time risk allowances on the programme is something that does not happen very often in practice. Unlike float on non-critical activities discussed above, where the Contractor shows time risk allowances on the programme, these cannot be used by the Client to cushion the effects of a delay event which is at the Client's risk.

---

[10] This applies in the UK and would be the case under standard JCT and ICC contracts following the judgement in *Henry Boot Construction (UK) Ltd v Malmaison Hotel* 70 ConLR 32 amongst several similar cases.

Within normal programming allowances, the Contractor will allow for events that are at its risk under the contract. The Contractor has to make such allowances in order to ensure that its programme is realistic. The type of matters that it allows for are such things as breakdowns of equipment, poor weather, late delivery of materials or absenteeism within its workforce. These are all types of risk normally absorbed by planned production rates. The ECC looks for these allowances to be clearly identified. This can be done by showing the time risk allowance associated with each activity as a separate bar (albeit that this can double the number of activities on a programme, which is not always desirable), by having an additional information column which states the allowance included in the overall duration for each activity, or by identifying the allowance in the supporting statement. Some software will allow the main activity bar to be annotated with the allowance.

The showing of these allowances has a practical purpose. When assessing compensation events, the Contractor is permitted by clause 63.8 to include risk allowances for matters which are at its risk under the contract. Providing that the Contractor has identified what time risks it has included on its programme then continuing to include such items on any revised programme, adjusted as appropriate for the changed circumstances, should make the analysis of the effect of any delay event more straightforward. Let me illustrate this with an example based on an actual project.[11]

The project in question for was the replacement of the waterproofing to several bridge decks along a section of dual carriageway trunk road.[12] The bridges all carried the road over secondary roads, watercourses and a railway line. In order to reduce the amount of traffic management in place at any one time, the Contractor was only allowed to work on one bridge at a time. The contract started in April of one year and was due to complete by the September of the following year. On its programme, the Contractor showed a time risk allowance for weather delays against the activity for installing the new waterproofing system on each deck. Readers will be aware that the installation of such systems is weather dependent which, in the UK (where rain is a frequent but unpredictable occurrence) is something we always have to consider. The amount of time allowed against each bridge varied with size and the planned season for the work. The Contractor made it clear how the allowances had been calculated.

Within the first few months a problem was found with the underlying concrete on the deck that was being worked on. This had to be repaired before the replacement waterproofing could be installed. It was clear that there would be a delay to the programme because of this event, which was notified and accepted as a compensation event.[13] The revision of the programme to show the effect was a straightforward exercise. The length of the delay was put into the critical path and all the following activities were re-scheduled accordingly. As the Contractor's time risk allowance for weather delays was clearly shown, the amount of time allocated to each waterproofing activity could easily be adjusted (longer or shorter) as appropriate. Where an activity was moved into

---

[11] This tale is based on a story told to me and is not therefore a totally accurate representation of the facts, nor is it intended to be. It is intended as an example only to illustrate time risk allowances.

[12] For readers not familiar with this description I am referring to two lanes of traffic in each direction with a central reservation between them and graded interchanges or roundabouts as the main form of ingress and egress from the road.

[13] See clauses 61.1, 61.3 and 61.4 and Sections 16.2, 16.3.1 and 16.3.2.

a season with a higher risk of rain the allowance was increased; when an activity was moved into a drier period, the allowance was decreased. The Contractor and Project Manager found that they could easily agree on the effects resulting in the quotation for the compensation event being promptly accepted. Everyone's efforts could then be concentrated on moving the project forward without distraction from a lingering issue. Without a clear understanding of the time risk allowances included, this was unlikely to have been achieved in such a straightforward manner.

12.4.7 A family of documents

It is important for users to recognise that the programme under the ECC is not just a Gantt chart; it is a family of documents which support each other. The most obvious example is the requirement at the eighth main bullet point of clause 31.2 requiring a statement for each operation. These statements cannot be shown on a Gantt chart. Although the necessity to produce a whole series of supporting statements can appear burdensome at the outset, these documents can assist when revising the programme. By giving each supporting document a unique reference number and inserting that reference number on the programme with the relevant activity, the need to re-issue every supporting document at each revision can be avoided. This approach reduces the administrative burden and the amount of information issued.

For instance, if an activity has been delayed for some reason and the effect of that delay is simply to put back the start time of that activity but the duration, method and resources otherwise remain unchanged, there is no need to amend the statement. A simple agreement between the Project Manager and the Contractor to the effect that supporting documents that have not been revised do not need to be re-issued at each programme submission thereby avoids that need. This approach has the advantage for the Project Manager that, at each submission of the revised programme, the only documents received are those that have been revised. This means that it is clear as to which documents need to be considered and thereby reduces the time taken to consider whether the programme can be accepted or not.

Depending on the complexity of the project as well as the Gantt chart, the family of documents that make up the Accepted Programme could also include:

- the supporting statements required by the eighth main bullet point at clause 31.2;
- a summary of the resources collected from all of the supporting statements;
- a schedule of information required by the Contractor from the Project Manager and/or Client (often referred to as an Information Required Schedule);
- a procurement schedule for subcontractors;
- a procurement schedule for major material purchases; and
- a separate design programme for the Contractor's design.

The above list is not exhaustive. Certain projects or market sectors may have other documents that they routinely find to be to the project's advantage to include. Neither will it be necessary for every project to provide all of the above. As with any project carried out under the ECC, the important factor is to use the framework to produce the optimum level of detail and support to provide a cost-effective programme that brings benefit to the management of the works.

12.4.8 Resourced programme

The fact that the programme (by means of the supporting statements) needs to show the resources to be employed by the Contractor in carrying out the works has been referred to several times previously in this chapter. In practice, evidence suggests that this requirement is one of the least observed. The majority of contractors appear not to want to give this information to the Project Manager, as they fear it will be used as a stick with which to chastise them if actual progress is not as good as planned. A commonly heard comment from contractors is that if they show, say, 10 gangs of bricklayers on the programme but they only employ 9 gangs on the site and a delay event occurs which is at the employer's risk, they are then refused an extension of time because the project manager says the delay was the contractor's fault for not having the planned resources. Notwithstanding that this is a common occurrence, the analysis by the project manager is of course incorrect.

In practice, contractors should not let such fears prevent them from showing resources on the programme. In the ECC showing this information is of significant benefit when it comes to assessing compensation events (see Section 17.6.1). Having the planned resource levels on the programme makes it significantly easier to forecast the changes in the resource level resulting from a compensation event. The forecast is always the difficult part of an assessment in such circumstances; once the forecast has been established, costing the resources involved is a relatively simple task.

When comparing the resources shown on a programme to those actually employed in doing the work, project managers should always remember that any aspect of a programme was only the best estimate of that factor when the programme was prepared. Should the resource levels fluctuate up or down, this will often be a function of the contractor adjusting its resources to suit actual production and in keeping the resource levels at their optimum in terms of cost versus value. When estimated, the resource levels can only be a best statement of intent and are usually an average of what is required. With many trades, there is a period at the start and finish of that trade's work where the resource level is steadily increased as the work faces open up and, conversely, reduced as the number of work faces reduce. This fluctuation also occurs as the available work varies because of the interaction with other trades. Showing this precise detail on a programme is difficult, so resource levels shown on programmes are often realistic but not entirely accurate. It is suggested that in practice, it is better to have a realistic average than nothing at all.

12.4.9 Using programming software

It is rare these days for anyone to produce a programme without the use of a computer. Some are drawn using a simple spreadsheet package which provides a pictorial view of the plan but makes revising such programmes tedious and inaccurate. Thankfully, the majority of programmes are now produced using specialist software packages of varying degrees of complexity. These packages will facilitate accurate production of the programme. Most of them allow the insertion of the necessary restraints and links that allows for automatic determination of the critical path. The practical difficultly here is that the restraints and links must all be included; the omission of just one link can result in the software providing an incorrect analysis of the critical path. When done correctly, the software will also allow the float (see Section 12.4.5) to be shown by selecting the appropriate function. The main advantage of using good software, providing that the input is done correctly, is that the production of revisions is made much simpler.

The downside of using software it is that there are several competing programmes, each of which operates differently from its competitors and all of which are complicated. Unless the user is familiar with the software, it can be difficult to get the best out of it. The complexity means that it is usually necessary to employ a specialist planner who is familiar with the software package in use. Even experienced planners sometimes find it difficult to switch to a software package that they are not familiar with. From the Project Manager's point of view, should the use of a particular software package be required, then that required use should be included in the Scope as a constraint setting out the particulars of the requirement. Without such a constraint, the Contractor would be free to use whichever software package it prefers, which would normally be the one which the Contractor's planners are most familiar with.

If the programmes are prepared using computer software, then anyone using such a method should ensure that when they revise the Accepted Programme they always save the current Accepted Programme with a new file name. By doing this, they will of course keep a copy of every programme in its electronic format. This point may seem obvious, but I have come across several examples where this simple rule has been forgotten. As a result, the history of the project in the form of programmes had been lost. This is generally only a problem if there is a need for forensic investigation of any area of the project in order to resolve any difference of opinion or interpretation between the parties. Remember this basic but important point.

12.5 Starting and finishing

12.5.1 Starting work

As commented on in Section 12.2.1, the Contractor often starts to perform its obligations under the contract before it accesses the Site. The date on which the performance of the Contractor's obligations begins is known as the starting date, the timing of which is included in Contract Data Part One. This can be problematic in practice, as all of the data included in Part One should be provided by the Client to the Contractor when the tender is issued. With many projects, this date is not known at the time tenders are invited from contractors. There can be a whole range of issues that make fixing the actual start date difficult, whether that date refers to starting work before going to the Site or actually on the Site at the time, when tenders are issued. When such difficulties exist, the only practical solution is to mark these dates as matters to be agreed. If this approach is taken, the parties must remember to record the dates they eventually agree and include them in Contract Data Part One before the contract is executed. Practicalities to one side, the 'starting date' is not referred to in core clause 3 (except for the requirement to show it on the programme at clause 31.2). This does not diminish its importance or relevance.

The time that the Contractor can go onto the Site and start physical work is restricted by clause 30.1 to the first 'access date'. Contract Data Part One can contain several access dates, each for a different part of the Site. This useful characteristic allows the Client to control access to parts of the Site and allows it to continue to use parts of its property until such time as the Contractor is allowed access. This is particularly useful where the project is for work on a client's existing site or for infrastructure projects where there may be complicated land issues in connection with pipe or cable runs.

12.5.2 Completion

Clause 30.1 also places the key obligation on the Contractor that it must achieve Completion by the Completion Date. This requirement makes reference to two defined terms. The definition of Completion Date at clause 11.2(3) is the most straightforward, being a reference to the date included in the Contract Data. This entry can be in either Part One, if specified by the Client, or in Part Two if the Client decides that the Contractor will be allowed to specify the date by which it will complete as part of the tender return. From the Client's point of view, the first option is the safe one if it needs the project completed by a certain date. However, imposing a date for completion upon the tendering contractors does not always result in the cheapest possible price. By electing to use the second option, the Client is allowing the tendering contractors to plan for and price the most cost-effective combination of resources and time. This option usually results in lower prices being offered by the tenderers than when they are obliged to price to a date decided by someone else.

The term Completion is defined at clause 11.2(2). The first part of the definition, as set out in two bullet points, provides by reference to the Scope an objective test as to whether the Contractor has achieved Completion or not. The use of this objective measure can only come into play when the Client has set out in the Scope the requirements in this respect.[14] The second bullet sets out a test that is more subjective by referring to the necessity to correct notified defects that would otherwise have hindered the Client from enjoying the works or may prevent Others from carrying out succeeding works.

The second part of the definition, which can only come into play if the Scope does not state what has to be completed, is subjective in a way that will be familiar to readers who have used other standard forms of contract which are popular in the United Kingdom. The default position within the definition refers simply to the Contractor having completed all the work that is necessary to allow the Client to use the works and for it to be possible to carry out any subsequent works to be carried by Others. The way the courts in the UK have interpreted this type of clause, referred to in other contracts as 'practical

---

[14] See Section 22.2.8 for comment on practical matters that should be considered for inclusion in the Scope.

completion' or 'substantial completion', confirms that such requirements do not mean completion of everything the Contractor is obliged to do. The UK courts have used the Latin term '*de minimis*' to denote that minor works that will not prevent the Client from enjoying the benefit of the works are not required to be completed where such wording is used in order to achieve the desired state that brings about the completion of the contractor's obligation. In practice, this approach to defining 'completion' is so subjective that it has spawned numerous disputes and should be avoided. By setting out defined requirements in the Scope, users can avoid this potential pitfall.

The responsibility of deciding when Completion has been achieved is placed in the hands of the Project Manager by clause 30.2. This clause does not require the Contractor to notify the Project Manager that the Contractor thinks that it has achieved Completion, or to give the Project Manager any advance warning that it considers that this state is about to be achieved. This means that, in practice, the Project Manager has to be alert to what is going on in order to determine the right time at which to confirm the position. This is one of the several matters that the Project Manager should seek to rely on the spirit of mutual trust and cooperation. Once the Project Manager decides that the Contractor has achieved Completion, then the same clause requires the Project Manager to certify (see clause 13.6 and Section 6.2) that this state has been reached within one week of the date upon which that condition had been achieved.

12.5.3 Key Dates

Section 12.4.4 includes comment on the significance, contractual term and practical issues surrounding Key Dates and their related Conditions. Clause 30.3 imposes the obligation on the Contractor to complete each Condition stated for a Key Date by that Key Date. There is no requirement in the ECC for the Project Manager to certify that the Contractor has achieved each Key Date that may be included in Contract Data Part One. Practically, it would be sensible for the parties to record that any such obligation had been successfully fulfilled, if for nothing more than administrative neatness.

What the ECC does provide is a provision which sets out the Client's remedy should the Condition for a Key Date not be achieved by the Key Date. This provision, at clause 25.3, is triggered by the Project Manager deciding that the Contractor has failed to fulfil its obligations by the stated date. Once the Project Manager has so decided then, within four weeks of when the Contractor does actually achieve the Condition, the Project Manager must assess the additional cost (if any) incurred by the Client on this project for stated reasons as a result of the Contractor's breach of its obligations. The Contractor then pays this amount, usually by deduction in the assessment of the amount due under clause 50.3 (see Section 13.4.2). While this is the Client's only remedy, it is one which the Contractor's only right of objection is via a reference to the dispute resolution procedures (see Chapter 19).

In legal terms, this clause gives rise to the imposition by the Client of unliquidated damages for delay on the Contractor. This type of provision is not liked by contractors as the unknown quantity of such damages does not allow them to assess the risk involved with failing to achieve the obligation. This is especially significant when a tendering contractor considers that meeting any Key Date stated in the Contract Data could be difficult or even impossible. Without knowledge of what the damages may be, which the contractor would have if they were liquidated, the contractor is unable to assess its risk accurately and make a suitable provision within the pricing. It must also be said that many employers do not like such provisions as the amounts claimed as damages are often hard to calculate and support. While the ECC only requires the assessment of the damages, which is a lesser test than having to ascertain the damages, even assessments can be overturned in the dispute resolution procedures if they cannot be shown to be reasonable. In practice, it appears that up to the time of writing, the UK is not seeing many examples of Key Dates being used in contracts.

12.6 Other matters

12.6.1 Access to the Site

As would be expected, the Client is obliged by clause 33.1 to allow the Contractor to access the Site. Unlike many other contracts, this does not refer to the Contractor being given possession of the Site, just that it has the right to access it. The provision of access must, in practice, also include allowing or arranging for access from the public highway into the Site itself when they are not immediately adjacent to each other. As an example, should the Site be part of an existing industrial complex and located away from the public highway, the Client would have to provide access across the necessary parts of that complex in order to allow the Contractor access to the Site. Assuming that the Client owns or operates the whole of this complex, the Client can set out in the Scope any constraints that apply to, say, the routes the Contractor can take through the complex and any provisions to be satisfied before people, equipment or Materials are allowed into the complex.

Should the Client fail to provide access or maintain the provision of access to the Site, then such a failure would be a breach of the contract, which would give rise to a compensation event.[15]

The way that the ECC is structured allows for more than one access date into predefined parts of the Site to be stated. This facility provides practical flexibility, which allows clients the opportunity to lessen the impact of a project on their existing operations. This flexibility is also used in the infrastructure sectors where differing access provisions and times are arranged with separate landowners along an easement or wayleave obtained to install a pipe or cable route. The secret to the proper application is the accuracy and completeness of the information included in Contract Data Part One by the Client before the tender is issued. In the event that, for practical reasons, the Client cannot provide the required information during the tender period, this will be a point that can be negotiated (possibly with commercial alterations to the Prices) before the contract is formed. In the event that the Client states more than one access date but, for whatever reason, is unable to give access on the stated date, a compensation event will occur as set out in the preceding paragraph.

---

[15] See clauses 60.1(2), 60.1(3) and 60.1(18) and Sections 15.3.2, 15.3.3 and 15.3.18, respectively.

This same clause also requires that the access to be provided is made available no later than the access date stated in Contract Data Part One or an alternative date shown on the Accepted Programme. An alternative date will only be of effect if it is later than the date given in Contract Data Part One. Clause 31.2 encourages the Contractor to show a later date than the access date on its programmes if the Contractor does not need access at the date stated in Contract Data Part One. This is perfectly reasonable but may bring practical issues if other events impact (discussed in Section 12.8.1).

12.6.2 Stopping and starting work

The Project Manager is given the discretionary power in clause 34.1 to instruct the Contractor to stop or not start any work. The Project Manager can also reverse this instruction at some future time or remove the affected work from the Scope. That is the extent of the clause. It does not contain any reference to the type of event or reason that would give rise to the Project Manager having the right to use this power. The Project Manager is therefore free to use this power under any circumstances or for any reason.

However, as with many things in the ECC, in practice it is not that straightforward. Because of the lack of cross-referencing within the ECC, this short and simple provision does not reveal the two associated provisions which act to provide some form of protection for the Contractor against abuse of this power. It must be added here that both of these two associated provisions also act to extend the Client's options related to such an instruction.

The first of the associated provisions to consider is the application of the compensation event mechanism (see Chapters 15–17). Here the user must consider two issues in tandem: the compensation event that arises when the Project Manager issues the instruction that the clause empowers the Project Manager to issue at clause 34.1 (see clause 60.1(4) and Section 15.3.4) and the first of the four bullets points at clause 61.4 (see Section 16.3.2) which determines that it is not a compensation event if the event arises from the fault of the Contractor. If the Project Manager issues such an instruction, and the reason for that instruction was not due to a fault of the Contractor, the Contractor is protected by the occurrence of a compensation event. Similarly, the Client is protected in the opposite manner if the instruction was given as a consequence of a fault of the Contractor, as a compensation event would not occur.

The second associated provision is found at clause 91.6 where either or both parties may find that they have a right to terminate the contract depending on the cause of the instruction given by the Project Manager. This right will only arise where an instruction to stop or not to start given by the Project Manager is not countermanded by an instruction to start or re-start within thirteen weeks of the first instruction. This process is covered in more detail in Chapter 18.

12.6.3 Take over

In order to understand the relevance of 'take over' in the ECC, the user must appreciate what the significance of the mechanism is in practice. The only other reference to 'take over' in the ECC is found at the sixth main bullet point in clause 80.1. This reference acts to place the liability for loss, wear, or damage to the works, or parts thereof, that have been taken over with the Client. The purpose of take over is therefore to transfer the responsibility for these liabilities from the Contractor to the Client. With this in mind, it is then easier to understand the provisions.

The first sentence of clause 35.1 makes reference to an entry in Contract Data Part One where the Client may state that it will not take over the works before the Completion Date (see clause 11.2(3) and Section 12.5.2). If the Client elects to include this statement, then it is telling the Contractor that the Client has no interest in the project being completed early. Should the Contractor complete before the Completion Date when this statement is part of the Contract Data, the Contractor would still be entitled to a Completion certificate (see clause 30.2 and Section 12.5.2) but the liability for loss or damage to the Works would remain with the Contractor until the Completion Date. If this statement has not been made in Contract Data Part One, by the second sentence in clause 35.1 the Client takes over the works within two weeks of Completion.

Clause 35.2 gives the Client the right to use any part of the works before Completion has been certified by the Project Manager (see clause 30.2 and Section 12.5.2) and goes on to confirm that, should this happen, then the Client takes over any part of the works when it begins to use that part. There are two exceptions to this rule set out in the clause. Should it be set out in the Scope that the Client will use part of the works before Completion, this will not constitute take over. This can be common in, say, highway works where specified traffic management sequencing involves putting the traffic over completed sections of the works prior to the whole works being complete. By making it clear that this is a requirement, when the Client starts to use those new parts of the highway (for example), it will not take over those parts. The second reason would be where the Contractor's chosen method of working requires the Client to use part of the complete works while the Contractor carries out other parts of the project.

The Project Manager is required by clause 35.3 to certify completion of any part of the works that the Client takes over within one week of the event.

12.6.4 Acceleration

At clause 36, the ECC provides a set of provisions that allow for the bringing forward of the Completion Date (see clause 11.2(3) and Section 12.5.2). This can be instigated by either the Contractor or the Project Manager pursuant to clause 36.1 (mostly likely by the Project Manager at the behest of the Client) proposing to the other that an acceleration could be used to achieve Completion before the Completion Date. If both the Contractor and the Project Manager are willing to consider the proposal then the Project Manager issues an instruction to the Contractor for the provision of a quotation to achieve the desired revised Completion Date. When issuing such an instruction, the Project Manager must state what associated changes to any Key Dates (see clause 11.2(11) and Section 12.5.3) are to be made. The Contractor then submits a quotation within three weeks of the instruction being issued. The quotation, as required by clause 36.2, sets out and provides details of the proposed change to the Completion Date, the Key Dates and the Prices. The Contractor includes details of the assessment with the quotation.

Users should note that these acceleration provisions do not refer to the Project Manager being able to instruct a quotation for a revised Key Date in isolation. Clause 14.3 allows the Project Manager to instruct a revision to a Key Date without any need to obtain a quotation from the Contractor first. Although this mechanism creates a compensation event (see clause 60.1(4) and Section 15.3.4) it also acts to bypass the principle that acceleration should not be instigated unless the parties have agreed the effects on both the affected time and cost elements.

In the event that a quotation for acceleration is accepted then clause 36.3 applies and confirms that when the Project Manager notifies the acceptance the Completion Date, any Key Dates and the Prices are changed in line with the quotation.

12.7 Secondary Options related to Time

The Secondary Options available within the ECC allow three matters to be added to the core and Main Option clauses concerned with Time. All three can be used as standalone provisions. The first can be used either alone or with either or both of the other two. The other two could be used at the same time, although only one could come into play unless the first was also in use, when all three could impact on the contract. If that seems a little confusing a simple explanation of the options below should clear things up.

12.7.1 Sectional Completion: Secondary Option X5

The actual provision at clause X5.1 is straightforward in its approach. All it does is make any reference (wherever it might appear in the clauses) to Completion, the Completion Date or the works a reference to the same things, but for each section of the works. By use of italics (see clause 11.1 and Section 6.6.1) the clause refers the user to Contract Data Part One, where it is necessary for the Client to set out a clear definition of what works for each Section of the project.

It is in Contract Data Part One that the important information required to make this Secondary Option effective is required to be entered. A description of each Section needs to be provided which contains enough detail to make it clear to the Contractor and any third party who might become involved (e.g. the Adjudicator; see Section 5.7) the precise amount, part or area of the works which forms each Section. Unless this detail is precise and definitive there can be problems with making these provisions work effectively; it is advantageous to make sure that these details are clear in practice. There is no limit to the number of Sections that can be created. The provision of Sections for Completion is not linked in any way to the number of access dates (see Section 12.5.1) that are included in a project.

The use of Sectional Completion is common in practice. It provides a mechanism whereby the Client can plan for the progressive use of the works in their completed state, in order that the Client can start to enjoy the benefit of its scheme without having to wait for all the works to be finished.

12.7.2 Bonus for early Completion: Secondary Option X6

The provision at clause X6.1, if selected, creates a situation whereby the Contractor is paid a bonus for every day it achieves Completion (see clause 11.2(2) and Section 12.5.2) before the Completion Date (see clause 11.2(3) and Section 12.5.2) or the date the Client takes over (see clause 35 and Section 12.6.3) the works, whichever is the earlier. The daily bonus rate must be stated in Contract Data Part One. If this provision is used in conjunction with Secondary Option X5 (see Section 12.7.1), the entry at Contract Data Part One needs to be in respect of each Section of the works.

Should Sectional Completion be in use and the Client only wants to reward the Contractor should it manage to finish early on one or some of the Sections rather than all of them, the Client simply needs to insert NIL as the bonus rate against those sections to which no bonus applies. In this way, it will limit those sections of the works for which a bonus could be earned.

In the UK, where secondary Option Y(UK)2 (see Section 13.9) is included in the contract, the operation of clause Y2.1 would mean that any day that is bank or public holiday would be excluded from the calculation of the number of days for which the bonus would be paid.

12.7.3 Delay damages: Secondary Option X7

The last of the three Secondary Options that are related to time concerns the provisions that enable the Client to recover a predetermined level of damages on a daily basis in respect of any delay by the Contractor in achieving Completion (see clause 11.2(2) and Section 12.5.2) by the Completion Date (see clause 11.2(3) and Section 12.5.2). In the UK, the common name for this type of provision is 'liquidated and ascertained damages' or LADs. The basic principle (contained in the ECC in clause X7.1) is that the Contractor, being in breach of its obligation to complete the works by a stated date (see clause 30.1 and Section 12.5.2), pays the Client a predetermined amount for each stated period of time that it continues to be in breach of this obligation. In the ECC, the chosen period of time is the 'day' so the penalty will continue to run for each calendar day. In the UK, where secondary Option Y(UK)2 (see Section 13.9) is included in the contract, the operation of clause Y2.1 would mean that any day that is a bank or public holiday would be excluded from the calculation of the number of days to which the damages would apply.

Should the Client take over part of the works before Completion, the Project Manager is required by clause X7.3 to assess the benefit to the Client of taking over that part in proportion to having taken over the whole of the works. The rate of delay damages is then reduced by that assessed proportion. This approach is different to most other standard forms of contract which work on the basis of using the value of the portion taken over against the value of the whole. The ECC's method in this respect is subjective as it is based on the Project Manager's assessment, whereas the other method is based on values which can be determined and therefore provide an objective measure. The use of an objective measure puts the Contractor in a position where if it is in a delay position it can calculate the benefit, if any, of concentrating its efforts in completing a portion of the works (which it knows the Client will take over) if at all possible and thereby reduce the damages for the remainder of the project. The use by the ECC of the subjective assessment prevents this approach.

In the event that the Completion Date is changed after delay damages have already been paid by the Contractor, clause X7.2 requires the Client to repay any overpayment plus interest. This provision is common in other forms and must be considered as fair and reasonable. In practice, this is a reasonably common occurrence as the agreement of compensation events and their impact on time often lags behind the progress of the works despite the ECC's best efforts and intentions to achieve the agreement of such matters within a short timescale (see Chapters 15–17).

The provisions of Secondary Option X7 can be used in conjunction with Secondary Option X5 by providing a level of delay damages for each section of the works set out under the provisions of X5. An entry is made in Contract Data Part One, separate from the list of sections and completion dates, which lists the sections and the associated rate of damages.

In the UK, until 2015, provisions such as these were only effective if the amount of damages stated was a genuine pre-estimate of the loss that the Client would suffer should the Contractor be in breach of its obligations. If the amount inserted was considered by a court to be a penalty, then the provision would be struck out and the Client would have to prove its actual loss. If, on the other hand, the amount is understated the Client is prevented from claiming any amount over the stated sum.

In 2015 the UK Supreme Court changed this position by its combined decision in *Cavendish Square Holding BV v El Makdessi* and *ParkingEye Ltd v Beavis* [2015] UKSC 67. In this judgement the Supreme Court held that the test is not just about whether the amount is a genuine pre-estimate of the loss. Equally, it was not sufficient to argue that a high figure was a deterrent because such a position can be legitimate in intending to affect how a party behaves. The Court determined that the test was whether the damage figure was unconscionable or extravagant, but in doing so the parties' freedom to contract would be respected. There would be a strong initial presumption that it was the parties who were best placed to determine what was legitimate. It was held that the true test is whether the damage liquidated damage provision is a secondary obligation, which imposes a detriment on the contract breaker out of all proportion to any legitimate interest of the innocent party in the enforcement of the primary obligation. This decision has made it harder to challenge liquidated damages provisions. The lesson is that the Contractor needs to avoid signing up to what it considers to be an unconscionable level of delay damages as it will be extremely difficult to overturn that provision at some later date.

12.8 Practical issues

12.8.1 Showing access as being later than the access date

In Sections 12.2.1 and 12.6.1 it was identified that the Contractor can show later access dates on its programme than stated in Contract Data Part One. In theory, there is nothing wrong with this approach; it does nothing more than reflect the principles of clause 10.1.

Consider that the Contractor shows a later access date than that stated in Contract Data Part One and some other occurrence at the Contractor's risk happens, which results in the Contractor being unable to achieve Completion (see clause 11.2(2) and Section 12.5.2) by the Completion Date (see clause 11.2(3) and Section 12.5.2). The question then arises of whether the Contractor could, in a revision of the programme, show an access date into a particular area earlier than previously shown on the Accepted Programme (but no earlier than that stated in Contract Data Part One). Clause 33.1 is quiet in this respect.

On the other hand, the provisions at clause 32.1 (see Section 12.3) allow the Contractor to make any other change which it proposes (third bullet point) and require him to show how it plans to deal with any delay (second bullet point). The problem for the Client in practice is that it may, being aware of the Contractor's original intentions, have made arrangements for other things to be done in that area until the later access date shown on the programme by the Contractor. Changing these arrangements may be difficult, inconvenient, or even impossible. Without any clear statement in the ECC, the parties may find themselves in a position where a dispute becomes unavoidable. Although the position is unclear, the provisions at clause 31.2 referred to at the start of this paragraph would appear to be in favour of the Contractor and it may prove difficult for the Project Manager not to accept the programme without creating a compensation event (see clause 13.8 and Section 6.2 and clause 60.1(9) and Section 15.3.9) by not operating within the reasons given at clause 31.3 (see Section 12.3).

Without a clear solution to this problem being provided by the ECC, the only sensible practical answer would appear to be to preach extreme caution should a Contractor decide that it can show a later access date than that stated in Contract Data Part One. This extreme caution should be exercised by the Contractor and the Client: the Contractor before showing such a position on the programme and the Client before arranging to use the area for some activity which is not easily stopped or reversed. This is one of those situations where the spirit of mutual trust and cooperation needs to operate before the programme is accepted. The Project Manager, as the person best placed between the Client and the Contractor, needs to act as the catalyst to ensure that both the Contractor and the Client fully understand the implications of the situation. A dispute resulting from the fact that the ECC does not provide a clear answer to the conundrum (which it creates in this circumstance) should thereby be avoided.

12.8.2 Instructions to stop or not start any work

On the practical side, the main issue that users have to consider in connection with any decision by the Project Manager to issue an instruction under clause 34.1 (see Section 12.6.2), whether it is by the Project Manager himself prior to issue or by any other affected party, is the cause behind the instruction. The consideration of this point will provide all the answers for what happens next.

The matter to be established is whether the reason for the instruction is:

- as a result of a fault of the Contractor;
- as a result of a fault or decision of the Client; or
- as a result of something outside the control of either the Client or the Contractor.

Once the reason for the instruction can be categorised into one of the three classes above, then the Project Manager and the Contractor should know how to proceed. If it falls into the first class, then a compensation event will not arise and the Client will have a reason to terminate if the instruction is not rescinded in thirteen weeks. In practice, the most common cause of such an instruction that would fall into this category would be a serious breach of a health and safety requirement, leading to the Project Manager instructing the Contractor to stop work.[16]

Should the reason fall into either the second or third classes, then a compensation event will arise. If it is the second class the Contractor will have a reason to terminate, and if it is the third class both the Client and the Contractor will have a right to terminate. In practice the second class may arise as, say, the result of the Client wanting to stop progress while a major redesign is considered. The third class may occur due to some external factor such as a decision by a government or local authority that results in the project having to stop.

12.8.3 Bonuses for early Completion

In practical terms the provision at Secondary Option X7 (see Section 12.7.3) provides the Client with a mechanism by which the Client can incentivise the Contractor to finish before the Completion Date (see clause 11.2(3) and Section 12.5.2) if it is in the interest of the Client to be able to have use of the project early. The way this provision is drafted means that the Contractor is rewarded for every day it finishes early. If the Client will only benefit if the Contractor finishes, say, three or more weeks before the Completion Date, then the way the provision is drafted will not serve the Client's needs. In such circumstances, the Client needs to consider amending the entry in Contract Data Part One in order to match the bonus payment to the time that it will be of benefit to him and to set the bonus at a suitable level. This could include inserting a lump sum to be paid if the Contractor managed to achieve the objective by or before a stated date.

In practice, this type of bonus has been used to provide an incentive in the private sector where the income stream from the development starts early if the project can be completed earlier. In the UK public sector, a similar provision was used for many years in connection with repair and re-construction works on motorways and trunk roads where the Contractor was paid a bonus for every day it managed to open a section of road, as this was seen to be of benefit to the community and the economy. Like all provisions in the ECC, this provides a framework for the Client to use to suit its own requirements.

12.8.4 Using the bonus for early Completion and delay damages Secondary Options together

Secondary Options X6 and X7 are in many ways diametrically opposed, as one rewards the Contractor for finishing early and the other charges it for being late. It is rare to see

---

[16] This is in addition to a serious breach of health and safety giving rise to a separate reason at clause 91.3 allowing the Client to terminate.

both provisions used on the same project. There is a good practical argument, however, in respect of some projects, that if there is a cost for overrunning, then there must be a saving if the project is finished early. Clients are encouraged to consider this position; they may find that by combining these two provisions they incentivise better performance from their contractors without the true cost actually increasing (any bonus paid is gained by the benefit of early use).

Chapter 13
Payment

13.1 Introduction

It is through the payment process that the overall financial risk profile between the Client and the Contractor is determined. This is achieved, in simple terms, by varying the meaning of the term Price for Work Done to Date (PWDD) between the six Main Options (four of them have the same definition for this term). Within those four Main Options, the financial risk profile is varied by reference to different definitions of other defined terms, to other documents and the inclusion or exclusion of a separate mechanism known as 'the Contractor's share'. This chapter considers all of these variations together with the common processes and the Secondary Options that can be applied to payment.

This is also one of the areas where UK law has to be provided for by separate provisions. This chapter looks firstly at how the common payment procedure for use anywhere in the world applies and then considers the UK-specific option as a separate subject. Users who are operating under jurisdictions other than those of England, Wales, Scotland and Northern Ireland should satisfy themselves that the common payment procedure complies with the statutory requirements in the state in which they are operating.[1] If not, then changes will need to be made; these are not considered within this guide.

13.2 The payment process

13.2.1 Payment intervals and applications

The payment process in the ECC is structured so that interim payments become due to the Contractor with the passage of time, i.e. it is what is often called a periodic payment procedure. The ECC does not provide any option whereby stage payments linked to milestones within the construction process trigger the right to be paid (although Main Option A could be considered to work in this way to some extent).

---

[1] By late 2009 all Australian states had payment legislation in place. At November 2017 similar legislation was in place in New Zealand, Singapore, Malaysia, Ireland and Ontario, Canada. It is possible that it will be introduced in Hong Kong at some future date. Other Commonwealth countries are expected to follow the trend.

A Practical Guide to the NEC4 Engineering and Construction Contract, First Edition. Michael Rowlinson.
© 2019 John Wiley & Sons Ltd. Published 2019 by John Wiley & Sons Ltd.

How the time interval is determined is set out at clause 50.1. To operate successfully, this requires a period to be stated in Contract Data Part One which determines the interval between interim payments. Most parties use monthly for this interval, resulting in 12 payments over each calendar year. Users should note that if they insert 'four weeks' in the Contract Data for this period, then, in any period, there will be 13 interim payments. Equally, if 'five weeks' is inserted in this entry, then there will only be 10 payments in a calendar year. The opening phrase of this clause makes the Project Manager responsible for carrying out the assessment of how much is to be paid.

The Project Manager also determines when the first assessment date shall be. The Project Manager is required to do this to suit the requirements of the Contractor and the Client and to make the first assessment date no later than the stated interval after the starting date (see Section 12.5.1). If the parties require the assessments to be on, say, the last working day of each month but the work started on the 20th of the month, the first assessment would be done at the end of that month and thereafter at the end of each subsequent month. This process then continues until the Supervisor has issued the Defects Certificate (see Section 9.6), i.e. throughout the period following Completion, or until the date of issue of a Termination Certificate (see Section 18.4 by the Project Manager).

Before the Project Manager carries out the assessment the Contractor is required by clause 50.2 to submit a payment application to the Project Manager. This application must set out how much the Contractor considers is due at the assessment date and include details as to how that sum has been assessed. The detail provided must comply with any requirements and form set out in the Scope. This step in the process will not be alien to Contractors. Most, if not all, Contractors will compile such an application and submit it as part of their own processes. As will be seen in Section 13.2.2 below a failure to submit a payment application will have consequences that many will consider to be against the spirit of the ECC (see clause 10.2 and Section 4.2.3).

13.2.2 Timing of payments

The Project Manager is obliged by clause 51.1 to issue a certificate for the amount assessed as being due within one week of the assessment date. This certificate includes details of how the amount due has been calculated. This transparency is of course fully consistent with clause 10.2 and, in the United Kingdom, a requirement of the statute referred to above and detailed in Section 13.9. Clause 51.1 then goes on to state that the amount to be paid is the change in the certified amount, subject to the first payment being the amount of the certificate. This provision caters for the possibility that the certified amount reduces, in which case it is clearly stated that the Contractor pays the Client. All other payments are of course paid by the Client to the Contractor.

The payments are made within three weeks of the assessment date or such other period, whether shorter or longer, stated in Contract Data Part One. This requirement is set out in the first sentence of clause 51.2 and is applicable unless national statute requires something different, as it does in the United Kingdom. In the UK, for projects captured by the relevant legislation, this provision is overridden by Secondary Option Y(UK)2 (see Section 13.9).

If the Contractor has not submitted a first programme to the Project Manager, clause 50.5 states that one-quarter of the PWDD shall be retained until such a submission is made. The wording of this clause is such that the programme submitted must contain all the information which the contract requires. As discussed in Chapter 12, the required contents of any programme submitted for acceptance are so detailed that it will be rare for any such submission to be totally compliant. Indeed, there is often no discernible benefit for the project to go to those lengths; if a programme that is suitable for the scheme and the input to produce it is proportionate to the gain from doing so, then the spirit of the programming requirements will have been satisfied. Unfortunately, the wording of this clause is such that the Project Manager has no option; the deduction is automatic. This provision is seen by many to be contrary to clause 10.2 and unhelpful. Further, as most tenders require a programme to be submitted (which is then referenced in the Contract Data), this provision is bypassed in any case. For clarity, many Clients/Project Managers delete clause 50.5 from the provisions thereby giving the Contractor a signal that they do not intend to use the carrot and stick approach to obtain information.

In the event that any error is made in the assessment of the amount to be certified, then the ECC is quite clear that this is corrected in a later payment certificate. This provision at clause 50.6 does not create the need for additional assessments to those carried out pursuant to clause 50.1 nor, by referring to a 'later' certificate, does it mean that the correction must be in the next certificate. The clause clearly leaves flexibility in this respect.

13.2.3 Interest

All but the first sentence of clause 51.2 plus all of clauses 51.3 and 51.4 are concerned with interest (see also Section 13.11.2). These clauses identify what situations result in the Contractor being entitled to receive interest payments, the periods for which the entitlement runs and how the amount is calculated. The situations resulting in the entitlement are:

- the paying of a certified payment late by the Client;
- a delayed payment because the Project Manager did not issue a certificate[2] which should have been issued;
- the correction of the amount due as a result of a mistake in an earlier certificate;
- the correction of the amount due in relation to a compensation event;
- a delayed payment due to an unnecessary delay to a test or inspection that was required to be done by the Supervisor;
- following a decision of the Adjudicator;
- following a decision of the tribunal; or
- following a recommendation of the Dispute Avoidance Board.

The period over which the interest is applied is either the period of late or delayed payment or the period between when the incorrect amount was certified and when the

---

[2] The wording does not limit the word certificate to payment certificates only. The failure to issue any other certificate that causes a delay to payment, e.g. the Completion certificate, results in the same entitlement.

changed amount is certified. In both cases, the interest payment is included in the next payment certificate or the one in which the correction is made, as applicable. The amount of interest is calculated on a simple daily interest basis, but then compounded annually at the rate of interest stated in Contract Data Part One.

13.2.4 Tax

Should the law of the country where the site is located requires any tax to be paid in addition to the amount due, then clause 51.5 prescribes that the law requires is added to the amount due (see Section 13.4) and paid under the Contract. What this tax will be will vary around the world. In the United Kingdom this will be VAT (Value-Added Tax), which must be added at the relevant rate, which itself will be dependent on the nature of the project. Such additions should not be controversial or create any issues between the Parties.

13.3 Payments in multiple currencies

The final sentence of clause 51.1 details that the currency in which the payment is made is that of the contract; this requires an entry in Contract Data Part One. The last phrase in this clause curiously suggests that a different currency in which payments are to be made may be stated in the contract. This is a rather oblique reference to Secondary Option X3 (Multiple currencies), which can only apply to Main Options A and B.

For Main Options C, D, E and F the question of multiple currencies is dealt with by clauses 50.7 (Main Options C and D) and 50.8 (Main Options E and F).

These three provisions essentially work using the same piece of additional information included in the Contract Data Part One: exchange rates. The standard entries in Contract Data Part One refer to exchange rates as published in a journal to be named and on a date to be stated.

When this facility is used under Main Options A and B with Secondary Option X3, the Client will take the risk created by the potential fluctuation of exchange rates. If the Secondary Option is not used, then the Contractor will carry that risk. Secondary Option X3 also requires that items or activities that are listed in Contract Data Part One are paid for in the other currency or currencies with a maximum amount to be paid for each item in the other currency. Under this option the exchange rates are used to convert the value to be paid for the listed items, as stated in either the Activity Schedule (Option A) or Bills of Quantities (Option B), into the relevant currency up to the stated maximum. These amounts are then paid to the Contractor by the Client in that currency. Should the maximum amount stated in Contract Data Part One be reached, then any surplus value is paid in the currency of the contract. This can be avoided by stating unlimited in the entry.

For Main Options C, D, E and F, under which the payments are all based on actual expenditure by the Contractor (see Section 13.4.5 and 13.4.6), the exchange rates are used the other way around. Here the exchange rates are used to convert the amounts paid in the other currency into the currency of the contract in order to calculate the Fee (see Section 13.4.7) to be paid to the Contractor in the currency of the contract. For Main Options C and D, this conversion is also used to provide a level field for calculating the Contractor's share under clause 54 (see Section 13.7).

13.4 Interim payments – The amount due and the Price for Work Done to Date

13.4.1 Introduction

The object of the assessment carried out by the Project Manager at each interim assessment date (see Section 13.2) is to determine the amount due and therefore the payment to be made, usually by the Client to the Contractor although the payment may be due the other way around. What Section 13.2 did not cover was how the amount due is determined.

Within core clause 5 the amount due depends on whether an application for payment, as required by clause 50.2, has been submitted by the Contractor. It is necessary to consider the amount to be assessed as due when either an application has been submitted or when an application has not been received.

13.4.2 The amount due when a payment application is submitted

When the Contractor has submitted a payment application as required by clause 50.2 then the amount due is simply set out at clause 50.3 as consisting of three elements, these being:

1. the Price for Work Done to Date (PWDD);
2. any other amounts payable by the Client to the Contractor; and
3. any amounts payable by the Contractor to the Client or to be retained from the Contractor by the Client.

The third element above is of course deducted the other two are both positive within the calculation.

Working backwards, the third element acts as a collection point for provisions found in several places throughout the ECC, where the Project Manager assesses amounts to be paid by the Contractor to the Client or where the contract allows the Client to retain amounts from the Contractor. By scrutinising the ECC, the clauses that may give rise to such a deduction are as listed below:

- clause 25.2;
- clause 25.3;
- clause 41.6;
- clause 46.1;

- clause 46.2;
- clause 54.2 (Main Option C only);
- clause 54.6 (Main Option D only);
- clause 82.1
- clause 85.1;
- clause 93.6 (Main Options C and D only);
- clause X1.3 (Main Options A and B only);
- clause X1.4 (Main Options C and D only);
- clause X7.1;
- clause X14.3;
- clause X16.1 (not used with Main Option F);
- clause X16.2 (not used with Main Option F); and
- clause X17.1.

In a similar way the second element acts to collect together other sums, not covered by the PWDD, which may be payable to the Contractor where the circumstances arise under the following clauses:

- clause 51.2;
- clause 51.3;
- clause 54.2 (Main Option C only);
- clause 54.6 (Main Option D only);
- clause 82.2
- clause 86.3;
- clause 93.6 (Main Options C and D only);
- clause X1.3 (Main Options A and B only);
- clause X1.4 (Main Options C and D only);
- clause X6.1;
- clause X7.2;
- clause X12.4(1);
- clause X14.1
- clause X16.3 (not used with Main Option F);
- clause X20.4; and
- clause X22.7(1)

The first element, which should always be the major element of the amount due, is the PWDD. The PWDD has three different definitions depending on the Main Option being used. There are only three definitions, as four of the main options (i.e. C, D, E and F) share the same definition. As we will see in Sections 13.4.3–13.4.6, the definition of the term PWDD will lead users to up to four further definitions (Prices, Defined Cost, Fee and Disallowed Cost) and, depending on the Main Option, one of three major supporting documents (see Sections 13.5.1, 13.5.2 and 14.2); all indicating that this is one of the major terms within the ECC. It can be said with certainty that it is the definition of this term which is the predominant factor in the balance of risk between the Client and the Contractor.

The rules as to how to calculate the amount due will also apply to the final assessment to be carried out under clause 53 (see Section 13.6).

13.4.3 Main Option A

Starting with Option A, the definition of PWDD is at clause 11.2(29) and is stated as being the total of the Prices for activities or groups of activities that have been completed. The definition makes it clear that a completed activity must be free of defects that would either be covered up by or delay immediately following work. In essence, this would appear to introduce a mini completion requirement for each activity as the work progresses, and suggests that the Contractor and the Supervisor need to be operating the testing, inspection and defects regime (see Chapter 9) from the very outset.

As this definition refers to the term 'Prices' with a capital P, it is also necessary to consider the definition found at clause 11.2(32). 'Prices' is defined by reference to the Activity Schedule and consists of the total of all the lump sums included therein for each of the stated activities. The content, preparation and standing of the Activity Schedule are considered in Section 13.5.1; all that needs to be detailed here is the relationship between its contents and the definition of PWDD.

The lump sums from the Activity Schedule which are included in the PWDD are only those for completed activities or groups thereof. As the payment process is carried out at the passing of intervals of time then, in order for the value of any work carried out by the Contractor to be included in the assessment, the relevant activities must be complete. In order to keep its cash flow in a reasonably healthy position, the Contractor needs an Activity Schedule which contains a large number of short-duration activities that are capable of being completed in isolation of other activities. By structuring the Activity Schedule in this way, the Contractor will be able to secure the value of work carried out at an earlier time in assessments; this would not be possible if activities were for a longer duration and span one or more assessment dates. If longer-duration activities are included, it will only be when the activity is complete that the value of work done in the opening days or weeks of the activity is included in the assessment. The Contractor would therefore have to finance the cost of carrying out such activities until they have been completed. On the other hand, it must be recognised that this approach also acts as an incentive for the Contractor to complete activities rather than leaving works at an almost-complete stage.

This definition of PWDD brings another problem for contractors to consider. The definition refers to completed activities but makes no reference to any payment for what are often referred to as 'materials on site', i.e. unfixed goods which have been brought to the Site but have not yet been permanently incorporated into the works. Again, without payment for such items the Contractor will find that its cash flow suffers. The way around this problem is to include the delivery of materials to the site as separate activities. When tackling this problem, careful thought must be given as to how to structure the activities with the Activity Schedule. It is something which is easier to resolve for large single items as opposed to high-volume materials that will be delivered over many weeks or months, e.g. bricks. A single activity for the delivery of say bricks would not be complete until all the bricks required for the scheme had been delivered. This could be a large element of the building so, in order to maintain cash flow, the Contractor needs to plan deliveries in line with the activities for laying the bricks and set up related activities so that the completion of the delivery and laying follow each other in stages. It must be recognised that for the small and everyday materials that are often used over many activities, it is very difficult to come up with a reasonable solution to cater for delivery only that would be acceptable to a project manager.

The reference to groups of activities provides a tool whereby individual activities relating to a larger stage of the project may be grouped together. By adopting this approach, the periodic nature of the payment mechanism can be converted into what will effectively become a stage payment mechanism. Further comment on achieving this position is given in Section 13.5.1.

13.4.4 Main Option B

In Main Option B, the definition of PWDD is found at clause 11.2(30) and is similar to that for Main Option A except that the reference here is to the Bill of Quantities rather than the Activity Schedule. However, the application of the rules is somewhat different. The two bullet points detailing how the PWDD is established refer to, first, the actual quantity of work that the Contractor has completed against each item to which the rate for that item is applied and, secondly, the proportion of any lump sum item included that the Contractor has carried out. The sum for all the items are then added together to establish the total PWDD. The requirement for the work to be free of Defects is the same as for Main Option A.

The way this definition is worded means that partly completed activities will be paid for at each assessment date; this is a proper periodic payment mechanism related to a Bill of Quantities. As the payment is assessed by reference to the actual quantity of work done by the Contractor, rather than the quantities in the Bill of Quantities, it also means that this Main Option creates a re-measurement contract. To that end, the Client takes the risk on the accuracy of the Bill of Quantities, a fact that is confirmed by the additional compensation events (see clauses 60.4, 60.5, 60.6 and 60.7 and Sections 15.3.22–15.3.24 and 13.5.2) which are introduced by selecting this Main Option. That the work needs to be re-measured as it progresses also acts to impose an additional burden on the Project Manager (although considerable assistance in this task will probably be received from the Contractor in practice).

Unlike Main Option A this definition of PWDD does not refer to the Prices. In Main Option B the term 'Prices' is defined at clause 11.2(33) as being the sum of the quantities times each rate and the lump sums. In many ways they are simply symbolic, having no practical relevance to the administration of the contract. As set out in Chapter 17, the significance of the prices is that they are adjusted by the financial consequences of a compensation event.

On the other hand, as for Main Option A, the definition of PWDD does not make provision for the payment of 'materials on site'. The problem with Main Option B is that as the calculation of the PWDD is based solely on items in the Bill of Quantities, this does not leave any mechanism available to the Contractor to obtain interim payment for such unfixed goods. Accordingly, the Contractor will need to include in its pricing structure the cost of financing unfixed materials and will need to develop management systems to reduce the time that materials are stored on site before being fixed. Only by taking these steps can the Contractor reduce the impact of this feature of Main Option B.

13.4.5 Main Options C, D and E

The definition of PWDD at clause 11.2(31) is used in Main Options C, D and E (and also F) and leads directly to two other definitions and indirectly to a third. In order to appreciate the definition of the PWDD under these Main Options, it is worth remembering that the payment is essentially based around what the Contractor spends in doing the works subject to the application of some formulae. In Main Options C and D this amount is then subject to variation as a result of the application of the Contractor's share (see Section 13.7), whereas in Main Option E the payment is simply the amount assessed as being due.

The definition determines that the PWDD consists of two elements – the Defined Cost and the Fee – which are themselves defined terms. The definition also states that the Defined Cost element of the PWDD is that which is forecast by the Project Manager to be paid by the Contractor before the next assessment date. This statement about the forecasting of what will have been paid deserves some attention. What this introduces is a situation whereby what the Project Manager assesses is not the cost of the work that has been done at the assessment date or the Contractor's liability to that date. What the Project Manager is required to assess is the amount (subject to the formulae) that the Contractor will have paid out by the next assessment date, that being the one after the current assessment date that is being considered by the Project Manager. In order to be able to make this forecast reasonably, the Project Manager will need to know the payment terms under every contract entered into by the Contractor in connection with the project. Visibility of such terms is, of course, available to the Project Manager in respect of subcontractors under clause 26.4 (see Sections 8.3 and 8.4). In respect of other payments, the clauses concerning the keeping of records (see Section 13.5.3) should provide this information.

As an example let us assume that the Project Manager is assessing the PWDD at the end of, say, June and that the assessment interval is monthly. The Project Manager therefore needs to forecast what the Contractor will have paid by the end of July. For the people directly employed by the Contractor who are paid either weekly or monthly, the monies forecast to be paid to them in July will be included in the assessment. For other parts of the Contractor's supply chain, such as suppliers of Plant, Materials and Equipment and Subcontractors (see Section 8.3), the Project Manager should include only those amounts that will be paid by (in this example) the end of July. If one of these suppliers submits an invoice at the end of June for materials delivered in June and the payment terms are, say, 28 days, then the amount will be included, as the payment will be before the end of July. However, if the payment terms for this supplier are, say, 35 days, then any goods or services supplied and invoiced at the end of June will not be payable by the Contractor by the end of July and do not qualify for inclusion in the assessment at the end of June; these goods or services would be included in the end of July assessment as the payment would be before the end of August. To take the example a stage further, if the payment terms were, say, 70 days, then the payment for work to the end of June would not be made by the Contractor until early September and these monies would not qualify for inclusion until the end of August assessment.

The statement that has been examined in the preceding two paragraphs was introduced into the third edition as a change to the previous wording in the second edition in order to improve the Contractor's cash flow. The intent of the wording is to make the Contractor's cash flow almost neutral, which is considered by most to be a good thing. A side effect of this process is to remove any benefit to the Contractor of having its supply chain on payment terms that are longer than the assessment interval; such terms simply act to delay the inclusion of an amount into the calculation of the PWDD. Indeed, payment terms that are just longer than the payment interval actually create negative cash flow for the Contractor.

Of the two primary components of the PWDD, the smaller of these, i.e. the Fee, is considered in Section 13.4.7. That leaves us to consider the term 'Defined Cost' which is defined, for these three main options,[3] in clause 11.2(24). This definition is in short referring to the cost of components in the Schedule of Cost Components less Disallowed Cost.

The first of the above elements covers all the work carried out by the Contractor, including that which it has subcontracted. Payment is made for all such work in accordance with the Schedule of Cost Components. This document and its companion, the Shorter Schedule of Cost Components, contain so much detail that they are covered on their own at Chapter 14 of this guide, to which the reader is referred for the necessary detail.

The second element takes users to the third subsidiary definition referred to at the start of this section: 'Disallowed Cost' at clause 11.2(26). This is the longest definition in the ECC, a fact that creates differing views among users. The definition of the term contains 10 classes of cost that can be disallowed. These are split into two broad categories: the first covers five classes which are disallowed on the decision of the Project Manager and the second includes five classes which are disallowed should the stated events apply.

Within the first category, the first class requires the Project Manager to disallow any cost which is not justified by the accounts and records kept by the Contractor and, by implication, made available to the Project Manager (see Section 13.5.3). Experience suggests that this class is likely to be the one that causes more cost to be disallowed than any other, and therefore presents the biggest area for potential differences between the parties. The second class refers to amounts that should not have been paid to a Subcontractor. Reference is made to the subcontract between the Contractor and the Subcontractor as being the baseline against which the payment should be established, which of course is perfectly right and proper. What this class creates is a position whereby the Contractor cannot enter into a commercial settlement of a subcontractor's account without fear of the amount of such settlement being disallowed by the Project Manager. The Project

---

[3] All six main options include a definition of this term. For Main Option F see Section 13.4.6. For Main Options A and B, the term only has significance in respect of the assessment of compensation events; see Chapter 17.

Manager would be entitled to take such a decision and, in doing so, would be acting as required by clause 10.1. However, it is suggested that in practice this is an area where clause 10.2 can prove to be more beneficial for the parties, including the Client.

It is common commercial practice for parties to a contract to reach a commercial settlement over the cost or value of minor items where there may be some differences of opinion. Such settlements are often made as it will cost both parties more than they could possibly gain or save to follow the subcontract to the letter. Under the three Main Options under consideration here, if the Contractor's staff are involved in protracted discussions with a Subcontractor following the letter of the subcontract, it is probable that the cost of this staff time will be paid by the Client as part of the Defined Cost. It is therefore in the Client's interest that the money expended in such discussions does not exceed the benefit it receives in the form of a reduced Defined Cost in respect of the payment made to that Subcontractor. By either keeping the Project Manager informed of the commercial position or by directly involving the Project Manager in the discussions about a possible settlement, the spirit of mutual trust and cooperation can be used to benefit the Client. Providing that the Project Manager is satisfied that such an agreement is to the Client's benefit, then there should be no fear of sanctioning such settlements and confirming that such amounts will not be disallowed under this class of the definition of Disallowed Cost.

The third, fourth and fifth classes share an introduction which refers to them all being costs that the Contractor incurred due to some failure. The failures themselves are the third, fourth and fifth classes of Disallowed Cost. The third class is where the Contractor has failed to follow an acceptance or procurement procedure set out in the Scope. This can only be applied when such procedures have been set out and the Contractor's actions cause additional cost to be incurred by failing to follow these procedures. The fourth class is one of two potential negatives[4] that the Contractor can incur as a result of failing to give an early warning pursuant to clause 15.1 (see Section 6.4). This class will require the Project Manager to decide what mitigating steps could have put in place had the early warning been given and, as a result, how much money would not have been spent in order that the cost could be classified as Disallowed Cost. The fifth class arises from a failure of the Contractor to notify the Project Manager of the preparation for or conduct of proceedings, either adjudication or of the tribunal, in respect of a dispute with a Subcontractor or supplier. Such costs will cover the cost of the Contractor's staff involved in such matters, plus the cost of any consultants or lawyers employed to assist. That such costs can be disallowed confirms that if they are notified then they are allowable.

The second category is introduced by a simple statement referring to the cost of the items within the five classes which, for clarity, are referred to herein as classes six to ten inclusive. Classes six and seven refer to the cost of correcting Defects. In the first instance the reference is to those corrected after Completion, which in itself is easy to determine but can be more difficult to apply. The problem of application arises when the Contractor, as well as correcting Defects at this time, is also carrying out permanent works which were not required to be completed by Completion but which form part of the Contractor's obligations. While disallowing the cost of correcting the Defects is in theory straightforward it is likely that, in such circumstances, there would be some resources which

---

[4] The other one forms part of the compensation event mechanism; see clauses 61.5 and 63.7 and Section 17.5.1.

would be employed in carrying out both types of work. The Project Manager would have to make some form of reasonable apportionment of such shared resources between the two types of activity in order to apply the disallowed cost fairly.

The seventh class concerns the correction of Defects that have been caused by a failure of the Contractor to comply with a constraint contained in the Scope as to how it is to do something. The definition carefully uses the word *constraint* in order to be consistent with the definition of Scope at clause 11.2(16) (see Section 22.2.6). This could open up numerous situations where the Project Manager might have to consider whether a defect has arisen because of such a failure and would apply whether the defect was corrected before or after Completion.

The eighth and ninth classes refer to wasted and unused resources, respectively. In the case of the eighth class it is wasted Plant and Materials beyond what is a reasonable wastage factor. This is relatively straightforward in theory, but the actual application can require a considerable amount of reconciliation between the quantities delivered and the quantities actually required and is then subject to the application of a subjective percentage allowance for what is reasonable wastage. It must be appreciated that the wastage factor is subjective as, for example, this factor applied to, say, facing bricks would vary depending on the facing brick in question. Some facing bricks are by the nature of their manufacture far more susceptible to breaking and chipping than others. Accordingly, the wastage factor for a brittle facing brick would reasonably be higher than the same factor for a tougher brick. All such factors would have to be taken into account when determining whether cost could be disallowed under this class. As a simple example, if 50 000 bricks were required to construct the works and a reasonable wastage factor was 5%, then the cost of any bricks bought by the Contractor in excess of 52 500 could be disallowed by the Project Manager. This class includes a further proviso that it does not apply where the excess has resulted from a change to the Scope. In such instances, if the excess could be returned to the supplier, then a credit would be given to the Client[5]; such occurrences would not be dealt with by use of the disallowed cost mechanism.

The ninth class uses the general term *resources* to identify the class of cost that the Project Manager could disallow. This term is wide and will capture everything used by the Contractor to carry out its obligations, both permanent and temporary, including Plant and Materials although the eighth class expressly deals with these resources. Careful reading of the two classes reveals a difference in the caveats applied, which may mean that circumstances arise, however rare, where it is the ninth rather than the eighth class that is used to disallow costs associated with Plant and Materials. This ninth class refers to availability and utilisation or a failure to take resources away when requested to do so by the Project Manager. A good indicator of forecast resource levels and the activities for which they are required that project managers can use in this respect is the Accepted Programme (see Section 12.2.4). The reference to the failure to take resources away when requested to do so is somewhat curious as there is no provision elsewhere in the ECC that allows a project manager to make such a request. The provision at clause 24.2 (see Section 7.5.1) allowing a project manager to instruct the removal of a person might apply in these circumstances, but that clause refers to an instruction rather than a request.

---

[5] Under item 32 of the Schedule of Cost Components; see Section 14.2.3.

The tenth and final class refers to costs incurred by the Contractor in preparation for and the conduct of any dispute resolution procedures, whether they are in front of an adjudicator or the tribunal. This class is self-explanatory and, subject to the separate rules regarding costs associated with disputes with subcontractors and suppliers (see fifth class above), should not present users with any difficulties.

It can be seen from the above review of the meaning of the term PWDD and its subsidiary defined terms that calculating the amount due under Main Options C, D and E requires consideration by the Project Manager of a significant number of factors. In practical terms, this will be very difficult for a project manager to do without the assistance from the Contractor. Contractors can provide such assistance by submitting an application, which they are required to do by clause 50.3 (see Section 13.4.2) and by assuring that the records they are required to keep are properly available to the Project Manager (see Section 13.5.3). Without such assistance, which commercially it is in the Contractor's interest to provide, the Project Manager will find the task extremely difficult to undertake with any reasonable accuracy.

For completeness, the matter of 'materials on site' as raised in respect of Main Options A and B above is of course not an issue under these three Main Options. If the materials have been delivered and the Project Manager forecasts that the Contractor will have paid for them by the next assessment date, then the Project Manager is obliged to include the value of such unfixed items in the assessment.

13.4.6 Main Option F[6]

The definition of PWDD in Main Option F is the same as in Main Options C, D and E (see Section 13.4.5). However, the definitions of 'Defined Cost' and 'Disallowed Cost', while containing many similarities, do differ as it is necessary to provide a mechanism for calculating the amount due under a Management Contract as opposed to a contract based on a 'cost reimbursable' formula. This form of contract is based around the Contractor managing the construction process and subcontracting all but a few limited operations that the Contractor itself may carry out. The definitions are structured in order to cater for that difference.

In order to prevent repetition, this section will refer to Section 13.4.5 whenever there is commonality between the terms referred to above and those under Main Option F.

The definition of Defined Cost is at clause 11.2(25) and differs from that above as it includes three elements, these being:

1. payments to Subcontractors;
2. the prices for work done by the Contractor; and
3. disallowed cost.

The first element is easy to understand and should constitute most of the value at any particular time. This is because under this procurement route it will be the Subcontractors (often called Work Contractors or Trade Contractors under other contracts covering

---

[6] Main Option F is not included in the NEC4 Engineering and Construction Subcontract.

this type of arrangement) who carry out the work. The Contractor will need to ensure that the Subcontractors are entitled to be paid the whole of the sum included under the terms of the relevant Subcontract, otherwise Disallowed Cost will come into play (see below).

The second of the three elements refers to the prices for the work done by the Contractor. This is different to Options C, D and E, which refer to 'other work' in the equivalent element. Under this Main Option, the Contractor includes its prices (which does not have a capital 'p'; see clause 11.1 and Section 6.6.1) in Contract Data Part Two. These are fixed prices for the parts of the operation that the Contractor is to do itself, including its management of the project.

The third element differs from Main Options C, D and E only as set out in the following paragraphs.

It is at clause 11.2(27) that the definition of Disallowed Cost is found in Main Option F. This contains eight classes of cost that can be disallowed, the first five of which are identical to the five classes in the first category of clause 11.2(26) (see Section 13.4.5 above). The sixth and seventh classes concerns payments made to Subcontractors in connection with, firstly, work which the Contract Data states the Contractor is to do and the cost of which is therefore included in the prices and, secondly, any of the Contractor's management functions or activities which are also included in the prices or the Fee (see Section 13.4.7). The eighth class is the same as the tenth class in clause 11.2(26) (see Section 13.4.5 above).

While it can still be complicated, this version of the PWDD is easier to calculate than under Main Options C, D and E; the majority of the work is subcontracted and that which is not is covered by quoted prices rather than being subject to the detail contained in the Schedule of Cost Components. That said, as with the other three cost-based options, the provision of an application by the Contractor and good access to the records will be necessary if the Project Manager is to produce a reasonably accurate assessment of the amount due.

Within this Main Option, the issue of 'materials on site' will usually depend on the terms and conditions of the subcontracts between the Contractor and the Subcontractors. If an NEC is used, then the rules in the preceding sections will apply as applicable.

13.4.7 Fee

The Fee, as we have seen in Sections 13.4.5 and 13.4.6 regarding Main Options C, D, E and F, forms part of the calculation of the PWDD in all four cases. The term is defined at clause 11.2(10) as being the amount calculated by applying the fee percentage stated in Contract Data Part Two to the Defined Cost.

Clause 52.1, which being a core clause, is common to all the Main Options, makes it clear that any of the Contractor's costs which cannot be allocated to one of the elements of the Defined Cost are included in the Fee. Contractors should therefore consider carefully what costs associated with the project can be recovered through that mechanism, and ensure that the percentages are sufficient to cover every cost that cannot be recovered under that term. This means that a contractor's standard head office overheads percentage may not be sufficient for many projects where costs that a contractor might allocate to a project do not qualify as Defined Cost under the Main Option in use. Many contractors have been caught out by this provision.

13.4.8 The amount due when a payment application is not submitted

Clause 50.4 sets out that, in the event that the Contractor does not submit an application before the assessment date, then the amount due is the lower of the amount the Project Manager assess to be due under the rules set out in clause 50.3 (see Section 13.4.2) (i.e. as if the Contractor has submitted an application) and the amount due at the previous assessment date.

The way that this is written means that if the Contractor omits to submit an application before the assessment date, then either the Contractor will not receive any payment from the Client or the Contractor will make a payment to the Client. The most likely position will be that, despite having carried out considerable work on Site during the period, the Contractor will not receive any payment. As in such circumstances the amount due will go up, then the previous assessment amount due will be lower and therefore the amount that applies.

It is theoretically possible that the amount due at the assessment date will be lower than certified at the previous assessment date. The author has seen such occurrences on a very small number of occasions in his 40-year-plus career. This has resulted from significant defects being found causing the value to be abated by significant sums, or the correction of an arithmetical error that existed in the calculation of the previous amount due, or the firming up of the value of compensation events against which an 'on account' amount had been included previously. In such rare but not unknown circumstances, then, the amount due would go down if the Project Manager assessed that this was the case.

Whilst considering clause 50.4 it is necessary to appreciate that the clause refers to the application received 'before' the assessment date. This would not include being received on the assessment date. Receipt of an application on the assessment date would not be 'before' the date and accordingly would not satisfy clause 50.3 (see Section 13.4.2). Contractors in particular need to be very aware of this distinction.

13.5 *Supporting documents and records*

13.5.1 Activity Schedule

The definition of PWDD (clause 11.2(29)) in Main Option A refers, via the separate definition of Prices (clause 11.2(32)), to a supporting document called the Activity Schedule. This term is defined at clause 11.2(21) by reference to the same term but in lower case and italics, leading the user to an entry in the Contract Data Part Two. As an aside, users will note that this entry in Contract Data Part Two refers to both Option A and Option C. The Activity Schedule plays a different role in Option C (see below) and, as a result, can be formatted differently.

Section 13.4.3 for Main Option A describes the relationship between the contents of the Activity Schedule and the definition of PWDD. What is to be considered here is the content, preparation and standing of the Activity Schedule. The ECC does not prescribe which party shall prepare the Activity Schedule or set any format for the document. This is left entirely to the parties in order that something suitable for their needs can be produced. However, what the ECC does provide is three subclauses that provide some parameters and standing for the Activity Schedule. Clauses 55.1 (Option A) and 55.2 (Option C) confirm that the contents of the Activity Schedule do not form part of either the Scope or the Site Information, which makes perfect sense. Clause 55.1 also requires the Contractor to correct the Activity Schedule if the activities in it do not relate to the Scope. The combination of clauses 55.3 and 55.4 determine that the contents of the Activity Schedule should relate to the activities on the Accepted Programme (see Section 12.2.2), relate to the Scope and that the total of all the activities should equal the total of the Prices (i.e. the tendered sum at the outset and the adjusted contract sum at the end[7]). This can be provided in as many or as few activities as required. What users of Option A should always remember is the practical application of the PWDD as covered in Section 13.4.3. This suggests that, under Option A, the duration of each activity should be short in order that it can be completed and included in the PWDD.

In practice, it appears to be common that the preparation of the Activity Schedule is left to the Contractor. It also appears to be common that the Activity Schedule is not provided with the tender but only called from and prepared by the successful contractor, thereby reducing the tendering costs of the unsuccessful contractors. Where this is the practice adopted, it is within the power of the Contractor to formulate the Activity Schedule in order to protect its cash flow as best as it is able within the meaning of PWDD.

If on the other hand the Client decides that it wants to predetermine the form of the Activity Schedule, it is suggested that the Client can do this by either providing a predetermined schedule for the tenderers to complete or include its requirements within the Scope. By using this technique or by requiring all the activities for stated stages of the works to be grouped together, the Client can change the periodic payment nature of the ECC into a stage payment mechanism as favoured by some design and build type contracts. This method impacts on the Contractor's cash flow and should therefore only be used after careful consideration and recognition that it may cause the tendered prices to increase to cater for the financing of the project.

The number of activities in the Activity Schedule for Main Option C is not an issue. Under this option the Activity Schedule is the total of the Prices at tender, which is then adjusted up or down for the effects of compensation events (see Section 17.2). The level of detail has no impact on the interim payment process. The only use of the total of the Prices (as adjusted) and therefore the Activity Schedule, once the Contract has been formed, is in the Contractor's share mechanism at clause 54 (see Section 13.6).

---

[7] This will not necessarily match the final amount due as the PWDD is subject to additions and deductions as set out at clause 50.3; see Section 13.4.2.

13.5.2 Bill of Quantities

Main Option B also makes reference to a supporting document called the Bill of Quantities, but this time directly in the definition of PWDD at clause 11.2(30). The term Bill of Quantities is itself defined at clause 11.2(22) by reference to the same term but in lower case and italics. In this instance, the user is led to an entry in Contract Data Part One (which identifies the standard method of measurement and any amendments used to produce the bill of quantities) and a second entry in Contract Data Part Two (which identifies the actual document itself). As an aside, users will notice that both these entries in the Contract Data refer to both Options B and D. While the role of the document in the two main options differs, the method of production and associated risk is exactly the same.

The ECC does not specify which party should prepare the bill of quantities but, as Contract Data Part One requires an entry in connection with the document and the Client will have selected one of the two main options that makes use of it, it would normally be the Client who arranged for its production. This would be consistent with the traditional way, certainly in the UK, that such documents have been used under various contract arrangements over many decades. However, it is perfectly feasible that the Contractor could be required to produce the bill of quantities and the document so prepared could be referenced in the Contract Data, even though this would be out of the ordinary.

While feasible, the second of these two options (i.e. the Contractor prepares the bill of quantities) is not advisable: assessing the PWDD under Main Option B involves remeasuring the quantities; under Main Option D the Total of the Prices (being a factor used in the calculation of the Contractor's share; see Section 13.7) is finalised from a re-measure of the quantities; and both Main Options B and D include compensation events (see clauses 60.4–60.6 and Sections 15.3.22–15.3.24), which clearly place the risk of the accuracy of the bills of quantities on the Client. Accordingly, that risk is probably easier to control and manage or lay off[8] if the Client makes the necessary arrangements to prepare the bill of quantities.

Both Main Options B and D include the single clause 56.1 that refers to the Bill of Quantities and confirms, as one would expect, that information in that document is neither Scope nor Site Information. However, users should note that in this context under clause 60.7, which applies to Main Options B and D, the Contractor is assumed to have taken the Bill of Quantities as being correct in respect of any compensation event that arises as a result of the correction of any inconsistency between any other document and the Bill of Quantities.

It is worth noting at this point that the use of a bill of quantities, under Main Option B or D, increases the administrative burden on the parties. In particular, the use of a bill of quantities through Main Option D, given the re-measurement aspect of that option, seems contrary to the principle of target contracting.

---

[8] In this context, 'lay off' means by subcontracting the preparation of the bill of quantities and therefore creating a contractual link between the Employer and the preparer under which the Employer may have a claim for errors or negligence as the terms of that subcontract permit. Such a subcontract would be outside the ECC contract between the Employer and the Contractor.

13.5.3 Records

Main Options C, D and E all contain the same two clauses, 52.2 and 52.4, which refer to the keeping of records and the making of such records available to the Project Manager. In Main Option F the two operative clauses, covering the same requirements are clauses 52.3 and 52.4. The keeping of records by the Contractor and the availability for inspection is vital under all of these Main Options in order to allow the Project Manager to fulfil the obligation to assess the amount due. Set out at the four bullet points in clause 52.2 is a requirement for the Contractor to keep accounts of payments made that qualify as Defined Cost (Subcontractors in Option F), together with proof that those payments have been made, records in connection with the assessment of compensation events for all of the Contractor's Subcontractors and any other records that are stated to be required by the Scope.

On the face of it, this is all straightforward and should not present any real issues. Note, however, that the way in which the fourth bullet point is worded implies that the other records to be kept, as required by the Scope, are not restricted to only accounting records. The wording is loose enough that, despite the clause being part of core clause 5 (Payment), it acts to catch all other records, of whatever type, required by the Scope. This is one of those little quirks of the ECC that does not appear to cause problems for parties in practice.

The second of these two clauses requires the Contractor to make all these records available to the Project Manager at any time during working hours. A statement as to what the working hours are is not something that is required to be made in the Contract Data; it is suggested that, in practice, it is a good idea to state these as a constraint in the Scope. Another quirk in the wording here is that clause 52.4 refers to the Project Manager being allowed to 'inspect' the records; it doesn't say anything about him being allowed to take or receive copies of the information. As this frequently becomes a point of contention, if the Project Manager wishes to take copies of any or all of the records, a simple amendment to clause 52.4 should suffice.

As Main Options A and B do not contain any equivalent clauses to these two, the only records that the Contractor will be obliged to provide under those Main Options will be those generated in respect of tests and inspections as a result of the requirements of clause 41.3 (see Section 9.3). Should the Client or Project Manager require other records to be provided, then it would be prudent to add clauses similar to clauses 52.2 and 52.4 to these main options via Option Z.

13.6 Final assessment

Clause 53.1 requires the Project Manager to carry out a final assessment of the amount due and to then certify a final payment, if anything is due, no later than two potential points in time, one of which must apply to every contract. The first of these points in time is four weeks after the issue of the Defects Certificate by the Supervisor (see clause

44.3 and Section 9.5). This will be the most common of the two points in time to occur. The second point in time is thirteen weeks after a termination certificate has been issued by the Project Manager (see clause 90.1 and Section 18.4). This will be a relatively rare occurrence but is not unknown.

Having made the assessment the Project Manager is required to give the Contractor details of how the amount due has been assessed. Clause 53 does not contain any rules as to how the amount due is to be calculated or what it consists of. Without any statement to the contrary then, reference to clause 50.3 (see Section 13.4.2) should be made to determine what should be taken into account. Clause 50.3 is the only clause in the ECC which details the components of the amount due. Clause 53.1 then continues by stating that the final payment is made within three weeks of the assessment or a different period if one is stated in Contract Data Part One. Clause 51.1 (see Section 13.2.2) will need to be applied in order to determine whether the payment is made by the Client to the Contractor or by the Contractor to the Client.

Unlike many other provisions in the ECC, clause 53.2 then sets out what can happen if the Project Manager does not make the final assessment within the time period allowed (i.e. four or thirteen weeks depending on which point in time triggers the requirement). This allows the Contractor to issue an assessment of the final amount due to the Client, not the Project Manager. The assessment so issued must give details as to how the amount due has been calculated. If the Client agrees with the assessment, then a payment is made by the resulting payee to the other party within the same timescale as set out in clause 53.1.

Clause 53.2 does not say what happens if the Client does not agree the assessment which is issued to the Client by the Contractor. The user needs to refer to other clauses to see what might happen. At clause 53.4 it is stated that the final amount due is changed based on any agreement of the Parties. This suggests that the Parties can endeavour to reach agreement over the final amount due. This is always the better option when compared to the other options available.

These other options will all involve referring the difference to dispute resolution under which ever of Options W1, W2 or W3 is in use. Clause 53.3 anticipates one of these Options being used once the final assessment has become final and conclusive. That the final assessment becomes final and conclusive is set out in the opening part of clause 53.3. This applies to any final assessment that is issued in the period stated in the contract. So if the Project Manager issues a final assessment with the period allowed for its issue under clause 53.1 (see above), then that becomes final and conclusive. What is not clear is if the Project Manager does not issue the assessment in that permitted time will any assessment issued by the Contractor, which does not include a required time period, also become conclusive? Volume 4 of the User Guide says that clause 53.3 applies to the final assessment whether produced under clause 53.1 by the Project Manager or under clause 53.2 by the Contractor. The author is not convinced of this position by reading the wording of the clauses alone.

Regardless of what the User Guide says or the author thinks, what happens next in either case will be a reference to dispute resolution under the relevant Option. Clause

	If Option W1 applies, a Party refers any dispute or issue:	If Option W2 applies, a Party refers any dispute or issue:	If Option W3 applies, a Party refers any dispute or issue:
Step 1	to the *Senior Representatives* within 4 weeks of the assessment being issued	to the *Senior Representatives* within 4 weeks of the assessment being issued OR to the *Adjudicator* within 4 weeks of the assessment being issued	to the Dispute Avoidance Board
Step 2	to the *Adjudicator* within 3 weeks of the list of issues not agreed being produced	if Step 1 was to the *Senior Representatives*, then to the *Adjudicator* within 3 weeks of the list of issues not agreed being produced	to the *tribunal* within 4 weeks of the recommendation being made
Step 3	to the *tribunal* within 4 weeks of the decision being made	to the *tribunal* within 4 weeks of the decision being made	

Figure 13.1 Steps under the Dispute Resolution Options.

53.3 sets out for each Option the steps that can or must be taken through the dispute resolution hierarchies. Each step comes with a time period, which if not complied with will result in the assessment becoming final and conclusive either as originally issued, or if a step or steps in the dispute process have been taken as changed by that step. Clause 53.4 also establishes that any decision or recommendation arising from dispute resolution also acts to change the final amount due. The steps required under the three available dispute Options are tabulated in Figure 13.1.

13.7 The Contractor's share

'The Contractor's share' is the term used in the ECC to refer to what is commonly called the 'pain/gain mechanism' by people who work with target cost contracts. Within the ECC Main Options C and D are referred to as 'target contracts', the first using an Activity Schedule and the second a Bill of Quantities (see Sections 13.5.1 and 13.5.2, respectively). The idea of such contracts is that a target is established and then the Contractor is paid its cost for carrying out the work. Should the cost be less than the target, then the Contractor is paid a share of the saving (the gain). Conversely, should the cost be more than the target, the Contractor pays its share of the overspend (the pain). In order to make them work, this type of contract requires some form of mechanism or formula to determine what the Contractor's share is at the end of the project.

In Main Option C this mechanism is set out at clauses 54.1–54.4 inclusive; for Main Option D clauses 54.5–54.8 apply. These two sets of four clauses are identical except for one minor difference which applies in all four clauses in each Option. This difference is that clauses 54.1–54.4 use the phrase 'total of the Prices', whereas clauses 54.5–54.8 use the term 'Total of the Prices'. Following the rule at clause 11.1, users will quickly see that the phrase at Option C includes the term 'Prices' as defined at clause 11.2(32) (see Section 13.4.3). On the other hand, at Option D the term 'Total of the Prices' is defined at clause 11.2(35), for which the material definition is exactly the same as the definition of PWDD for Option B at clause 11.2(30) (see Section 13.4.4). This phrase for Option C and term for Option D are, as a matter of practice, the target against which the spend will be compared. In simple terms, for Option C the target is the tender sum adjusted up and/or down for compensation events. For Option D, the target is a re-measure of the actual work carried out against the Bill of Quantities. Option D therefore carries more risk for the Client.

Under both Options (see clauses 54.1 and 54.5 for Options C and D, respectively) the Project Manager has the task of assessing the Contractor's share of the difference between the target and the PWDD.[9] The Project Manager is only required to carry out this task on two occasions. The first is in the payment immediately following Completion (see clauses 54.3 and 54.7 for Options C and D, respectively) based on forecasts of the target and the PWDD. The second is in the final assessment (see clause 53.1 and Section 13.6), which should be four weeks after the Defects Certificate (see Section 9.5) has been issued by the Supervisor (see clauses 54.4 and 54.8 for Options C and D, respectively) at which time the final target and final PWDD will have to be known. In the event that termination has occurred (see core clause 9 and Chapter 18) then the final assessment will be no later than thirteen weeks after the termination certificate has been issued (see clause 53.1 and Section 13.6).

Clauses 54.1 and 54.5 (for Options C and D, respectively) set out in words how the Contractor's share is calculated. In order for this exercise to be possible it is necessary to have the share ranges and share percentages set out in Contract Data Part One. The share ranges are expressed as percentage bands or increments, e.g. '90%–110%'. The percentage is established by dividing the PWDD by the target. If there is more than one share range, which in practice there sometimes is, then the saving or overspend is allocated into the relevant share ranges to suit the percentages. The share percentage is then applied to the amount allocated to each share range to calculate the Contractor's share for each share range. The Contractor's share for each of the share ranges is then summed to establish the total of the Contractor's share. In accordance with clauses 54.2 and 54.6 (for Options C and D, respectively) where there is a saving (i.e. the PWDD is less than the target), the Client pays the Contractor the Contractor's share. Where there is an overspend (i.e. the PWDD is more than the target), the Contractor pays the Client the Contractor's share.

These two groups of four clauses are far easier to understand when demonstrated by worked examples. Such worked examples were provided in the NEC3 ECC Guidance Notes (NEC Panel, 2013k) (at page 66) but these are absent from Volume 4 of the NEC4 ECC User Guide (NEC Panel, 2017e). I refer readers back to the NEC3 ECC Guidance Notes for these examples.

---

[9] See Section 13.4.5 for how the PWDD is established for both Main Options.

There are, however, some serious practical implications that can arise from this mechanism of which novice users should be aware; these are covered in Section 13.8 below.

13.8 The Contractor's share: Practical issues

13.8.1 Principles and accuracy

The principles behind the Contractor's share are not difficult to understand. In simple terms, it presents the users with an incentive mechanism to encourage the Contractor to use all of its skills and experience to reduce the cost of constructing the project in return for a share of the saving made. Equally, it provides a mechanism whereby if the Contractor does not perform in line with the assumptions used to arrive at the target, then it will pay its share of such inefficiencies. The key to this mechanism being fair to both parties is that the target is both reasonable and robust at the outset. It is therefore necessary that the detail available to prepare the target is in an advanced state. Preparing a target from sketchy details will inevitably lead to the inclusion of risk factors and result in a proliferation of compensation events that adjust the target as the Scope is changed to reflect exactly what the Client requires as the project progresses.

The procurement and contract strategy guidance at Volume 1 of the User Guide actually states that target contracts are typically used where the detail is not fully defined. This is indeed common and acceptable. However, users who select this method of contracting where the detail is not developed must realise that one or both parties will carry greater risk. The Contractor will price that risk element, and it is likely that the project will require more resources in order to manage all the compensation events that will arise. To this end, it should be recognised that one of the aims of the ECC is that, no matter which Main Option is being used, the financial aspects are resolved as the job proceeds and that the final figures are known either at Completion or very soon after, subject to the impact of any works carried out after Completion but before the defects date.

13.8.2 'Win–win' outcome

What the Contractor's share mechanism will show is that if those involved work effectively together and reduce the actual cost of carrying out a project through efficiencies, thereby reducing the PWDD, then the Contractor's share (which will usually be bonus profit) will increase as the amount the Client pays decreases. This is a win–win situation where both parties benefit from working together to construct the project efficiently. In such situations, the Contractor's share mechanism only benefits the Client and the Contractor. In practice, the Project Manager and the Supervisor can both have a significant influence over the success of a project. It is worthwhile for the Client to consider how the separate contracts between it and the Project Manager and Supervisor, in which roles external consultants are employed, can be structured to incentivise these two agents to achieve the goal of lowering the Client's overall expenditure without compromising quality.

13.8.3 Guaranteed Maximum Price

The variation in the entries made in Contract Data Part One, in respect of the share percentages and share ranges, has no bounds. One particular combination, which is used by several clients that the author is aware of, creates a position where a target contract becomes subject to what is effectively a Guaranteed Maximum Price (GMP). Under these arrangements, such a GMP is always subject to adjustment, up or down, by compensation events; if exceeded, it still has severe consequences for the Contractor. The GMP is achieved by using a share range where the percentage is stated as being 'greater than X%', where X will be 100% or more, and a corresponding share percentage of 100%. This means that once the PWDD exceeds the X% of the target the Contractor's share of that overspend is 100%, i.e. all of the overspend from that point on, hence the GMP. Many contractors who have been novice users of the ECC have inadvertently entered into such arrangements without being aware of the risk that they have taken on. On the other hand, users regularly and knowingly enter into such arrangements but, being aware of the risk, price the target accordingly. This is just another of those risk elements that contractors take account of and manage.

13.8.4 Target exceeded before Completion

Clients who enter into target contract arrangements under the ECC must be aware of the effect of the combination of the payment process (see Section 13.2) and the timing of the application of the Contractor's share mechanism (see Section 13.7), should the target be exceeded before Completion: the Client will be required to pay the Contractor a greater sum in the interim payments between the time that the target is exceeded and the payment immediately after Completion than would otherwise be due to the Contractor at the payment immediately after Completion. The net result of the payment immediately after Completion will, in such circumstances, require a payment from the Contractor to the Client.[10] This obviously presents a risk to clients who may pay more money than would be due to a contractor who is possibly in a loss-making situation on that project. Faced with this situation, many clients have simply refused to make any further payments. Such an action is a breach of the contract that will give rise to a right to interest under clause 51.2 (see Section 13.2.3). This is a potential reason for the Contractor to terminate the Contract at clause 91.4 (see Section 18.2.2) and, in the United Kingdom, the ability for the Contractor to suspend performance under the Act (see clause Y2.5 and Section 13.9).

13.9 Special provisions for the United Kingdom

In the United Kingdom, any project that qualifies as a 'construction contract' as defined by Sections 104 and 105 of The Housing Grants, Construction and Regeneration Act 1996 as amended by the Local Democracy, Economic Development and Construction Act

---

[10] That such a position could arise is envisaged by clause 51.1; see Section 13.2.

2009 is required to incorporate payment provisions that comply with Sections 109–113 of that Act.[11] These are statutory provisions that cannot be avoided by any means whatsoever. In the event that the necessary provisions are not incorporated into a qualifying contract, then the Act implies the necessary terms from a supporting Statutory Instrument called The Scheme for Construction Contracts (England and Wales) Regulations 1998 SI 649, amended by The Scheme for Construction Contracts (England and Wales) Regulations 1998 (Amendment) (England) Regulations 2011 SI 2333. When using the ECC and where the statute requires these provisions to be included in the contract, then the use of Secondary Option Y(UK)2 partially satisfies this requirement (see below).

Clause Y2.1 adopts the approach set out in Section 116 of the Act in respect of how periods of time stated in days are calculated. This is an important provision in practice as it determines that where periods are stated in days, which is the norm throughout this Secondary Option, then those days are calendar days except for days that are a bank or public holiday[12] are not included in such a calculation. Day one of such periods starts the day after the event that triggers the period, which is stated in the Act but not the contract.

Clause Y2.2 sets out that interim payments become due (a term used by the Act) 7 days after the assessment date and that the final date for payment (another term used by the Act) is 14 days after the due date unless an alternative period is stated in Contract Data Part One. The clause also provides that the Project Manager's certificate serves as the payment notice to the Contractor specifying what the Act calls the 'notified sum' and stating how that sum is arrived at, thereby satisfying that further requirement of the Act.

Clause Y2.2 also sets the due date for the final payment following the issue of the final assessment under clause 53 (see Section 13.6). If the Project Manager has made the assessment after issue of the Defects Certificate, then the due date is five weeks after the Defects Certificate was issued. If the Project Manager fails to do the assessment as required by clause 53.1 (see Section 13.6) and the Contractor then issues a final assessment under clause 53.2 (see Section 13.6), then the due date is one week after the Contractor issues the assessment. If the Project Manager has made the assessment after issue of a termination certificate as required by clause 53.1 (see Section 13.6), then the due date is fourteen weeks after the termination certificate was issued. Two of these variants create the position where the due date for the final assessment is one week after the last date for the issue of the final assessment by the Project Manager. Where the Contractor issues the final assessment, then it is one week from that issue.

The requirement at section 111 of the Act, that a notice must be given a specified number of days before the final date for payment if a party intends to pay less than the notified sum that is due to the other party, is satisfied by clause Y2.3. Any party that issues such a pay less notice must be aware that the courts have found the timing of this notice to be

---

[11] The Housing, Grants and Construction Act 1996 as amended by the Local Democracy, Economic Development and Contraction Act 2009 applies to all new contracts entered into from 1 October 2011. (Readers should note that the 1996 Act applied to all new contracts entered into from 1 May 1998).

[12] At the time of writing these are 1 January (New Year's Day), Good Friday, Easter Monday, Early Spring (first Monday in May), Late Spring (last Monday in May), Summer (last Monday in August), Christmas Day and Boxing Day, giving 8 in total. Additional public holidays are sometimes given for special events such as the wedding of Prince William (2011) and the Queen's Diamond Jubilee (2012).

critical. If it is late, then it is ineffective; there are no grey areas as far as this interpretation is concerned.

A further provision allowed by the Act is included at clause Y2.4. This addresses the position where a payment has been certified but not paid at the date of a termination certificate (see clause 90.1 and Section 18.4). The Client must make the payment unless a pay less notice has been issued under clause Y2.3 or the reason for the termination was one of Reason R1 to R10 (see Section 18.2.1), i.e. the Contractor has become insolvent. In the case of insolvency if the termination certificate was issued before the last day on which a pay less notice could be issued, then that pay less notice must be issued.

As stated at the end of the introductory paragraph to this section, not all the provisions of the Act are provided for in Y(UK)2. Section 112 of the Act (which allow a payee to suspend performance of part of all of its obligations after the expiry of a seven-day notice that a sum due has not been paid by the final date for payment) has not been included as an express provision within this Secondary Option. Clause Y2.5 states, however, that if the Contractor exercises its statutory right, then a compensation event arises. This provision means that the ECC satisfies the Act as far as the required compensation to the payee is concerned.

The amendments introduced by the Local Democracy, Economic Development and Construction 2009 also provide for a default system should the Client or the person designated to act in its stead (the Project Manager in this instance) fail to issue the payment due notice specifying the 'notified sum'. The Act now allows for two alternatives. First, if the payee (i.e. Contractor) has issued an application for payment (see clause 50.4 and Section 13.2), then in the absence of a payment due notice the amount applied for by the payee becomes the 'notified sum'. Subject always to the issue of a pay less notice pursuant to clause Y2.3, the amount applied for would then have to be paid by the final date for payment. Second, if the payee has not submitted an application, then at any time after the payment due notice should have been issued the payee may submit a default notice stating how much it considers it is due to be paid. This sum becomes the 'notified sum'. The final date for payment is delayed by the number of days between the date when the payment due notice should have been issued and the date when the default notice was received by the payer. Subject always to the issue of a pay less notice pursuant to clause Y2.3, the amount stated in the default notice would then have to be paid by the final date for payment. The ECC does not include any such default provisions in respect of interim payments (see Section 13.4); nonetheless, the payee will always be able to rely on the default mechanisms stated in The Scheme for Construction Contracts (England and Wales) Regulations 1998 (Amendment) (England) Regulations 2011 SI 2333.

Why the default provisions provided by the Act have not been incorporated in connection with interim payments is not explained in the User Guide.

This process, excluding the default mechanisms described in the preceding two paragraphs, is shown at Figure 13.2, for interim payments.

Should the Project Manager fail to issue the final assessment as required by clause 53.1 (see Section 13.6), then clause 53.2 (see Section 13.6) allows the Contractor to submit its own final assessment. This would become due one week later as determined by clause Y2.2. This position satisfies the default requirements in the Act.

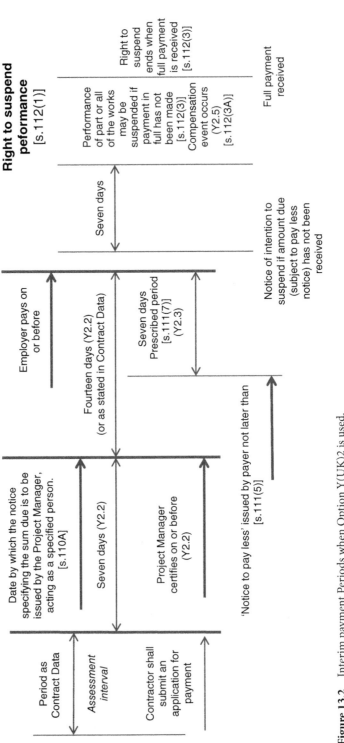

Figure 13.2 Interim payment Periods when Option Y(UK)2 is used.

The odd thing about this Secondary Option is that it does not omit the payment provisions and periods contained in clauses 51.1 and 51.2. That leaves the contract in a position that, when the Secondary Option is used, two conflicting periods for payment will be included. Volume 4 of the User Guide says that the Secondary Option 'supplements the core clause', a statement which must be held in doubt. The likely outcome of the scrutiny of this conflict by a court would probably be that, as the parties had elected to incorporate optional terms, then those optional terms would take precedence over the standard (core clause) terms. As no party has yet elected to waste legal fees exploring this point this view, this is of course speculation which might be incorrect. I have not yet encountered this duplication as being a problem in practice.

13.10 Related Secondary Options

13.10.1 X1: Price adjustment for inflation (used only with Options A, B, C and D)

For the fixed price and target contract options, users can choose to provide provisions for price inflation. In the UK these types of provisions were common in times of high inflation but have not been included in the majority of contracts over the past 25 or so years of low inflation. By selecting this Secondary Option, the Client will assume the risk for any price inflation that occurs over the duration of the project.

The mechanism set out in the option works by means of an index which is used to account for changes from a base date to the date of the assessment. This is the common way that all such indexes work. The definition of the terms used are in clause X1.1. Clause X1.2 determines that any change in an index after it has been used is not corrected.[13] It also sets out the Price Adjustment Factor calculated for the last assessment before Completion (see clause 30.2 and Section 12.5.2) is used for all assessments after that date. Details of the index to be used, any subdivision of the Prices or Defined Cost into proportions of the work and the base date all require entries in Contract Data Part One. The amount for price inflation to be included in each assessment is set out at clause X1.3 for Options A and B and at clause X1.4 for Options C and D. Simple worked examples for both clauses are included on pages 85 and 86 of Volume 4 of the User Guide for anyone who is not familiar with this methodology.

In respect of compensation events clause X1.5 states that Defined Cost (see clause 11.2(23) and Section 17.2.4 for Options A and B, or clause 11.2(24) and Section 13.4.5 for Options C and D) at base rate levels is used for assessing people and Equipment for compensation events. Should any cost forming part of Defined Cost in the assessment of a compensation event be at current rates then the indices are used to discount the current cost back to the base date; this allows the compensation event to be included in the main calculation for price inflation at the time the work is included in the assessment.

---

[13] It is common practice for the compilers of such indices to issue a provisional value for the current month which is then firmed up, say, three months later when the compilers have verified all the data. It will be the provisional value which is used in Option X1.

13.10.2 X14: Advanced payment to the Contractor

The Client can elect to make an advanced payment to the Contractor. This decision is usually only made in situations where the Contractor is required to make a large initial investment in either Equipment or by paying significant funds to secure Plant or Materials that have a long lead-in period. Secondary Option X14 provides the necessary conditions to facilitate such a payment and for its repayment by instalments via the periodic interim payment mechanism (see Section 13.4). Provision is also included for a bond to protect the Client if required.

Clause X14.1 imposes an obligation on the Client to pay the amount of the advance payment to the Contractor either as part of the first assessment (see clause 50.1 and Section 13.2.1) or if an advance payment bond is required in the next assessment after the bond has been received by the Client. Only if a bond is required, but has not been provided, will the situation arise where there can be any legitimate delay in making this payment. Clause X14.2 contains requirements for the provision of an advanced payment bond if one is required by the Client. The Project Manager is responsible for accepting the issuer of the bond and for accepting its issue on behalf of the Client. The amount of the bond is the amount of the advance payment that is outstanding and therefore reduces as repayments are made. This clause is quite clear that, if the reason for not making the payment is not legitimate, then a compensation event will arise.

The repayment of the advance payment is set out in clause X14.3 and is done by deducting instalments from the assessment of the amount due in accordance with details included in Contract Data Part One. Whether it is one or more instalments is something for the parties to decide. When setting up this mechanism, the Client (or Project Manager on the Client's behalf) should ensure that there is no duplication between the matters the advance payment is intended to fund and, in Main Option A in particular, between any activities included in the Activity Schedule that have been framed to also provide advanced payment. It should also be made clear as to whether the advance payment amount is inclusive or exclusive of any tax that the Client must pay the Contractor.[14]

13.10.3 X16: Retention

If the Client decides that it wants to retain a portion of the PWDD from each interim assessment as both security and as motivation for the Contractor to complete the works, then it needs to select Secondary Option X16. Many view such mechanisms as contrary to the spirit of trust that the ECC attempts to engender between the parties. What it certainly does do is have a negative effect on the Contractor's cash flow and increases the administration, albeit marginally. Clients should also consider whether this provision is required if they also request a performance bond by using Secondary Option X13 (see Section 20.5).

---

[14] See clause 51.5 and, for example, VAT in the UK.

The provisions at Secondary Option X16 work in much the same way as similar provisions in other construction contracts. Clause X16.1 provides that the retention percentage is applied to and retained from the PWDD in all interim assessments. It also includes, as for some other construction contracts, a principle that a stated amount can be retention-free. In this clause the retention percentage is not applied until after the retention-free amount has been exceeded and then only to the portion of the PWDD in excess of that retention-free amount. If the Client does not want any amount to be free of retention, then the entry in Contract Data Part One should be completed as 'nil'.

Under clause X16.2, the amount retained is halved in the assessment made immediately after the Client has taken over the whole of the works. This level of deduction is then maintained until the Defects Certificate (see Section 9.6) is issued, after which no amount is retained.

Unlike some other construction contracts, there are no requirements in this Secondary Option requiring the Client to either deposit the retention funds in a separate bank account and/or to act as a fiduciary trustee. The Contractor is therefore at risk of the retention monies becoming an unsecured debt should the Client become insolvent.

13.10.4 Y(UK)1: Project Bank Account

Introduced in the 2013 reprint of NEC3, Option Y(UK)1 allows for the implementation of a Project Bank Account through which some or even all of the payments associated with the project will be made. In simple terms, if this option is selected then the Contractor sets up a bank account, approved by the Project Manager, into which Client and on occasions the Contractor will make payments. Both the Contractor and a class of people or organisations known as 'Named Suppliers' will receive payments from the bank account after completion and approval of a document known as the Authorisation.

In more detail, by deciding to use Option Y(UK)1 it will be the Client who makes the decision that these provisions will apply to the Project. If the Client makes that decision, then it will need to ensure that the Trust Deed and Joining Deed forms are included in the Contract. The Client will also need to ensure that the Scope contains an entry stipulating any restrictions on the use of the Project Bank Account (see clause Y1.6 and Section 22.2.44).

As part of its tender the Contractor will need to complete the required entries in Contract Data Part Two (see Section 21.4.9).

Once the contract comes into being, then it is the Contractor, under clause Y1.2 who sets up the Project Bank Account. The project bank will have already been named by the Contractor in Contract Data Part Two (see Section 21.4.9). As a result, there is no need for any approval of the bank itself but the Contractor must provide details of the actual account to the Project Manager (clause Y1.4). The Project Manager is given a reason for not accepting the project bank account, that being that the arrangements for the account will not allow the payments required by the Contract to be made. This requirement would include any timings required for the payment of deposits by the Client and the Contractor and the subsequent payments out to the Named Suppliers and the Contractor, and must be within the times required by clause 51 (see Section 13.2) or Y(UK)2 (see Section 13.9) as applicable. Should the account not be able to satisfy these contractual timescales such a failure would be fundamental and make the account unfit for purpose. Clause Y1.3 provides that the Contractor will pay any charges in connection with the administration of the bank account and have the benefit of any interest paid by the bank on the deposits in the account unless the Contract Data states otherwise. To facilitate this position there is an optional statement in Contract Data Part One (see Section 21.3.24), which if included transfers this responsibility and benefit to the Client.

Those Named Suppliers listed in Contract Data Part Two, together with the Client and the Contractor, all sign the Trust Deed before the first assessment date (see clause 50.1 and Section 13.2.1) for the project. The Trust Deed, as defined at clause Y1.1(5), is an agreement that contains the provisions for administering the project bank account. It will be in the form set out in the contract. A template is provided and needs to be included in the contract when formed. It is recommended that the Trust Deed template and Joining Deed template (see below) are included in Contract Data Part One after the entries regarding the Option (see Section 21.3.25).

Within the contracts between the Contractor and the Named Suppliers listed in Contract Data Part Two (see Section 21.4.9) the Contractor includes terms identical to Option Y(UK)1 in those subcontracts and notifies the Named Suppliers of all the relevant details of the account including the payment arrangements (clause Y1.5). Where the Contractor is using the NEC4 Engineering and Construction Subcontract (ECS) (NEC Panel, 2017i) as the form of subcontract a suitable version of Y(UK)1 is available within that subcontract. The ECS version also allows for 'Subcontractor Named Suppliers' thereby allowing the benefits of this payment arrangement to be stepped down into the next tier of the supply chain. If the NEC4 Engineering and Construction Short Contract (NEC Panel, 2017m) is used, this does not contain suitable wording, which would therefore have to be added as an additional condition of subcontract.

Should the Contractor wish to add a Supplier (term defined at clause Y1.1(4)) to the list of Named Suppliers and thereby add that Supplier to those listed in Contract Data Part Two then the Contractor is required by clause Y1.6 to submit proposals to that effect to the Project Manager. The Project Manager's only reason (in addition to that given at clause 13.4 (see Section 6.2)) for not accepting the proposal from the Contractor is that the Supplier proposed does not satisfy the requirements of the Scope. If the Supplier proposed by the Contractor is accepted, then the Client, Contractor and new Named Supplier are all required to sign the Joining Deed. The template for the Joining Deed is provided at the end of the text for Option Y(UK)1 and must be included in the Contract (see above).

As is appropriate, the longest part of this Option concerns the process for making the payment. This is covered in clauses Y1.7–Y1.12. The process starts by the Contractor, before the assessment date (clause 50.2 and Section 13.2.1) for the payment in question, submitting an application for payment to the Project Manager. The application that the Contractor submits must not only show the amount due (see clause 50.3 and Section 13.4.2) to the Contractor but also the amount due to each Named Supplier under the subcontract between the Contractor and each Named Supplier. This requirement means that the application will need to contain more detail than one submitted under a contract in which Option Y(UK)1 did not apply. It is likely that the exercise will reveal to the Project Manager the margin (positive or negative) that the Contractor is making on each Named Supplier. Whilst this information would be known where Options C or D apply, it is not information that would normally be revealed where Options A or B applied.

Clause Y1.8 then requires the Client, and if applicable the Contractor, to pay whatever sums are due under the Contract into the project bank account within the time required by the arrangements for the project bank account that will enable the payments that are to be made to the Named Suppliers and the Contractor to be made from the project bank account. The sum to be paid by the Client will be the amount certified by the Project Manager under clause 51.1 (see Section 13.2.2). The amount to be paid by the Contractor, if any, will be the sum withheld by the Client and notified to the Contractor in a pay less notice issued under clause Y2.3[15] and which is required to enable the payments to be made to the Named Suppliers. This creates a position whereby if the amount to be withheld by the Client is less than the Contractor's share of the payment to be made from the project bank account, then there would be no need for the Contractor to pay anything into the account as sufficient funds would be available from the payment in by the Client. Where the Contractor's share of the payment would be insufficient to cover the amount withheld, then the Contractor is only required to make up the difference so as to ensure that there are sufficient funds to pay the Named Suppliers.

In order for the payments to be made, it is necessary for a document called the Authorisation (definition at clause Y1.1(1)) to be prepared. Under clause Y1.9 this is prepared by the Contractor and sets out the sums to be paid to the Named Suppliers as assessed by the Contractor together with the amount to be paid to the Contractor, which will be the balance of the payment made in by the Client. Depending on the amount of any withholding (see previous paragraph), it is possible that the payment to the Contractor will be nil. The Contractor is required to sign the Authorisation and submit it to the Project Manager no later than four days before the final date for payment. The clause then requires the Client to sign the Authorisation and for that document to then be submitted to the project bank no later than one day before the final date for payment.

Clause Y1.10 confirms that the Contractor and the Named Suppliers receive payment of the sums set out in the Authorisation from the project bank account as soon as is practicable after the payment or payments into the account have been received. Related to this clause Y1.12 confirms that the payments made from the project bank account are treated as being the payment from the Client to the Contractor as required by and in accordance with the Contract. The way this is worded suggests that all payments from the account are treated as payments to the Contractor. As the total of the payments should total the amount certified to be paid, less any sums withheld as set out in any pay less notice issued under clause Y2.3, to the Contractor had Option Y(UK)1 not been used this would make sense. Some of the monies will go straight to the Named Suppliers then, also confirmed by clause Y1.12, and those payments are treated as being payments from the Contractor to the Named Supplier under the individual contracts with each Named Supplier.

---

[15] This position can only apply where secondary Option Y(UK)2 applies to the Contract (see Section 13.9).

In addition, clause Y1.12 also states that should the payment be delayed (i.e. is not made until later than the date required by the contract) and the reason for the delay is due to a failure of the Contractor to comply with the requirements of Option Y1.12, then that delay is not treated as late payment under the contract between the Client and the Contractor. That being the case, the Contractor would have no right to claim interest (see clause 51.3 and Section 13.2.3) or to suspend performance under the Housing Grants, Construction and Regeneration Act 1996, as amended by the Local Democracy, Economic Development and Construction Act 2009 (see Section 13.9), providing of course that the Act applied to the contract.

However, if the payments to the Named Suppliers are late when measured against the requirements of the individual contracts between each Named Supplier and the Contractor, those individual contracts would treat the payment as being late. Such a position would entitle the Named Supplier to claim interest on the late payment and also issue the required notice regarding suspension of performance as above, again providing that the Act applied to that contract. Whether actual suspension could take place would depend on how late the payment actually was once paid.

The payment requirements discussed above all concern payments to be made from the Client to the Contractor. Clause Y1.11 records that should a payment be due from the Contractor to the Client (see clause 51.1 and Section 13.2.2), then such a payment is not made through the project bank account. It would be made directly from the Contractor to the Client.

In the event that the Project Manager issues a termination certificate (see clause 90.1 and Section 18.4), then clause Y1.14 provides that no further payments are to be made into the project bank account. The contract is silent on what happens to any funds remaining in that account. That being the case, the banking arrangements should cover that position. The sensible outcome would appear to be that if any balance comes from interest receipts and there are any remaining charges, then the net of these two amounts should be borne by either the Client or the Contractor depending on who has the benefit of and liability for these two aspects. Any money in the account destined to be paid to the Contractor or Named Suppliers should be paid to the party identified in the Authorisation. In the event of insolvency of the Contractor being the cause of the termination (see clause 91.1 and Section 18.2.1) the Named Suppliers would continue to be paid as per the terms of the Trust Deed. However, under the Housing Grants, Construction and Regeneration Act 1996, as amended by the Local Democracy, Economic Development and Construction Act 2009 (see Section 13.9) the Client would be able to withhold payment to the Contractor.

13.11 Practical issues

13.11.1 Applications for payment are vital

Whilst passing judgement in the case of *Gilbert-Ash (Northern) Ltd. v Modern Engineering (Bristol) Ltd.* [1974] AC 689, the then Master of the Rolls, Lord Denning, said, 'There must be a "cash flow" in the building trade. It is the very lifeblood of the enterprise'. This statement is as true today as when it was made.

The author remembers his first day of employment with a contractor and the Chief Quantity Surveyor saying, 'Just remember lad, it is our money and it is your job to get it'. In other words make sure that the monthly payment cycle works and that the company would receive fair payment for what it had done each month. This might be by meeting the client's QS on site every month and carefully taking him through everything that had been done to ensure that the value was all included in the valuation that was issued by the client's team. Alternatively, if the contract required a payment application to be submitted, then prepare one and submit it.

Clause 50.3 of the ECC is very specific in making it clear that the Contractor must submit an application before the assessment date. If the Contractor does not make such a submission before the assessment date, not on or after the assessment date but before, then clause 50.4 makes it clear the most the Contractor can get paid for that assessment period is nil. It could be that the Contractor has to pay the Client but under no circumstances will a payment be due to the Contractor.

This stark position, which can hardly be considered to be one of mutual trust and cooperation, can be avoided by submitting applications before every assessment date. It is a discipline that most contractors comply with for most periods. Given the above position then, it is vital that it is complied with for every assessment period. Whilst the position that it creates is only temporary, it could still create major problems for the Contractor. Contractors: you have been warned.

13.11.2 Interest

In Section 13.2.3 the contractual provisions for interest were covered. The way the ECC payment provisions are structured, it is the Project Manager's obligation to calculate any interest payments that are due to the Contractor as a result of any of the triggers to that entitlement occurring. Given that clause 50.3 requires an application from the Contractor, should that application include an item for interest due under the contract it is difficult to see how the Project Manager could avoid including it in the certified sum. This would include assessing interest when the Project Manager had made a mistake or failed to carry out one of the Project Manager's own duties properly. In practice, it would be unusual to find project managers who would take such a step (especially if they were an external consultant and, as a result, could expose themselves to a claim from their client). Further, in many situations contractors would value the commercial relationship between themselves, their client and the project manager more than what may possibly be fairly small sums of money in the context of the whole project. This all means that, in practice, it is rare to see interest added to assessments. Users should, however, be aware that (certainly in the UK), on the rare occasions that a project goes to some form of formal dispute resolution, interest will inevitably be claimed. Providing that the Contractor is awarded some payment, interest will be applied to that sum.

For this reason, commercial users in the UK must be aware of and cater to the provisions of The Late Payment of Commercial Interest (Debts) Act 1998. This statute requires commercial contracts, a term which would cover most construction contracts,[16] to include contractual provisions that provide for the payment of interest when a debt that has arisen under that contract is paid late. This includes a requirement that the contract includes what the statute refers to as a 'substantial remedy' in the case of default by the payer. Where a 'substantial remedy' is not included then, at the time of writing, the statute and its supporting Statutory Instruments provide that the rate of interest shall be 8% over the base rate of the Bank of England, unless the contract refers to another bank or banks. Users in the UK should be aware of this requirement and ensure that the interest rate inserted in Contract Data Part One constitutes a substantial remedy.[17]

13.11.3 Taxes

In the UK construction industry, it is standard practice that rates and prices quoted and used in contracts are exclusive of Value-Added Tax (VAT). Similar conventions appear to operate in other jurisdictions. Clause 51.5 requires that any such taxes which the relevant law requires the Client to pay to the Contractor are included in the amount due, which is simple enough. What the ECC does not state is whether the Prices (i.e. tendered sum) as stated in Contract Data Part Two or any of the rates that may also be stated in Contract Data Part Two are inclusive or exclusive of such taxes; it is therefore a good idea to clarify this point. Many users do this by adding a sentence to the end of clause 51.5 to state that all rates, prices and other monetary sums stated in the Contract Data and other supporting documents referred to therein are exclusive of the relevant taxes.

13.11.4 Option B and 'Materials on Site'

When discussing the PWDD for Option B in Section 13.4.4, it was identified that there is no contractual route to enable the Contractor to be paid in interim assessments for unfixed Plant and Materials that have been delivered to site and are awaiting incorporation into the works. That is generally the case, but readers of this guide may be interested in a situation that actually happened in practice some years ago.

The project, under Option B, was for the construction of a pipeline along an easement that ran across fields from X to Y. Given the restrictive access to and width of the easement on this project, the Contractor needed to bring all the pipes to site and then

---

[16] Contracts with someone classed as a consumer, especially residential occupiers, are likely to be excluded.
[17] While the courts have not stated exactly what rate above base constitutes a 'substantial remedy', users in the UK should be aware of the comments and findings by J. Edwards-Stuart in *Yuanda (UK) Co Ltd. v WW Gear Construction Ltd.* [2010] EWHC 720 (TCC) (13 April 2010). This gives certainty that 0.5% over base is not a substantial remedy, while 5% over is and also that 3–4% might qualify as a substantial remedy. The judge makes it clear in his judgement that he was not deciding this point as he was only required to consider the rate in the contract between the parties. Having determined that it did not constitute a substantial remedy, then the statutory rate applied.

string them out along the route of the trench before it started to excavate and lay. If this course had not been taken, the Contractor would have struggled to get the materials past the completed work without running heavy vehicles over the completed pipeline with a possibility of damage being caused. This method of working formed part of the statements supporting the Accepted Programme (see Section 12.2.4). By the time of the first assessment the Contractor had taken delivery of and had strung out approximately £250,000 of pipes. and then discovered, not being familiar with the ECC, that there was no mechanism to include this sum in the assessment. It was suggested to the Contractor that as it had physically done something with the pipes (i.e. string them along the route of the pipeline) which formed part of the operations covered by the measured rate, then the Project Manager should assess the quantity by a portion of the rate in the Bill of Quantities. As the Project Manager and the Client were both sympathetic to the Contractor's position, this was agreed. An interim payment based on the quantity, and a rate that covered the cost of the pipe per metre and a nominal amount per metre for the people and Equipment used in stringing them out, was certified. In addition to being a practical example of overcoming what would have been a serious problem for the Contractor, it is also an example of mutual trust and cooperation at work.

13.11.5 Paying for Defects corrected before Completion

The definition of Disallowed Costs at clause 11.2(26), as used in Options C, D and E, results in the position that the cost of correcting Defects before Completion is not disallowed unless the Defect has resulted from a failure of the Contractor to comply with a constraint on how it is to Provide the Works. This is a situation that many clients, project managers and supervisors find concerning; some express the view that this must be wrong. That, of course, is not the case.

Our industry routinely builds prototypes in muddy fields rather than controlled factory environments. The prototypes concerned involve using a considerable number of differing trades, many who will not have worked together before, in this uncontrolled environment. To expect perfection the first time in such circumstances is unrealistic; even prototypes built in controlled environments need amendments to get them right. The consequence of all this is that it is inevitable that defects will occur when operating in such a manner.

Given that contractors as a collective body make a profit from the sum total of all their activities, which includes correcting defects that occur in the work they carry out, it follows that clients, collectively, are paying for defects to be corrected regardless of the contractual arrangements in place. We can expand on this piece of business logic as follows. When a contractor prices a fixed price tender then, either by its production outputs or within the subcontract prices it uses, the Contractor is consciously making an allowance to cover the cost of correcting defects. The client then pays this cost regardless of how few or many defects actually occur. The position created by the definition of Disallowed Costs means that the Client only pays for the correction of those Defects that actually occur, making the overall payment mechanism relevant to the actual cost of carrying out the work.

13.11.6 The Fee

What is not clear from either the definition of Fee or the definitions of PWDD is whether these percentages are added before or after the deduction of any Disallowed Cost. Common sense would suggest that the percentages should not be added until after the deduction of any such Disallowed Cost. If this were not the case, it would seem unfair that the Contractor would be paid a fee on non-qualifying cost.

13.11.7 Using Secondary Option Y(UK)1 without Secondary Option Y(UK)2

Some of the terminology in Option Y(UK)1, in particular the term 'final date for payment' in clause Y1.9, only appears elsewhere in the contract within Option Y(UK)2. This is a term taken from the legislation covered by Option Y(UK)2. It does not appear in the standard payment terms set out in core clause 5. In practice this means that Option Y(UK)1 should only be used when Option Y(UK)2 is also used, otherwise the terminology will not be consistent and difficulties would occur in respect of the interpretation of Option Y(UK)1.

As both of these Options are provided to be used in the UK only that Option Y(UK)2 would not be used when Option Y(UK)1 is used might seem inconceivable. However as the ambit of the relevant legislation, i.e. the Housing Grants, Construction and Regeneration Act 1996, as amended by the Local Democracy, Economic Development and Construction Act 2009 (see Section 13.9) is restricted to what the legislation classifies as a 'construction contract' some construction projects will fall outside of the definition (for example a water treatment works – dirty or clean). For such projects it would not be necessary to include Option Y(UK)2 in the contract, but the Client may wish to include Option Y(UK)1.

In any such instance it is suggested that the Client has two choices of how to proceed. First, it could include Option Y(UK)2 in the contract even though there is no legal requirement to do so; equally there is nothing preventing a non-qualifying contract from being subject to the provisions. The second option is to make an amendment to clause Y1.9 to remove the phrase 'final date for payment' (note that the term is used twice in this clause) and replace it with terminology consistent with core clause, and in particular clause 51.2 (see Sections 13.2 and 13.2.1). This second option could also be achieved by making clause 51.2 consistent in terminology with clause Y1.9. It is as broad as it is long.

13.11.8 Avoiding a dispute arising about the final assessment

The way that clause 53 is written leads users to the potential for a dispute. Clause 53.3 goes as far as setting out the steps through each of the three dispute resolution options (i.e. Options W1, W2 and W3). As it is far better to avoid disputes as opposed to getting dragged into one and all the uncertainty and costs that go with those processes, then the Parties and the Project Manager are better advised to work together to consult and negotiate to get to an agreed final assessment which the Project Manager can then issue under clause 53.1.

If this position cannot be achieved within the timescale allowed by clause 53.1 to the Project Manager, i.e. four weeks after the issue of the Defects Certificate, then as there is no fixed period in clause 53.2 in which the Contractor is required to submit an assessment to the Employer, the Parties have whatever time is needed to reach agreement and avoid the dispute. What the Parties must remember if making use of this lack of a submission period is that if the period does drag on, then eventually one of the Parties (usually the Contractor) will lose patience and a dispute will become inevitable. There is a balance between constructive negotiations which can be seen to progress and stalling or reluctance to progress shown by one or other party (usually the Employer). The latter is a rich breeding ground for disputes.

Recognising these processes and the advantages to all of reaching a settlement should enable the Parties and the Project Manager to avoid the harmful dispute.

Chapter 14
The Schedules of Cost Components

14.1 Introduction

The ECC contains a document called the Schedule of Cost Components (SCC) together with an abbreviated version known as the Short Schedule of Cost Components (SSCC). These documents do not form part of the terms and conditions, but one of them is referred to depending on the Main Option selected, except for Option F, which does not include any such reference. Their role is to provide a basis for the assessment of cost, whether in relation to the calculation of the Price of Work Done to Date (PWDD) (see Sections 13.4.5 and 13.4.6) or the effect of a compensation event (see Section 17.2).

The SCC (i.e. the full version) only applies to Main Options C, D and E, as referred to by the definition of Defined Cost (see Section 13.4.5) at clause 11.2(24). In this respect, it must be used when calculating the Defined Cost as part of the PWDD and is also the reference point for the assessment of compensation events (see clause 63.1 and Section 17.2.1).

The SSCC applies to Main Options A and B for the assessment of compensation events, as determined by the definition of Defined Cost at clause 11.2(23).

Neither version of the Schedules is used with Main Option F.

Throughout both versions of the Schedule, extensive reference is made to it applying to resources either within or outside the Working Areas. Readers are referred to Sections 6.6.4 and 6.7.2 where comment on this term was given and related practical issues were discussed.

14.2 The Schedule of Cost Components

The SCC contains rules for the reimbursement of people employed and not employed within the Working Areas, including wages and salaries, payments to people for bonuses, overtime, sickness, holidays, travelling and relocation together with Equipment, Plant and Materials purchases, Subcontractors, off-site manufacture and fabrication and design. These costs do not necessarily reflect the true cost incurred by the Contractor, so any difference between Defined Cost and true cost needs to be allowed for in the Fee (see clause 11.2(10) and Section 13.4.7).

A Practical Guide to the NEC4 Engineering and Construction Contract, First Edition. Michael Rowlinson.
© 2019 John Wiley & Sons Ltd. Published 2019 by John Wiley & Sons Ltd.

In order to achieve this objective, the SCC is divided into eight sections, the first seven of which cover costs while the eighth provides for credits in respect of insurance receipts. The seven cost sections, on the whole, provide for elements of cost that are reasonably recognisable to those used to working in construction and in costing works. Four of these sections use either defined terms or recognised principles as their heading. All eight sections are considered in Sections 14.2.1–14.2.8, which use the name from the Schedule as their title. Any references under these headings to item numbers or to bullet points are to those indicators in the actual Schedule.

14.2.1 People (item 1)

This section concerns the people that the Contractor uses to carry out all the activities in order to Provide the Works as it is obliged to do by the ECC. The people will include not just tradesmen and operatives who are physically involved in the construction, but all those involved in the management of the project who qualify for inclusion. The people that are recoverable are limited to three categories:

(i) people the Contractor employs directly and who for the period concerned are based in the Working Areas (see first bullet point);
(ii) the time spent working in the Working Areas by people the Contractor employs directly and who are not based in the Working Areas at that time (see second bullet point); and
(iii) people that the Contractor does not employ directly but who are paid by the Contractor for the time worked while they are within the Working Areas (see item 14).

The first and third categories are straightforward and simple to monitor for both the Contractor and the Project Manager. The second category is more difficult to monitor. This category would cover the Contractor's visiting staff and other workers brought in to carry out specific tasks or additional work. From the definition, it is clear that any time spent within the Working Areas carrying out work in connection with the contract would be paid for as part of Defined Cost. However, if a staff member, for example, took work back to head office (assuming that head office was outside the Working Areas, which should be the case in most circumstances) then the work carried out at the head office would not form part of Defined Cost; it would be covered by the fee.

The costs to be included in respect of people covered by categories (i) and (ii) above are listed at items 11, 12 and 13, the full list of which is drafted to allow for the cost of having people work anywhere in the world including relocating with their families. For projects carried out in the Contractor's home territory, many of these aspects are simply disregarded. For the remainder it is necessary to carefully go through each item and allocate the appropriate payment to the cost of each person. The majority of these cost components are easily understood, but reference to Volume 4, Chapter 5 of the User Guide comments specifically on item 12(f) and gives some general comments on other aspects of the SCC.

For the people who qualify under category (iii) above, being those normally referred to as labour-only subcontractors but also freelance or contract staff, the cost is simply the amount paid by the Contractor to or for these people.

14.2.2 Equipment (item 2)

As defined at clause 11.2(9), Equipment covers everything that the Contractor uses to Provide the Works but that is not incorporated into the works. The SCC classifies Equipment into five classes,[1] each of which has its own rules for calculating the cost.

Under item 21, the cost of any Equipment hired in from a company outside the group of companies to which the Contractor belongs is included at the cost charged by the hirer. Where the Equipment is owned by any part of the group of companies to which the Contractor belongs, then under item 22 such Equipment is included in Defined Cost at open market rates multiplied by the time for which the Equipment is required. This is done to prevent any distortion from the market cost caused by the internal charging policy with the Contractor's group.

Where the Contractor has purchased Equipment for work included in the contract, then under item 23 the amount paid for such Equipment has two components. The first is the change in value over the period for which the Equipment is required (i.e. the depreciation in value) and the second is a time-related cost charge as stated in Contract Data Part Two. A useful proviso within this item is that should Equipment within this category that is not listed in the Contract Data be required as the result of a compensation event, then the Project Manager is permitted to agree to assess such Equipment as if it had been listed. It is suggested that in such circumstances there is no other viable alternative unless the parties depart from the intent of the SCC.

At item 24 provision is made for payment for special Equipment at rates listed in Contract Data Part Two multiplied by the time for which the Equipment is required. As with item 23, the Project Manager can agree to add Equipment to this list. Little guidance is given in the Schedule as to what constitutes special Equipment. However, Volume 4 of the User Guide suggests that special Equipment will be some item of Equipment that the Contractor owns but for which open market rates (item 22) or a purchase/resale value (item 23) cannot be established. The example given by the User Guide is 'an old but effective crane barge for offshore work'.

Item 25 covers the cost of any consumables that are treated as Equipment, i.e. that qualify under the definition of the term Equipment, and states that these shall be included at cost. Examples given in the User Guide cover items such as fuel and welding rods, to which one would add drill bits, cutting blades, etc. Item 27 appears similar in its coverage except that it refers to materials used to construct or fabricate Equipment rather than being consumed. As temporary shuttering constitutes Equipment, then the timber and plywood used in the construction of such temporary works would be captured by this item or, arguably, where its use was exhausted by the project under item 25. Either way, the cost constitutes a cost component.

---

[1] Referred to by the item numbers in the Schedule (21, 22, 23, 24 and 25).

Under item 26, provision is made for the payment for transport, erection, dismantling and removal where the hire rates used elsewhere exclude such costs. These are usually clearly itemised on the relevant suppliers' invoices.

Finally, at item 28 it states that the cost of operators of Equipment is included in the people section unless included in the hire rate. Attention is required as to whether the equipment supplier or the Contractor supplies the driver or operator for each piece of operated Equipment.

14.2.3 Plant and Materials (item 3)

This item within the SCC is the most straightforward. In simple terms, it covers the cost of buying the Plant and Materials (see definition at clause 11.2(14)) together with any delivery (or removal) costs, the costs of any sampling or testing and any payments in relation to packaging (including the removal thereof).[2] Provision is also included for credits where items are returned to the supplier subject to a caveat to prevent the Client receiving a credit where the Project Manager has disallowed the costs from the calculation of Defined Cost (see clauses 11.2(24) and 11.2(26)).

14.2.4 Subcontractors (item 4)

The statement as to what is included in respect of Subcontractors is contained within a single sentence, the application of which needs more consideration than is apparent at first. Who constitutes a Subcontractor is determined by the definition in clause 11.2(19) (see Section 8.2); this definition will have to be applied rigidly in order to properly categorise the costs covered by this item.

The amount included for any person or organisation who qualifies as a Subcontractor under the definition referred to above is the amount paid to the Subcontractor. Any amount paid to or retained from the Subcontractor by the Contractor is not deducted should the deduction result in the Client either paying or retaining that amount twice. For example, any retention to be kept by the Contractor from the Subcontractor under Option X16 (see Section 13.10.3) would not be deducted in the cost exercise, as the Client would already be retaining that same money by retaining it from the Contractor under the main contract.

When considering the amount to be paid to the Subcontractor, the Project Manager will need to consider whether any of the cost paid to should be classified as Disallowed Cost under clause 11.2(26) (see Section 13.4.5). In particular, the first and second classes of Disallowed Cost could apply to payments to Subcontractors. It is also possible to see that some of the other eight classes from this definition could also apply in the appropriate circumstances.

---

[2] This element would cover any cost associated with packaging management from the growing volume of regulations affecting the industry in the UK and other countries.

14.2.5 Charges (item 5)

Item 5 is subdivided into four sub-items, all of which are fairly straightforward. The list of Charges at items 41, 42 and 43 is clear and easy to follow.

Item 44 on the other hand could be either a cost or a credit. This item refers to payments made or received, hence the previous sentence. The charges referred to will be in connection with the disposal (cost) or sale (credit) of materials that have arisen from either excavation or demolition on the Site as part of the works. This cost or credit ties in with the provision at clause 73.2 (see Section 10.2) concerning title to such materials.

14.2.6 Manufacture and fabrication (item 6)

Experience suggests that this is a little-used part of the SCC. That is because the opening part of the item limits its application to Plant and Materials that are 'wholly or partly designed specifically' for the project. This immediately rules out all off-the-shelf items and any products assembled from a menu of standard components or factors.[3] The design would have to result in something that would be exclusively for the project concerned. If that item were also to be manufactured and fitted by the designer, it is more likely that a subcontract would be placed for the complete package thereby resulting in the cost being allocated to the subcontractor costs (item 4).

Should that not be the case and the manufacture or fabrication of this unique item was done outside the Working Areas, then the hours taken are applied to the rates included at Contract Data Part Two. Should either the Project Manager or the Contractor foresee any element of the project falling into this category, it is vital, in practice, that they ensure that the necessary rates are included in the Contract Data before the contract is formed.

14.2.7 Design (item 7)

Where design is carried out outside the Working Areas, the cost of that design is assessed in accordance with item 7 of the SCC. It is quite clear that this only refers to such work done outside the Working Areas; where the designers are working within the Working Areas they are treated as people and covered by item 1 (Section 14.2.1). Users must also recognise that on many occasions the designers employed by a contractor may also qualify as a subcontractor (see clause 11.2(19) and Section 8.2). The cost of this design work can only be recovered under one class within the definition of Defined Cost.

Where design work outside the Working Areas is assessed under this item, then rates stated in Contract Data Part Two are applied to the hours worked by the designers in the categories listed. This reference to the categories listed is something that, in practice, the Contractor needs to consider. The Contractor must ensure that it has covered all categories of designers that may work on the project in order that it does not face

---

[3] For example, fire doors selected from a catalogue that allows purchasers to mix any combination of layout, size, finish, fire rating, etc.

any contentions that, as there is no category for a certain designer, then the cost of that individual designer is not recoverable.

Should it be necessary for any category of designer to travel from their place of work to the Working Areas in order to fulfil their duties, then the cost of travel is recoverable. As with the hourly rates it is necessary, in practice, to list every category of designer that may be involved in travel. There is a sustainable point of view that if the category of designer is not listed, then the travel expenses are included in the Fee (see clause 52.1 and Section 13.4.7).

14.2.8 Insurance (item 8)

Item 8 includes two deductions that are made from the Defined Cost in connection with insurable events. Firstly, where an event has occurred that the Contractor has insured against, the costs related to that event and covered by other parts of the Defined Cost are deducted. The Contractor recovers these monies under its insurance policy and therefore, in theory, the costs of the event are covered once. The second deduction prevents the Contractor from benefiting from any additional cover it may have in excess of that required by the contract, a position that in practice will probably be extremely rare.

14.3 *The Short Schedule of Cost Components*

The SSCC is approximately two-thirds the length of the SCC. It contains the same eight items (see Sections 14.2.1–14.2.8) as the SCC; indeed items 3 (Section 14.2.3), 5 (Section 14.2.5), 6 (Section 14.2.6), 7 (Section 14.2.7) and 8 (Section 14.2.8) are identical in both versions. The other three items are revised to some extent or completely.

In item 1 the class of people as listed at item 14 in the SCC is added as a third bullet point to the two classes of employed people at item 1. The total to be paid for each of the people is then calculated by multiplying the hours spent working in the Working Areas by the People Rates. The People Rates are defined at clause 11.2(28) in both Option A and B by reference to the same term but in italics, without the capital initials (see clause 11.1 and Section 6.6.1). This takes the user to Contract Data Part two where entries should have been completed with the category of person (e.g. bricklayer, joiner, general operative, site manager, foreman and the like), the charging unit (e.g. hour, day, shift or the like) and the unit rate per unit of charge. In both of Options A and B, clause 63.16 provides for the addition of new people rates if the current list does not include a suitable rate for a person who will or has been employed in relation to a compensation event. The new rate is either agreed by the Project Manager and the Contractor, or if not agreed then the Project Manager assesses the new rate based on the other rates in the Contract Data. Once the Project Manager has made such an assessment the only way that the new people rate concerned could be changed would be by reference to the dispute resolution procedures, as Option W1, W2 or W3.

For the cost of Equipment (item 2), the approach is changed from the SCC to one that revolves around a published list as referred to in Contract Data Part Two. In the

United Kingdom, the two primary lists are the 'BCIS Basic Schedule of Plant Charges' published by the Royal Institution of Chartered Surveyors (RICS) and the 'Schedules of Dayworks Carried Out Incidental to Contract Work' published by the Civil Engineering Contractors Association (CECA). A percentage adjustment[4] quoted by the Contractor and also stated in Contract Data Part Two is applied to these rates. The remainder of the provisions in item 2 concern the hours to be claimed, additional payments for transport, etc. and the treatment of operators.

Item 4, covering Subcontractors, simply states that it is the payment for work carried out by Subcontractors. As the SSCC only applies in the assessment of compensation events, the Contractor will need to show either the sum it expects to pay to the Subcontractor, if forecasting Defined Cost, or what it has paid to the Subcontractor when using actual Defined Cost as the basis of the assessment. As neither Option A nor Option B includes any rules for disallowing cost, the Project Manager would appear to be limited as to the amount of investigation that may be made into such payments.

14.4 Application to Subcontractors

The second sentence of clause 26.1 (see Section 8.3) states that a Subcontractor's people and equipment are treated as if they were the Contractor's. Earlier editions of the ECC followed this statement in clause 26.1 in the SSCC by requiring that all costs for subcontractor's resources as if they were owned, hired or bought by the Contractor. A separate component of the Fee was then added to just the element of the overall Defined Cost which applied to subcontractors. As the Short Schedule applied to Options A and B for compensation event assessments and could, by agreement or election by the Project Manager, be applied to the same task in Options C, D and E then this aspect was important but did cause a significant amount of confusion. However, that approach has been abandoned in NEC4.

As the SCC and the SSCC in NEC4 now include the amounts paid to Subcontractors as a component of cost at item 5 in both Schedules, the link with clause 26.1 has been broken in terms of its application to cost.

As Options C, D and E all include provisions for both Disallowed Cost (see clause 11.2(26) and Section 13.4.5) and for the keeping of records (see clause 52.2 and Section 13.5.3) the Project Manager will be able to get visibility of the Subcontractor costs.

On the other hand, as Options A and B do not include such provisions, then users may need to consider how the Contractor will be required to demonstrate the costs of Subcontractors and how deeply the Project Manager should be able to investigate the costs from this class of organisations. If the Contractor subcontracts on NEC4 Subcontract terms, then the Contractor will have visibility of the Subcontractor's cost components to the same degree as required by the ECC. That information can then be made available to the Project Manager.

---

[4] The use of the term 'adjustment' allows for either a negative or a positive adjustment to be quoted.

14.5 Practical issues

14.5.1 Visiting employees

There has always been a problem for Contractors in estimating what portion of the true cost of their visiting employees they will be able to recover under both versions of the Schedule. This only tends to be an issue on smaller projects that do not justify the full-time involvement of certain specialisations such as quantity surveyors, planners and senior contracts management staff. It is not unusual for such individuals to be involved in the management of several projects at the same time. In practice, these people tend to carry out their duties both in the Working Areas and at the head or regional office from which they are based. A strict interpretation of the classes of people at item 1 of both versions of the Schedule results in such people being included in the cost when they are working in the Working Areas, but falling within the Fee (see clause 52.1 and Section 13.4.7) when they are working outside the Working Areas. There a number of practical solutions to this problem which users may wish to consider.

Contractors can unilaterally overcome this problem by arranging for such visiting employees to only work on projects that are being carried out under the ECC when the employees are in the Working Areas. This method is effective in as much as it makes the recovery clear. In practice, however, it produces inefficiencies as it effectively takes away important people who can contribute to the success of the project for that time that those individuals are working outside the Working Areas on other projects.

Another alternative that contractors can instigate is to base such individuals in the Working Areas so that they use them as the base from which to carry out their work on other projects, which may well be under other contractual arrangements. While enabling any contractor who uses this method to cost the people concerned within Defined Cost, it also creates the potential for a difference where the Project Manager considers that more time is being booked than the individual is devoting to the project. This does not help foster a spirit of mutual trust and cooperation.

Other solutions require the agreement of the Client and/or Project Manager and involve a reasonable-to-high level of trust. There are two solutions to this position used by parties working together on a reasonable volume of repeat work, where the volume is sufficient that the types of specialist being considered here can be allocated to the programme of projects on a full-time basis. The Client can either pay for a core team of essential staff through a separate contractual arrangement (and not book them to the individual projects) or for these individuals to complete time sheets allocating their time among the individual projects that are active at any one time. The location of the individual's work is then no longer an issue, as the individual will be costed whether working inside or outside the Working Areas. This method can be used where there is only one project, but often leads to differences over the reasonableness of hours booked.

A variation on the previous method is to extend the Working Areas to include the Contractor's offices but to limit the people working in those extended areas whose time will be allocated to the Defined Cost via the Schedule. For example, the agreement could be that the Contractor's head office constitutes Working Areas but only for certain named individuals. By this approach, work done in connection with the project in the Contractor's office (by, say, a director or by accounts department staff) would not be taken into account and would therefore be deemed part of the Fee. This approach has been used successfully but does require trust.

14.5.2 Other ways of calculating the cost of people

Calculating the cost of people, given the variety of matters to be considered as set out at items 11, 12 and 13 of the full Schedule can become a time-consuming and complicated process. In the United Kingdom, concern is also expressed whether some of the information required as part of this calculation can actually be revealed to the Project Manager because of restrictions imposed by The Data Protection Act 1998. For no other reason than to reduce the amount of administration required, many employers and project managers change item 1 of the SCC.

The two most common changes seen in the United Kingdom are, first, that a schedule of job titles is inserted in place of the standard wording. Second, a schedule of basic salary ranges is given. In both cases the tendering contractors are then required to insert hourly rates against these schedules. These hourly rates are then used in calculating the Defined Cost no matter what the purpose (i.e. assessing the PWDD or assessing a compensation event). Both approaches have advantages and disadvantages.

The advantage of both approaches is that it simplifies the process of arriving at the cost of people. For Main Options C, D and E it also allows the Project Manager to feed the rates into a cost model (possibly based on the resources shown on the tendering contractors programme if one has been required) and thereby obtain a more accurate tender stage assessment of the outturn costs of the project rather than just relying on the tendered target Prices.

The primary disadvantage of the first approach is where, as a result of changes to the type of job descriptions and grades being employed on the project, someone whose job description is not in the schedule actually works on the project. This situation has resulted in differences arising between project managers and contractors, ranging from contentions about what rate should be applied to claims that the people concerned should not be paid for at all as they must be covered in the Fee as stated in clause 52.1 (see Section 13.4.7). With the second approach the previous situation cannot occur, but this does not prevent disputes arising about whether the salary level of a person is commensurate with the job that person is doing. Further, where professional staff in particular are promoted during a project and given an increase in salary, issues arise about whether that promotion and the increase that the Client is then asked to pay is reasonable or correct. By recognising these potential areas of dispute, the disadvantages can be avoided by careful wording and agreement before the contract is entered into.

It is the case, of course, that NEC4 has addressed this point in the SSCC by using People Rates (see clause 11.2(28) and Section 14.3), but these provisions are only included in main Options A and B. It seems likely that many users will adopt the People Rates clauses and included them in Options C to F by means of the Z clauses.

14.5.3 Managing actual cost against the Schedules

A good practical approach for a contracting organisation to take in order to ensure that its actual or real cost will be recovered by the operation of the Schedule (either version) is for it to allocate each of the cost categories within its cost capture or accounting system against the items within each Schedule. This will show which actual costs it needs to consider when calculating the Fee percentage and which (if any) of the costs it allocates as part of its normal accounting procedure to projects that are not covered by the Schedule. Those actual costs not covered then need to be added into the calculation for the Fee; otherwise they will be lost. This is not a time-consuming exercise. For the best results, it needs to be carried out by a small team made up of a representative from each of the estimating, commercial and accounting departments; at least one member of this team must have an in-depth understanding of the Schedule and its workings.

14.5.4 Completing Contract Data Part Two

A detailed practical review of completing the Contract Data is included in Chapter 21. What is not covered in that chapter is that, in practice, the blank Contract Data Part Two for completion by the tendering party is usually supplied as part of the tender documents. Often this has been edited in order to appear to require only the entries that are applicable to the Main and Secondary Options that apply to the tender in question. It is common in practice to find that entries that need to be completed have not been included in such a draft. Accordingly, tendering contractors and subcontractors are advised to check the blanks that they have been supplied with to ensure that all the necessary information to make the relevant version or versions of the Schedule work has been requested. If it has not, then either a question should be raised during the tender period (or the additional information should be supplied with the tender) or the contracting party should ensure that the oversight is corrected before any contract is formed.

Chapter 15
Compensation Events: Theory and Events

15.1 Introduction

This is the first of three chapters that consider the contents of core clause 6: compensation events. The fact that this subject is broken down into three chapters gives an indication of how big it is in relation to the other project management procedures contained in the other core clauses. The basic underlying theory will be explained in this chapter together with a guide to all the compensation events contained in the contract, whether in core clause 6 or elsewhere. This guide will concentrate on the practical aspects of the events and, in particular, the obligations that each event is linked to.

Chapter 16 will consider the requirements and practicalities of the procedures surrounding the notification of compensation events and the submission/acceptance of quotations. This is an area that in practice is often misunderstood or simply ignored. An improved understanding of this aspect can help users avoid disputes and the need for costly formal dispute resolution procedures.

The final part of the trilogy, Chapter 17, will consider how to assess a compensation event in terms of both time and money. Again this seems to be a much misunderstood area that many consider to be a black art. In reality, the money aspect is very basic and the time issues are not complicated if the programme is up to date and comprehensive.

15.2 The theory

15.2.1 Underlying principles

The compensation event procedure is radically different to the procedures for dealing with change in traditional forms of construction contract used in the United Kingdom and elsewhere in the world. This radical difference has caused and continues to cause what can only be described as a mixed bag of reactions from commentators and the industry as a whole.

The original intent of the drafting body was to create a mechanism where the Contractor carried no financial risk as a result of change and was therefore unaffected by the decisions of the Project Manager and motivated to minimise time and cost overruns. However, the Guidance Notes to the third edition admitted that the Contractor carries the potential risk or reward if its forecast (see Sections 17.2 and 17.3) of the impact of a

---

A Practical Guide to the NEC4 Engineering and Construction Contract, First Edition. Michael Rowlinson.
© 2019 John Wiley & Sons Ltd. Published 2019 by John Wiley & Sons Ltd.

compensation event is wrong. Depending on the Main Option in use, the risk taken can be nullified or reduced. The fourth edition follows the same approach.

As we will see, it is debatable whether the minimal or no risk objective has been achieved and, indeed, many say that the Contractor carries far more risk under this mechanism than it would under a more traditional arrangement. In addition, given the number and variety of events for which the Contractor is entitled to compensation, it may also be considered, from the Client's point of view, that the mechanism fails to provide any certainty over the final price of the project.

Further, the drafting body initially considered that compensation events would only arise infrequently; in practice, however, users find that the rate of occurrence can be high. As a result, some consider that the strict administrative procedures then serve to impose a high burden on the parties which is unmanageable. Chapter 16 will describe how this can be overcome, to everyone's benefit, by the application of practical steps.

15.2.2 What is a compensation event?

The term *compensation event* is not defined in the ECC.[1] It can only be determined by examining the provisions that make up the procedure. The principle of the definition is found at clause 62.2, which states that quotations for compensation events comprise proposed changes to the Prices and any delay to the Completion Date. From this we can conclude that compensation events, as detailed under the ECC, are those events for which the risk in terms of both time and money is transferred away from the Contractor and onto the Client.

It should also be noted that not all events that fall within the bounds of the compensation events detailed will result in a change to the Prices or the Completion Date. Further, clause 61.4 says that if the compensation event arises from a fault of the Contractor or if it has no effect on either Defined Cost or time then there will be no change to the Prices or the time for completion. Therefore events that do not cause a change are not compensation events. A compensation event only arises where there is an actual effect on the Prices and/or the Completion Date.

A feature of the compensation event mechanism is that the occurrence of any of the events gives rise to an entitlement to the Contractor that entails the potential to change both the Prices and the Completion Date. This adjustment is made in respect of matters that are treated as neutral events under other contractual arrangements widely used in construction generally. This feature is one that many users, especially on the employer's side of the industry, find difficult to accept.

15.2.3 What other contractual entitlements apply?

A question that commonly arises is whether the compensation event procedure at core clause 6 provides an exclusive scheme for dealing with all claims for time and money

---

[1] The term is not used with upper case initial letters; see clause 11.1 and Section 6.6.1.

under the ECC. This question includes consideration as to whether the procedure acts to exclude the Contractor's entitlement to seek damages at common law for breach of contract.

In England and Wales, the common law position[2] is clear in that the right to claim damages for breach of contract can only be excluded by express words to that effect. The presence of a contractual scheme that enabled the contractor to recover loss and expense (i.e. a form of damages) was found to be supplementary to the contractor's common law rights, rather than a substitution and/or replacement of such rights. It must be noted that some legal commentators will take issue with this position, but the decision of the court referred to above should, in most instances, be persuasive.

When the first edition of the ECC was issued, the promoters hoped that they had devised a scheme to remove the word *claim* from the vocabulary of those who used the form. Use of the first edition soon revealed that the list of compensation events did not provide protection for the Contractor against the Client's possible defaults. The second edition expanded the list to 18 events, which included at clause 60.1(18) an event covering any breach which is not one of the other compensation events listed in the Contract. The intent of the authors was clearly to widen the coverage of compensation events so that the procedure would provide a remedy for all breaches without any need to go 'outside' the Contract. With the inclusion of compensation event number 18, this intent would appear to have been achieved. However, what the second edition did not achieve was to make the compensation event provisions a condition precedent or a time bar and therefore an exclusive remedy.

In the third edition, provisions were introduced to act as a time bar where the Contractor fails to notify a compensation event that it should have notified. These are maintained in the fourth edition. The time bar works to stop the Contractor's entitlement to a change in the Prices, Completion Date or a Key Date. This time bar is coupled with clause 63.6, which acts to limit either party's rights in the event of a compensation event to one of these changes. Although this matter has not been tested in the UK courts at the time of writing, many believe that in practice it will be effective and prevent a party from making a claim outside of the contract for any breach covered by the compensation event mechanism. There are some commentators who disagree with this proposition; only a judge can have the final word.

15.3 The events

There are 21 compensation events that are common to the ECC no matter which Main Option is selected and with whatever combination of Secondary Options. These are all listed at subclauses 60.1(1)–(21). For ease of reference, these 21 events are identified as Event 1 to Event 21 and described in Sections 15.3.1–15.3.21. Sections 15.3.22–15.3.24 use clause numbers to identify three compensation events included in core clause 6 in relation to Main Options B and D, followed by a further six included within Secondary Options described in Sections 15.3.25–15.3.29.

---

[2] See *Hancock v Brazier* [1966] 2 All ER 901.

15.3.1 Event 1

There is little doubt that this will be the most commonly occurring compensation event. Feedback suggests that, on most projects, over 90% of all compensation events will fall within this classification. It appears to be simply the equivalent of a standard variation clause as found under most standard forms of contract used in construction. However, the actual wording refers to changes to the Scope (see clause 11.2(16) and Section 22.2) and not just the works. As well as changes to the works, this compensation event will therefore also include any changes made by the Project Manager to any condition, requirements, restriction or stipulation[3] contained in the Scope which governs the way in which the Contractor performs its obligations. Examples of such matters include, but are not limited to:

- working hours;
- access restrictions, either to the Site as a whole or to part of the Site;
- order of the works;
- works stipulated to be carried out by Others; or
- the scope of the Contractor's design responsibility.

Others reasons for changing the Scope, such as removing an illegality or impossibility (clause 17.2) or resolving an ambiguity or inconsistency (clause 17.1) may also constitute a compensation event under this classification.

The event includes two exceptions, which are set out in the two bullet points that form part of the drafting. The first of these excludes from the classification any change to the Scope resulting from any acceptance of a Defect (see clause 11.2(6) and Section 9.4) that arises as a result of the application of clauses 45.1 and 45.2 (see Section 9.7). Users should note that this exception does not apply to clauses 46.1 and 46.2 as the Scope is not changed under these clauses; it is merely treated as having been changed.

The second exception makes it clear that if the change is one to the Contractor's design made at the Contractor's request, then such a change is not a compensation event. Such changes are often referred to as 'design development' and are seen, where the basic principles of design and build contracting apply, as being at the contractor's risk. The way this exception is worded also acts to confirm that any savings that can be made by the Contractor to its design are to the Contractor's benefit, as there is no mechanism in the contract to value such a change. The second bullet point also makes it clear that should the Contractor change its design to comply with the Scope, then that change does not constitute a compensation event.

---

[3] Collectively referred to as 'constraints'; see clause 11.2(16).

15.3.2 Event 2

The wording of this event is clear: such a compensation event will arise if the Client fails in its obligation at clause 33.1 to provide access to and use of the Site or part of the Site by the required date. This requirement and the potential for the Contractor to show a different access date on any programme it submits for acceptance are covered in detail in Section 12.5.1, together with comments on the practical issues created by them in Section 12.8.1.

15.3.3 Event 3

This compensation event is worded in a way that is very wide in its application. The words 'does not provide something' can easily be interpreted to include all actions of the Client that it is required to carry out either by the Scope or some other provision of the contract. The event is linked to dates on the Accepted Programme and will only apply to dates that have been inserted on the programme by the Contractor. This again emphasises the importance of the programme and the completeness of the information that the Contractor should show. The obligation of the Client to provide things generally is stated in clause 25.2 (see Sections 7.5.2 and 7.6.5).

While this event refers to something to be provided by the Client, it makes no reference to anything that should be provided by either the Project Manager or the Supervisor. As the Project Manager and the Supervisor are both appointed by and responsible to the Client, it will be no defence to argue that the Client was not responsible for providing something that in practice is handed over by one or the other of the Client's agents. For example, if the Client is responsible for providing design information but in practice the Project Manager issues such information under cover of an instruction, any failure to provide the information by the time shown on the Accepted Programme would be a failure of the Client (despite the fact that it was the Project Manager who had not issued it). With this example, users must also be aware that should the information issued by the Project Manager under an instruction (whether that issue is by the required time or not) change the existing Scope, then that change would constitute a compensation event under Event 1 (see Section 15.3.1).

15.3.4 Event 4

This event contains the Contractor's remedy for the impact of two separate instructions that can be issued by the Project Manager. The first provides compensation where the Project Manager has issued an instruction to stop or not to start any work as the Project Manager is empowered to do by clause 34.1. This is straightforward by remembering that, in accordance with clause 61.4, if such an instruction arises from the fault of the Contractor neither the Prices nor the Completion Date will be changed.

The second instruction is that allowed by clause 14.3 to change a Key Date, and this compensation event merely caters for that change. However, the definition of Key Date at clause 11.2(11) includes a separate definition of the Condition to which a Key Date must relate. Clause 14.3 and this compensation event make no mention of a change to the Condition related to a Key Date, although the definition includes the words 'unless later changed in accordance with this contract' apparently in connection with the Condition. As the ECC does not expressly give the Project Manager the right to change the Condition related to a Key Date, it is suggested that the definition is at fault and that the Condition cannot be changed. It can, however, be corrected (see clause 63.11 and Section 17.5.3) where a change to the Scope has resulted in the description being incorrect. This correction would be taken into account as part of any assessment arising under Event 1.

15.3.5 Event 5

Event 5 covers two omissions and one act of the Client and the class of people known as Others (see clause 11.2(12) and Section 5.11). The inclusion of Others in this event means that the Client is accepting responsibility for the performance of all those people that fall within the definition of this term.

The first omission is that work is not done within the time shown on the Accepted Programme. The order and timing of such work should be given in the Scope. If the order and timing is not included in the Scope, then in accordance with clause 31.2, the Contractor should agree the order and timing with the Client and/or Others in order to prepare the first programme.

The second omission arises where the Client or the Others do not work within whatever conditions are set out in the Scope governing how they will do their work. Effectively, the conditions stated for such work in the Scope act as a constraint on the Contractor who must comply with such constraints and allow the necessary coordination on the programme. Similarly, the Client and/or the Others must work as stated otherwise the compensation event will arise.

The act, at the third bullet point in the event, is where either the Client or Others do work which is not described in the Scope on the Site. It could be argued that the introduction of additional work by the Client or Others constitutes a change to the Scope and that the Contractor would therefore be compensated under Event 1. The inclusion of the third bullet point in Event 5 serves to remove any doubt and create a clear right for compensation in such circumstances.

15.3.6 Event 6

Many consider that the provision of an express remedy for circumstances where an agent[4] of the Client fails to reply within a stated period[5] is unusual when they compare it to other forms of contract. However, when considered with the objectives of the ECC and

---

[4] For example, the Project Manager or the Supervisor (both of who are appointed by the Client and may be replaced by the Client).
[5] Known as the 'period for reply'; see clause 13.3 and Section 6.2.

the provisions it provides in order to achieve those objectives, this event is not only sensible but a vital part of that machinery. Clause 13.3 requires named people to reply to a communication within either the period stated in Contract Data Part One or whatever other period is stated in a particular clause. This requirement is there to ensure that the flow of information is not delayed. Being aware of these reply periods allows the Contractor to coordinate the various stages of the construction procedure, including matters such as design and procurement, with each other and to make submissions to the Project Manager for acceptance. This event is therefore needed to prevent this mechanism from breaking down or to provide security for the Contractor should it not be complied with.

In practice, it is not unknown for clients using in-house project managers and/or supervisors to delete this event from the list. This does not help the parties and greatly increases the risk for the Contractor. Otherwise it acts as a powerful incentive for the Project Manager and Supervisor to administer the Contract and reply to every communication within the relevant period for reply. This brings the opposite position where clients using external project managers and/or supervisors insist that this event remains in the list as it ensures that their representatives comply with the contract or face a claim under their own terms of appointment for a failure to carry out their duties.

There is no corresponding remedy for the Client should the Contractor fail to reply to a communication; however, it may prejudice the Contractor's ability to comply with its obligations, which in turn may give rise to delay damages.

15.3.7 Event 7

This event arises from clause 73.1 and does nothing more than provide compensation to the Contractor should antiquities be discovered on the Site. In practice, to determine whether an item is of historical interest or if it is of value is reasonably straightforward. Reference to whatever local body has jurisdiction over archaeological finds and/or the local museum will soon determine whether either of these factors applies.

On the other hand, it has been speculated by some commentators that the reference to items of 'other interest' is open to abuse by a Contractor who gives notices in respect of trivialities but claiming that the item it has found is of interest. Although I have never come across a reported incident of this type occurring in practice, it is of course not out of the bounds of possibility and must remain something that project managers keep in the back of their mind should such an occurrence arise.

15.3.8 Event 8

The application of this event depends on the interpretation given to the word 'decision'. The ECC uses the word 'decision' sparingly; most instances of its use are actually within core clause 6 and relate to the notification and assessment of compensation events. If a narrow meaning of the word 'decision' is taken, then there are very few applications for this event. On the other hand, if a wide interpretation is given, then this event would apply to any change of a previous communication.

15.3.9 Event 9

This event provides the remedy should the Project Manager withhold acceptance of something for a reason other than those given in clauses 13.4, 16.3, 21.2, 23.1, 24.1, 26.2, 26.3, 31.3, 40.2, 55.4, X4.2, X10.4(2), X13.1, X14.2, X16.3, X22.2(2), X22.3(3), Y1.4 and Y1.6. The ECC does not prevent the Project Manager from withholding an acceptance for another reason, but in the event that another reason is used then the Project Manager may subject the Client to additional cost and possibly time implications. This event is consistent with clause 13.8. That the reasons provided at the clauses listed are generally so wide in their application means that this event rarely arises.

15.3.10 Event 10

This event is analogous with other standard forms of contract and provides the remedy for the Contractor where the Supervisor has instructed him to search for a Defect pursuant to clause 43.1, but no Defect has been found as a result of that search. In practice, the principle of clause 43.1 and this corresponding compensation event does not appear to cause any significant issues between parties.

15.3.11 Event 11

The Supervisor is obliged by clause 41.5 to do tests without causing unnecessary delay. The only difficulty with this event is deciding the difference between necessary and unnecessary delay. Contractors would be advised in practice to include the Supervisor's tests on the Accepted Programme. This would provide a suitable benchmark against which to measure any delay caused by such tests.

15.3.12 Event 12

Event 12 is probably the event that, in practice, causes the most problems and may well be the one that is deleted more often than any other (except possibly for Event 13; see next section). This event passes the risk of physical conditions occurring that should not have been allowed for to the Client, subject to certain provisions about what should have been expected by the Contractor. It is easy to see that the principle of this clause is that, by the Client taking the risk for unforeseen conditions on site, the Client pays for what actually happens rather than for what might happen (assuming, of course, that the Contractor will price risk if it is left to carry it). From a client's point of view, this can make commercial sense, subject of course on the approach taken by any financial backers.

In order to make this clause work in practice, it is necessary to understand what the term 'physical conditions' means and how the benchmark against which those conditions that are actually encountered is established. The term 'physical conditions' has been interpreted by an English court[6] to have a much wider meaning than the term 'ground conditions' (the term which is found in similar provisions in other contracts). The matter being covered by this event is that concerning anything physical which has an effect on the works, not just conditions in the ground. It is suggested that any reader who wishes to obtain a deeper understanding of this interpretation should read the court decision referred to in the footnote below.

In order for a physical condition to become a compensation event, it must occur within the boundaries of the Site (see clause 11.2(17) and Section 6.6.4) and not be a weather condition (see below). It must also be something that an experienced contractor would have judged to have a small chance of actually being encountered. In accordance with clause 60.2, the Contractor is required to judge what actual physical conditions it may encounter by reference to information as listed in that clause. The first item of information referred to is a document called Site Information (see Section 22.3) which is a term defined at clause 11.2(18) as being a description of the Site and its surroundings. The second item of information is any information referred to in the Site Information that is publicly available, followed by anything obtainable by a visual inspection of the Site at the third bullet point. The fourth bullet point refers to other information which an experienced contractor could be expected to have or obtain, although no guidance is provided as to what such information might be in practice (see Section 15.4.2).

Clause 60.3 provides rules for resolving any inconsistency within the Site Information. This clause follows the legal principle for the construction of contracts known as the *contra proferentem* rule and states that, in the event of an inconsistency, the information is interpreted least favourably for the party who prepared the information i.e. the Client. With closer consideration, it appears that clause 60.3 may overrule clause 60.2 in the event that the Site Information contains an inconsistency. The wording suggests that the interpretation arising from the resolution of the inconsistency will override the four factors listed in clause 60.2 as being the test to be applied.

The wording of the second bullet point in this event refers to 'physical conditions' that do not come under the description of 'weather conditions'. This provision only acts to exclude specific weather conditions; it does not exclude physical conditions that are 'due to' weather conditions. Accordingly, a physical condition such as a flood, which in itself is not a weather condition but may be due to a weather condition,[7] will qualify as a compensation event under this event. This is an important point to note as, in the physical world, it is possible for a flood to occur in an area that has not been subject to heavy rain but suffers as a result of heavy rain in a catchment area somewhere upstream of the site of the flood.

The final paragraph of the event confirms that it is only the difference between what should have reasonably been allowed and the actual conditions encountered that is taken into account when assessing the compensation event. This sentence was added in the third edition as users of the second edition had reported that some users were claiming for the total effect of the consequences of an event, rather than discounting for what

---

[6] See the decision in *Humber Oil Trustees Ltd. v Harbour and General Public Works (Stevin) Ltd.* (1991) 59 BLR 1.
[7] In practice, a flood can occur from another event such as a burst main or the failure of a dam.

they should have reasonably allowed for. As this position was clearly incorrect, the third edition was strengthened to confirm that the event only gives rise to compensation for the difference. The fourth edition maintains this position.

15.3.13 Event 13

Event 13 concerns weather conditions and is drafted in such a way that the determination as to whether a weather condition is a compensation event or not has become an objective test. What is unusual is that the Client carries the financial risk of the event rather than the more common situation in other contracts, where weather delays are treated as neutral with time granted to protect the Contractor from delay damages and with the Contractor standing the costs. From this background, rather than examining each line of the clause, what follows is a practical look at the weather compensation event. What information the Client needs to provide, what the Contractor needs to consider when tendering, how the actual weather conditions encountered are considered and how the extent of any compensation event is determined must all be examined.

When preparing Contract Data Part One prior to the issue of the tender, the Client needs to consider the four main entries to be made in that part of the document against Section 6 (compensation events). These items concern the weather in two periods: (i) the past and (ii) the period during which the works will be carried out. For both these periods of time, the Client needs to decide at which location the necessary weather information will be collected. For practical purposes, it is sensible that the location at which the information is collected should be the same for both the historic information ('weather data') and the project period information ('weather measurement'). Contract Data Part One requires entries for both sets of information together with statements as to where the information is available from and/or is to be supplied by (e.g. the Met Office in the UK). These entries cover three of the four main segments of information required.

The fourth of the main entries required concerns the weather measurements that are to be taken each calendar month. Four standard measurements are given: two concerning rainfall, one regarding the minimum air temperature being less than freezing and the fourth regarding snowfall (which requires an entry by the user as to the hour of the day when the measurement will be taken). Subsidiary to this main entry is a box that provides a prompt for the Client to add any other weather measurements that it considers should be measured for comparison with the historic records. The problem with adding further measurements here is that the recording authority may not have historic data available. If this is the case, a further entry has been provided where the Client can insert an assumed baseline against which to compare any such additional measurements. This facility can also be used if (unlike the UK) there is no organisation which has records of historic weather patterns and data in the vicinity of the Site.

When preparing its tender, the Contractor needs to consider the historic weather data for the place stated in Contract Data Part One. All necessary allowances in terms of time and money for all weather conditions stated, to be measured up to and including a one-in-ten-year level of occurrence, must be included in the Prices and programme. For example, if the one-in-ten-year historic level for the cumulative rainfall in a particular calendar month is 150 mm, then the Contractor must decide how much to allow for up to 150 mm of rainfall for that calendar month. Similarly, the Contractor needs to assess the necessary allowance for each of the measurements in each of the calendar months that the works will be carried out. This will involve the Contractor in a certain amount of risk judgement and management. It is unlikely that the one-in-ten-year level will be encountered in every month, so to make allowances consistently at that level is likely to render the Contractor uncompetitive. The Contractor must therefore judge what level of weather disruption to include in its pricing.

During the currency of the works, the Contractor – should it have any intention of trying to obtain a compensation event as a result of bad weather – needs to obtain the weather measurements for each calendar month from the source stated in Contract Data Part One. The Contractor then needs to compare each of the actual measurements for that month with the historical weather data for each of the categories. Should any of the actual measurements exceed the historical level for a one-in-ten-year occurrence, then the Contractor will be entitled to a compensation event for the difference between the actual and the historic. For example, if using the 150 mm of cumulative rainfall from the preceding paragraph as the historic one-in-ten-year level if the actual was, say, 175 mm, the result is that the Contractor would be entitled to a compensation event for the additional 25 mm. The last sentence of the event clearly states this to be the case.

Having examined the clause as above, it is necessary to refer to the second of the three bullet points which confirms that the event refers to weather conditions that are encountered before the Completion Date (see clause 11.2(3) and Section 12.5.2). This acts to confirm that, should the Contractor overrun, then it cannot claim for compensation due to weather conditions in the overrun period. The Completion Date itself can of course be extended by the occurrence of any of the compensation events. In the event that the Contractor has not finished by the Completion Date current at that time, but the parties are still resolving other compensation events that may act to extend the Completion Date and weather conditions occur that would otherwise entitle the Contractor to a compensation event, common sense is required. The parties will need to record the position and then analyse any entitlement to a weather-related compensation event once the assessment of the other compensation events has been resolved and agreed.

It is worth stating here that, in practice, compensation events for weather do not occur as frequently as users expect them to. That does not mean that they do not occur, or that the frequency will not increase if the climate continues to change as some say it will.

15.3.14 Event 14

This event refers to the Client's liabilities. There is a list of such liabilities at clause 80.1 (see Section 11.2). These are insurable liabilities that are the Client's responsibility under the division of liability set up in core clause 8. Once this has been appreciated, then the practical application of this event becomes straightforward.

In simple terms, if an event arises that is insured by the Client, then the Contractor receives a compensation event to cover the effects of that event and the Client recovers its losses from the insurance company. There would, however, appear to be a conflict between this event and clause 82.2. Under clause 82.2, the Client pays the Contractor any costs the Contractor pays to Others due to an event which is the Client's risk. However, under this compensation event the Contractor is entitled to compensation based on the contractual definition of Defined Cost which (see Sections 13.4.3–13.4.6) may be more or less than the true cost actually expended by the Contractor. Users in the United Kingdom have not reported this apparent conflict to be an issue in practice.

15.3.15 Event 15

The trigger for this event is the certification of takeover by the Project Manager in accordance with the provisions of clause 35. The event would not be triggered if the Client uses part of the works before Completion, but that use did not constitute takeover by being one of the two exceptions at clause 35.2. It must also be noted that the event only arises if the takeover is certified as having occurred before both Completion and the Completion Date. In practical terms, if the Completion Date has passed but the Contractor has not achieved Completion (i.e. the Contractor is in culpable delay), the Client can use and take over parts of the works without triggering this event and, in all probability, claim damages from the Contractor.

15.3.16 Event 16

The description of this event is self-explanatory and is in relation to the Client's obligation to provide materials, facilities and samples for tests as it is obliged to do by clause 41.2. The key is the extent of the Client's obligation as stated in the Scope.

15.3.17 Event 17

This event is unique among the other events in ECC because it does not refer to some act, omission, default or breach related to the physical works being carried out under the contract. Instead, this event acts to reconcile two other provisions in core clause 6. These are the Project Manager's power at clause 61.6 to state assumptions on which assessments for compensation events are to be based and clause 66.3, which determines that assessments are only revised if the contract provides that such a revision shall occur.

This event acts as the mechanism by which to correct any assumptions given by the Project Manager in accordance with clause 61.6 (see Section 17.4). It must be remembered that the assumptions stated by the Project Manager are entirely different from any forecasts made either by the Contractor when preparing its quotation or by the Project Manager carrying out assessments of a compensation event under clause 64 (see Section 16.7). In practice, it is essential that any assumptions given under the provisions of clause 61.6 need to be clearly stated as such by the Project Manager in order to avoid confusion and dispute in the future.

15.3.18 Event 18

This event is intended as a 'catch-all' for any breach of the contract by the Client not covered by other compensation events. The provision of such 'catch-all' provisions is common in construction contracts; it is necessary to ensure that the remedy for such breaches stays within the bounds of the contract rather than having to be dealt with as a common law claim for damages.

In theory, this event provides a conflict with other clauses in the contract that provide remedies for the Contractor in circumstances where the Client has committed a breach for which a remedy that is not a compensation event is provided. The two main potential breaches by the Client for which a separate remedy is provided are a failure to pay on time, for which the remedy is interest under clauses 51.2 and 51.3, and a payment made on termination (see clauses 93.1 and 93.2 and Section 18.6.1) where the reason for the termination was a breach by the Client. In practice, these conflicts should not cause a problem; providing the remedies at clauses 51.4 and 93 are paid to the Contractor, there should be no need to consider a compensation event as there would be nothing to pay under such events.

15.3.19 Event 19

This event covers matters that are referred to as a *force majeure* in other standard forms of contract, but as 'Prevention' in the ECC (see clause 19.1 and Section 6.6.7). When considering this event, the starting point must be the fifth bullet point, which makes it clear that this event can only arise if the event is not one of the other compensation events. If this condition applies, then this event cannot arise and the user must rely on the other compensation event.

Providing the event is not one of the other compensation events, then the first two bullet points are the next to consider. In respect of the first of these, that the matter in question stops the Contractor from completing the whole of the works is an absolute test: it either does or does not.

In relation to the second, the User Guide, Volume 4 states that 'this is a strict test – it is not simply a matter of delay'. The dates must be impossible to meet no matter what resources are used to try to catch up the time'. This leads to the position that if the Contractor cannot demonstrate that there is no reasonable means by which it can complete on time, then this event will not arise. This approach by the User Guide appears to put the Contractor under the burden of proving a negative, i.e. that it cannot complete as a result of the event. This is contrary to the general rules of law where the burden of proof falls on the party asserting the positive, i.e. that it is possible to complete. Although the occurrence of something that would qualify as one of these events will (in all likelihood) be few and far between, it appears that the interpretation of whether the Contractor could complete by planned Completion as shown on the Accepted Programme will be subjective and therefore grounds for dispute between the parties.

15.3.20 Event 20

This event acts to provide compensation to the Contractor when a quotation it has been instructed to provide under clause 65 (see Section 16.8) for a proposed instruction is not accepted. The inclusion of this compensation event recognises that the Contractor can incur significant costs in preparing such a quotation, especially in a design and build type arrangement.

As this event will only arise retrospectively, i.e. after the quotation has been prepared and submitted, then Contractors will need to keep records of all the resources used to prepare the quotation for use as the basis of the assessment.

15.3.21 Event 21

The final core compensation event acts to create a link between the list of compensation events and an optional entry in Contract Data Part One. This entry, which is located in Section 6 (compensation events), allows the users to detail other events that they intend to be ones for which the Contractor will be compensated. There is no limit as to what could be included in this space.

An example that the author has seen in several contracts has been to provide that any change in the minimum wage or the London Living Wage will entitle the Contractor to compensation. It could be said that such increases would be covered by Option X2 (see Section 20.2), but if that Option is not going to form part of the Contract, these matters could be isolated and included in this entry.

15.3.22 Clause 60.4 (Main Options B and D)

Main Options B and D both contain the same three compensation events which are additional to the 21 events included at clause 60.1. The three events are all connected with the use of a Bill of Quantities (see Section 13.5.2) in these Main Options. None of the other Main Options contain any compensation events that are additional to the core list.

In practical terms, the compensation event at clause 60.4 provides a mechanism for re-rating a bill of quantities rate when the quantity changes, but only where the change is not the result of a change to the Scope. If the change in quantity results from a change to the Scope, then the affect should be assessed in the compensation event for that change (clause 60.1(1) as covered in Section 15.3.1). This event is concerned with the situation where there was an error in the Bill of Quantities regarding the quantity for a particular item. Should the two qualifying tests at the second and third bullet points in the clause be satisfied, then the rate for the item is adjusted up or down. This is to permit the correction of the rate to take into account the economies of scale in respect of both output and purchasing.

Consider the example where an item is stated as being 1,000 units in the Bill of Quantities, but should only have been 10. The purchase price for the materials is £10 per unit if 1,000 units or more are purchased but £25 per unit if less than 20 are purchased. The rate for the item would therefore be adjusted by this compensation event to reflect that each unit would cost £15 more to buy than the price that would have been obtained for the quantity in the Bill of Quantities. Through the operation of the definition of the Price of Work Done to Date (PWDD) (see clause 11.2(30) and Section 13.4.4) in Main Option B and the definition of the Total of the Prices (see clause 11.2(35) and Section 13.7) in Main Option D, the actual quantity carried out by the Contractor (i.e. 10 units) would be included in these calculations at the revised rate (i.e. the original rate increased by £15 plus the Fee).

The way this event is structured and, in particular, the test at the third bullet point means it can only apply to the unit rates for those items in the Bill of Quantities which represent more than 1/200th of the total value of the Bill of Quantities at the Contract Date (see clause 11.2(4) and Section 12.3). This is probably meant to pick up on the old maxim regarding Bills of Quantities, which states that 80% of the value is in 20% of the items (known as the 80/20 rule). This approach acts to stop the parties expending effort carrying out such exercises on items of small value and therefore negligible worth.

15.3.23 Clause 60.5 (Main Options B and D)

The second of the three events concerns the position where the change in quantity between that actually carried out by the Contractor and that stated in the Bill of Quantities has the effect of delaying Completion or the meeting of a Key Date. The inclusion of this compensation event means that, in practice, the Contractor can prepare its programme on the basis of carrying out the quantity of work included in the Bill of Quantities. Should any of those quantities be understated (to the extent that the difference between the quantity stated in the Bill of Quantities and the quantity actually done means that a longer period would have been included in the programme and that a longer period would have impacted, via the critical path, on Completion) then the Contractor would be entitled to a compensation event. As with clause 60.4 in the previous section, this event will not arise when the reason for the change in quantity is because of another compensation event (e.g. a change to the Scope).

15.3.24 Clause 60.6 (Main Options B and D)

In practice, users are far more likely to use the third of the events related to the Bill of Quantities as opposed to the other two. This event provides for the correction of mistakes in the Bill of Quantities. The event is limited to mistakes that are the result of a departure from the rules in the method of measurement identified in Contract Data Part One, in respect of either the description in an item or the division of work into the items required by the method of measurement or due to an ambiguity or inconsistency. This event creates the position that is common in contracts which use bills of quantities that the Contractor prices the bill items for the purposes of the tender and that the Client accepts responsibility for any mistakes in the preparation of the bills.

The way the event is drafted restricts its application to departures from the method of measurement, and thereby excludes mistakes in the Contractor's rates which are the result of errors by the Contractor or its misreading of other documents. The correcting of errors in the rates because of errors in the items within the bills is straightforward and would be covered by this event. All errors that arise as a result of the application of this event can lead to a reduction in the Prices as well as an increase.

The way in which the event is worded, the obligation to correct the mistakes lies with the Project Manager; it could be argued that it is the Project Manager's responsibility to check the Bill of Quantities for mistakes and make all the necessary corrections. However, in practice, it will probably be the Contractor who first notices such mistakes and who will no doubt, following the spirit of mutual trust of cooperation required by clause 10.2, assist the Project Manager in this task. Once a mistake is found, the Project Manager issues an instruction to correct the mistake.

15.3.25 Secondary Option X2

(See Section 20.2 for details of this Secondary Option.) This event could be far reaching as, in many countries including the United Kingdom, changes in the law include matters such as taxation, employment rights, import and export duties, currency movements and exchange rates. By including this event, the Client accepts the risk of all such changes, thereby removing the need for the Contractor to anticipate what might happen and make allowances for such changes in its pricing. The way in which the event is worded, the Client takes the risk of all changes that occur after the Contract Date (see clause 11.2(4) and Section 12.3) even if it was known that the change in the law would occur before that date.

In practice, this can be changed by amending the wording so that the event only arises if the change had not been promulgated before the Contract Date. Whether this amendment is included or not, the parties simply need to appreciate which one of them is taking the responsibility for what. That said, it is found in practice in the UK that the inclusion of this event in contracts is exceedingly common and is becoming the norm in many market sectors.

15.3.26 Secondary Option X12

(See Section 20.4 for details of this Secondary Option.) This Secondary Option includes two events that essentially provide the same protection for the Contractor but in respect of two different documents created by the use of this Secondary Option. The two documents concerned are the Partnering Information (clause X12.1(6) and Section 20.4) and the partnering timetable (which is not a defined term). If either of these documents is changed by the Core Group (clause X12.1(4) and Section 20.4) and such a change has an impact under the Contractor's Own Contract (clause X12.1(3) and Section 20.4), then the event is identified at clause X12.3(6) in the case of the Partnering Information and at clause X12.3(7) in the case of the partnering timetable. It must be emphasised that the changes made to these two documents by the Core Group are not enough to trigger the event; the changes made must result in an impact on the Prices and/or the Accepted Programme.

15.3.27 Secondary Option X14

(See Section 13.10.2 for details of this Secondary Option.) Where this Secondary Option is in use, a delay in making the advance payment constitutes an event. The effect of such a delay may at first instance appear to be simply a matter of financing. However, if the advanced payment was going to be used by the Contractor to secure orders for vital components, then being denied use of the funds could also cause delay to Completion. This event therefore arises instead of any right to only receive interest payments on the late payment to which the Contractor would otherwise have been entitled.

15.3.28 Secondary Option X15

(See Section 11.3 for details of this Secondary Option.) Where this Secondary Option is in use providing that the Contractor used reasonable skill and care in respect of a defect arising from its design, then the Contractor is not liable for that defect and will be paid. Providing the Contractor used reasonable skill and care when carrying out its design, then it will not have been negligent and therefore the Client would have no claim against the Contractor in respect of a simple error. Without the presence of negligence, the Contractor does not carry any liability. The Contractor will still be required to correct the defect but a compensation event will arise. This provision does not relieve the Contractor of its liability if the defect is as a result of poor workmanship or materials.

15.3.29 Secondary Option Y(UK)2

(See Section 13.9 for details of this Secondary Option.) This event only applies in the UK to contracts to which the Housing Grants, Construction and Regeneration Act 1996, as amended by the Local Democracy, Economic Development and Construction Act 2009, applies. It provides compensation for the Contractor in circumstances where the Contractor has exercised its statutory right under that Act to suspend the performance of the whole or part of its obligations as a consequence of the Client's continuing failure to pay a sum certified as being due, but which has not been fully made by the final date for payment (see Section 13.2). The contract does not actually give the Contractor the right to suspend the performance of the whole or part of its obligations in these circumstances. This right is implied into the ECC by the Act and the Scheme. Because of the inclusion of this event, the Contractor can recover both time and money without having to rely on the Act and the Scheme which would, in any event, give it that remedy in such circumstances.

15.4 Practical issues

As this chapter concerns the general theory of what a compensation event is and has then considered the events that are set out in the ECC which constitute compensation events, the number of practical issues attached to these matters are limited. Readers of this guide will find far more practical issues associated with compensation events in the following two chapters concerning the Procedures associated with compensation events and the Assessment of the impact of such occurrences.

15.4.1 Events not in the list

Having gone through the compensation events in the ECC, some users comment that there are matters which they would expect to see which are missing from the contract. However, on careful consideration, these matters are covered by other listed events. The main matters that tend to be mentioned in this respect are as follows:

- Any failure by the Project Manager or the Supervisor to act in accordance with the requirements of the contract. As the Client has taken full responsibility for the performance of the Project Manager and the Supervisor under the ECC, then any failure by either of them to act in accordance with the requirements of the contract would be classified as a breach of contract by the Client and fall under clause 60.1(18).
- The late supply of information to the Contractor by the Client or Project Manager. Providing that the Contractor has included a milestone on the Accepted Programme showing the date by which the information is required, then such failures would (in all probability) fall under clause 60.1(3).

15.4.2 Site Information

(For more details about the contents of the Site Information, see Section 22.3.) Within clause 60.2, reference is made at the second bullet point to publicly available information. In order to be considered by the Contractor when determining what it can expect to encounter, the Contractor must take account of such information as is referred to in the Site Information. On its own, this is a straightforward provision that should not cause users any issues in practice.

However, the fourth bullet point in clause 60.2 makes reference to other information obtainable by an experienced contractor. It could be argued that this other information could include publicly available information not referred to in the Site Information. This argument is unlikely to succeed in the UK as the attitude of the courts has tended to be that, as the client had a significant time to prepare the documentation in comparison to the time the contractor had to tender the works, then the consequences of the client not finding information and making reference to it cannot be passed to the contractor without clear express words to that effect. In this respect, the fourth bullet point is not clear enough.

The third bullet point makes reference to a visual inspection of the Site made by the Contractor. This is a reasonable condition to apply; however, it must be appreciated that such an inspection is limited in practical terms to what can be seen without any intrusions into the structure and to the areas can be accessed safely given the circumstances of the Site. In practical terms, if there is asbestos in an existing building that is not otherwise identified in the Site Information but is hidden behind some cladding or even above removable suspended ceiling tiles, a visual inspection cannot be expected to pick this condition up. Further, if there is some condition that would influence the Contractor's pricing in an area that would require special health and safety provisions to access (e.g. an enclosed space or inaccessible roof area) then, unless the necessary access provisions were provided at the time of the inspection, the Contractor cannot be expected to pick up the condition as part of the visual inspection.

15.4.3 Using the Met Office for weather information in the United Kingdom

The UK Met Office provides an excellent source of weather-related information, both in terms of the historical data and the measurements of the actual weather during the contract period. However, there are some practical aspects that users should consider when using this service.

The Met Office operates many weather stations around the country, but not all of them collect data of the type required by the ECC and not all those that do have done so for long enough to be able to provide the statistics for the one-in-ten-year frequency. Before selecting the weather station to be used, it is advisable to check with the Met Office and ascertain that the weather station chosen can provide the information required. If it cannot, then an alternative needs to be identified. The Met Office will assist users of their service in selecting a suitable weather station nearest to the Site.

Another factor that users will need to consider will arise if they decide to add additional measurements to the four standard data types listed in Contract Data Part One. Before making such additions, users need to ascertain whether the Met Office can provide the measurement required during the project and whether any historical data (see following section) exists against which to compare the actual. If actual measurements are not going to be available, then the user will need to consider how to overcome this problem. The only answer may be to install appropriate measuring equipment on the Site to provide the relevant information.

15.4.4 The problems with compensation events for weather

While there is no doubt that the objective test as to whether a weather event has occurred or not is preferable to the subjective alternative provided in most other standard forms of construction contract, it must be recognised that the set of measures provided by the ECC brings with it a number of problems. We consider the practicalities of these problems here.

In the event that the historical data for any weather measurement is not available, then the ECC provides the solution by providing an entry in Contract Data Part One in which the user can insert assumed data for each such measurement. In providing this assumed historical data, the user will need to consider the variance throughout the year and either provide a single assumption or up to 12 such assumptions depending on the likely variance. There is no fixed proposition here, as it will be a factor that varies depending on the weather condition being considered and the geographical location in the world that it is being applied too. If any such weather measurement is used, then, as discussed in the previous section in respect of the UK, the problem exists everywhere in the world of how to obtain the measurement for each calendar month during the project for any weather condition that is not normally measured by any local weather agency.

Within Event 13, the first bullet point presents users, especially contractors, with a risk that they must determine and allow for. The bullet point states that the weather measurement is recorded in a calendar month. This conveniently divides the objective test into 12 distinct slices per year, over which there can be no doubt. If the actual weather conditions follow the anticipated pattern, then this does not generally pose any problem. However, consider the case of total rainfall, which, instead of falling in small amounts across two adjacent months, falls in a different pattern. In the first of our two months, assume that it stays dry until a few days before the end of the period when it then rains continuously with the total rainfall being just less than the one-in-ten-year average for that month. At the start of the second of our two months, it continues to rain for the first few days, again with the total rainfall being just less than the one-in-ten-year average for that month, before stopping and remaining dry for the rest of the month. As the total rainfall in both months is less than the benchmark, a compensation event associated with total rainfall does not arise.[8] However, because of the sustained rainfall over a number of days, which as a single continuous period caused substantial delay and other problems, the contractor has suffered the consequences of weather for which it would otherwise have been compensated but for the timing of the event. Contractors must be aware of this issue of timing of the weather occurrences and make appropriate allowances as necessary.

The third bullet point also acts to present a practical issue for the Contractor. Comparative weather measurements and data recorded at the place stated in the Contract Data are used to determine if a weather-related compensation event arises. This location may be several miles or more from the actual Site and may be in a position that does not accurately reflect the conditions being encountered on the Site. The Contractor must therefore always consider the relative location of the weather station and the Site when assessing its risk for weather. For example, if the weather station is in a sheltered valley but the Site is on exposed high ground some miles away, this difference must be allowed for by the Contractor in the pricing and programme allowances.

---

[8] It is also probable that an event related to the number of days with rainfall in excess of 5 mm does not occur either.

Chapter 16
Compensation Events: Procedures

16.1 Introduction

Consistent with the ECC's general principles of bringing all matters out into the open as soon as they occur the notification procedures within core clause 6 follow the same ethos. Clause 61 is headed 'notifying compensation events' and the first sections of this chapter will deal with the majority[1] of these provisions.

The second part of the compensation event procedures concerns the submission by the Contractor of quotations for changes to the Prices, Completion Date and Key Dates resulting from the occurrence of an event followed by the Project Manager's reply thereto. These later sections will also consider what happens if matters in connection with quotations don't go smoothly and what the participants can do in various scenarios.

The last section of this chapter considers various practical issues related to these procedures, which include ways in which parties can use the spirit of mutual trust and cooperation to avoid difficulties and to ensure that the goal of the procedures is achieved with the minimum expenditure of time and effort.

The procedures discussed in the chapter are those that users appear to spend time debating among themselves more than any others. It is these procedures which, in the second edition of the ECC, created the largest amount of comment from users. This comment resulted in these procedures being considerably strengthened in the third edition of the ECC and maintained in the fourth edition to the extent that they now contain both an effective condition precedent and 'deemed acceptance' procedures where the Project Manager fails to act as required.

Conversely, many teams have successfully operated these provisions and been in a position where the changes resulting from all notified events have been assessed and agreed by the time that the Project Manager certified Completion (see Section 12.5.2). The practical steps that these teams have used are incorporated into the final section of this chapter.

---

[1] While clause 61.6 will be mentioned, the detail and reasoning behind this clause and the practical application and implications are covered in Section 17.4.

A Practical Guide to the NEC4 Engineering and Construction Contract, First Edition. Michael Rowlinson.
© 2019 John Wiley & Sons Ltd. Published 2019 by John Wiley & Sons Ltd.

16.1.1 The basic principles

In simple terms, the ECC requires the Project Manager to notify the Contractor of compensation events arising from giving instructions or notifications, the issue of a certificate or from changed decisions of the Project Manager or Supervisor. The Contractor is required to notify the Project Manager of all other compensation events and of those that the Project Manager fails to notify to the Contractor. The Contractor is required to give all such notices within eight weeks of becoming aware of the compensation event, albeit that in certain instances late notification would still be effective.

The notification procedure briefly set out above and described in detail in Sections 16.2 and 16.3 below can, once the principles are understood, be reduced to a relatively straightforward flow diagram as set out Figure 16.1. Figure 16.1 is based on a sketch that the author has sometimes used when giving training sessions on how to use this procedure and aims to take the procedures down to its basic steps. Most delegates at the training sessions have found that this simple version enables understanding, especially for those who are relatively new to the concepts.

From this simple summary, the question that arises immediately is whether the notification of a compensation event is a condition precedent to the Contractor's entitlement to payment or a revision of the Completion Date. In the first and second editions of the ECC, it was clear that the aim of making the notification procedure a condition precedent had not been achieved.

In the third edition, the stated intent was to create a condition precedent. Changes to clause 61.3 and a new clause 63.4 were introduced to make notification of those compensation events, which the Contractor is responsible for notifying in the first instance, a condition precedent. This has been maintained in the fourth edition. It is interesting to note that the words 'condition precedent' are not used in these clauses. Instead, the contract states, on the one hand, that the Contractor is not entitled to a change in the Prices, the Completion Date or a Key Date and, on the other hand, that such changes are its only right. A poll of commentators concludes that the majority think the provisions will be effective, but there is a feeling that the wording of the contract is such that arguments can be formulated against this proposition. These provisions do not apply to a compensation event that should, in the first instance, have been notified by the Project Manager to the Contractor.

As an aside in respect of condition precedents, on numerous occasions the UK Courts have found against requirements being construed as condition precedents unless the particular contract has clear words to that effect. The UK Courts have resisted attempts to imply such a requirement from language that has left itself open to interpretation. If this matter is referred to the UK Courts, only time will tell if they find that the words constitute a condition precedent or not.

16.2 Notification by the Project Manager

Whenever the Project Manager initiates a compensation event, then the Project Manager is required by clause 61.1 to notify that event to the Contractor as being a compensation

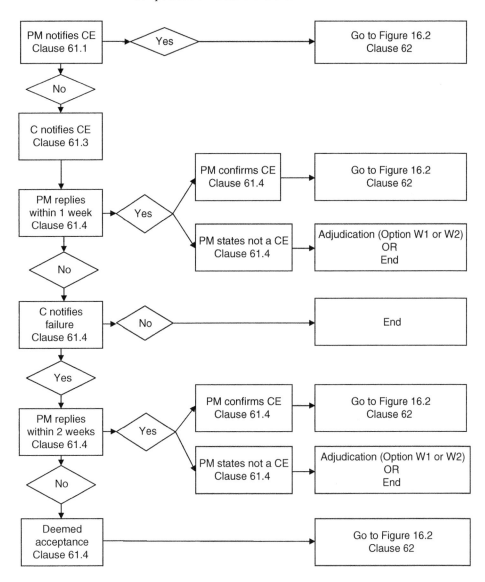

Figure 16.1 Clause 61 – Notification Procedure.

event at the time that the event was initiated. The clause expresses such events as those that arise from either the Project Manager himself or the Supervisor either giving an instruction or notification, changing an earlier decision that either of them had made or issuing a certificate (as required by the Contract). The way the clause is drafted, where the Supervisor is the originator of the instruction or changed decision, it is still the Project Manager who notifies the Contractor that a compensation event has arisen. This is something that the Project Manager will have to monitor carefully, although the only power that the Supervisor has to issue an instruction is that at clause 43.1 in respect of a search for defects (see Section 9.5).

In respect of the 21 compensation events at clause 60.1, the User Guide, Volume 4 states that the procedure at clause 61.1 would normally apply to compensation event numbers:

- 1: changes to the Scope;
- 4: instructions to stop or not to start work;
- 7: objects of interest;
- 8: changing a decision;
- 10: searches for defects;
- 15: take over before completion;
- 17: correction of an assumption;
- 20: a proposed instruction is not required; and
- potentially 21; an additional event or events.

The event at clause 60.6 is also identified as one that would be initiated by the Project Manager.

These compensation events are those that will arise as a result of the Project Manager or the Supervisor giving an instruction that they are empowered to give under the Contract or making some other decision required of them. As the User Guide states, that these are the events that would 'normally' be notified by the Project Manager, they seem to suggest that an event could occur that would be a compensation event but which would not arise as a result of a power exercised by either the Project Manager or the Supervisor. Given the nature and wording of the 10 events listed above, it is difficult to see how in practice any of them could arise in any circumstance other than where initiated by either the Project Manager or the Supervisor.

Clause 61.2 requires the Project Manager to also instruct the Contractor to submit quotations at the same time as the compensation event is notified. When all these requirements are considered in tandem with the requirement at clause 13.7 (see Section 6.2) that every notification is communicated separately, it can be seen that the Project Manager needs to issue up to three communications every time a compensation event is notified (see Section 16.9.1):

1. the instruction to change the Scope, for example;
2. a notification that a compensation event has occurred; and
3. an instruction to submit a quotation.

Clause 61.2 also tells the Project Manager that the instruction to submit a quotation would not be required if either:

- the event arises from a fault of the Contractor; or
- the event has no effect on Defined Cost, Completion or a Key Date.

A small number of commentators have created a fallacy about the ECC stating that, when a compensation event occurs, the Contractor is not obliged to and should not carry out any work in connection with that compensation event until the changes to the Prices and Completion Date have been agreed. If this fallacy was correct, which it is not, the effect would be that many compensation events would cause an additional and unwanted delay to the project while the implications were assessed. Clause 27.3 clearly obliges the Contractor to get on with the work. If the Project Manager issued an instruction one afternoon to change the Scope, which was a compensation event as clause 60.1(1), and that change needed to be physically actioned and carried out the morning after in order to fit the planned sequence of works and to avoid delay, then the Contractor would be obliged to carry out that change the following morning. The assessment of the compensation event would follow in due course after the work had been completed, unless the parties had agreed the effects on the same afternoon that the instruction was given.

16.3 Notification by the Contractor and the Project Manager's reply

16.3.1 The notification

Consistent with other provisions[2] in the ECC, there is a clear requirement on the Contractor to notify compensation events promptly. This requirement is stated at clause 61.3 and arises either when the Contractor believes that the event has happened or which is expected to happen. The obligation is therefore one that covers both the past and the future. In order for the obligation to impose itself on the Contractor, it must be an event which it believes is both a compensation event and which has not been notified by the Project Manager (see Section 16.2). The latter part of this obligation does mean that the Contractor is acting as the safety net, should the Project Manager fail to properly comply with the requirements of clause 61.1.

Should the Contractor's obligation to notify a compensation event arise, then the Contractor is obliged to notify the event to the Project Manager within eight weeks of becoming aware of the event. The clause then sets out that, unless the Contractor has notified the event within the eight weeks, it is not entitled to a change in the Prices, Completion Date or a Key Date. When read in tandem with clause 63.6, which acts to limit both the Client's rights and the Contractor's rights in respect to any matter which is a compensation event to a change in the Prices, Completion Date or a Key Date, these two provisions act to create what commentators consider will be an effective condition precedent. There is a caveat to this position that provides protection for the Contractor, which will be returned to later in this section.

Arguably, the wording used makes this a lesser test than if the clause included the words 'within eight weeks of the time that it should have become aware of the event'. The potential weakness in the wording used in the last part of clause 61.3 seems to present the Contractor with a potential 'get-out-of-jail-free' card if it doesn't notify an event within eight weeks by simply claiming that it was not aware of the events occurrence. Contractors must, however, be concerned that, to plead ignorance of an event having arisen (when any experienced contractor in charge of the Site would have known that the event had arisen) may not stand as a defence to the failure to issue a notice. Only a referral of this point to a court will provide the answer, but it will be a brave contractor who risks the costs of making such a referral.

---

[2] For example, at clause 15.1 (see Section 6.4) and clauses 17.1 and 17.2 (see Section 6.6.5).

The caveat to the provision requiring notification within eight weeks is contained in the last two lines of clause 61.3 where the rule is clearly excepted where the event came about as a result of the Project Manager or Supervisor issuing one of the instructions that they are empowered to give, either of them changing an earlier decision which they had made or the issue of a certificate required by the Contract by one of them. As covered in the first paragraph of this section, the Contractor has an obligation to notify any event which the Project Manager should have notified but did not. The caveat stops the condition precedent from applying when the Contractor fails to notify something that the Project Manager failed to notify. This prevents the Client from benefiting from a failure, in the first instance, of the Project Manager, a position that can only be considered as being fair and balanced.

The events that the Project Manager should normally notify were identified in Section 16.2. By reference to the Volume 4 of the User Guide, users are told that the Contractor would normally be expected to notify compensation events numbers:

- 2: failure to give access;
- 3: failure to provide something;
- 5: failure to work within times;
- 6: failure to reply to a communication;
- 9: withholding an acceptance;
- 11: tests or inspections causing delay;
- 12: physical conditions;
- 13: weather conditions;
- 14: event at Client's risk;
- 16: failure to provide materials, etc.;
- 18: breach of contract by the Client;
- 19: happening not caused by any party; and
- potentially 21; an additional event or events.

From a study of the compensation events within the Main and Secondary Options, it is also suggested that, in practice, the Contractor would normally be the person who notified the following events:

- X14.2: delay in making an advanced payment;
- X15.2: correction of a Defect for which the Contractor is not liable; or
- Y2.4: suspension of performance under the Housing Grants, Construction and Regeneration Act 1996.

Between the above two lists and their equivalents in Section 16.2 the compensation events at clauses 60.4 and 60.5, as applicable to Main Options B and D as well as those at clauses, X2, X12.3(6) and X12.3(7), have not been identified as being normally notified by either the Contractor or the Project Manager. While it is arguable that the Project Manager should notify the first three of these five events as a consequence of the Project Manager's obligation under clause 50.1 to assess the amount due, in practice, it would be safer for the Contractor to consider these matters and notify a compensation event each time it considers that either of these three clauses applies. As the latter two are initiated by a Core Group, of which neither the Project Manager nor the Contractor may be a member, it is difficult to determine who should notify these.

16.3.2 The reply

In every instance that the Contractor has notified a compensation event as set out above, the Project Manager is obliged by clause 61.4 to respond to that notification.

The clause starts by setting out the period for the reply to be made. In this case, the given period is one week from receipt of the Contractor's notification or such longer period to which the Contractor has agreed.

It then continues by setting out five reasons why the notified event may not qualify as a compensation event. The interpretation of these five grounds is straightforward and should not cause users any problem. That said, there will be disagreements about whether certain matters are compensation events; these will be based on differences of opinion about the event itself rather than interpretation of these five grounds.

From these five grounds, users will determine from the first bullet point that if the event has been caused by the Contractor,[3] then it will not qualify as a compensation event. The second and fifth bullet points, which state that the event has not happened, is not expected to happen or does not constitute one of the listed events, would also cause it to fail to be considered under core clause 6. The third bullet point refers to the timescales required by the Contract, namely that at clause 61.3 (see Section 16.3.1), for the notification of an event. These four grounds are all simple and straightforward to understand, as is the fourth, but it is the fourth which many users do not adopt. The fourth bullet point leads to the disqualification from being a compensation event of any event that actually qualifies under one of the compensation event descriptions and was notified in time, but that does not impact on the Prices, the Completion Date or a Key Date. In order to make this determination, the Project Manager is actually carrying out two steps in the overall process without consultation with the Contractor, which potentially acts against the spirit of mutual trust and cooperation.

In order for this fourth bullet point to be the only reason for rejecting the Contractor's notification, the Project Manager must first of all determine that the event itself is a compensation event. After making that determination, the Project Manager must then, in effect, carry out an assessment (see Sections 16.7.1–16.7.3 and Chapter 17) of the impact of the event on the Prices, the Completion Date or the Key Dates. Making such an assessment at this time is premature and contrary to the rest of the process that will unfold in the remainder of this chapter. It is therefore suggested that, in practice, if the Project Manager considers that an event notified by the Contractor is a compensation event then this fourth bullet point should not be considered as a ground for rejecting the Contractor's notification. The Project Manager should instead say that the event is a compensation event and ask for a quotation. Once the quotation has been submitted, the Project Manager can then determine whether there is any effect or not. Following this approach also

---

[3] The clause actually refers to it being due to the 'fault' of the Contractor; to be a 'fault' it must be caused by the Contractor.

means that any compensation events that do not have any effect can be entered into a log; consequently, there is an agreed record for audit purposes which records that the parties have agreed that there is no effect, thereby avoiding any attempts to re-visit the event in the future.

The conclusion of this part of the clause is that if any of the five bullet points apply then, notwithstanding the comments in the preceding paragraphs, the Project Manager should reply to the Contractor's notification stating that the decision is that the Prices, Completion Dates and the Key Dates will not be changed (in other words, that the Project Manager does not accept that the matter notified constitutes a compensation event). In order to comply with clause 13.4, the Project Manager should state reasons for making this decision.

In the event that the Project Manager decides that none of the five bullet points listed at the start of the clause apply, then the Contractor is informed that the matter notified to the Project Manager by the Contractor is a compensation event and at the same time the Contractor is also instructed to submit a quotation for the notified event.

In the event that the Project Manager fails to reply within the required timescale, then a procedure that was introduced into the third edition could be invoked at any time thereafter[4] by the Contractor. This procedure is initiated by the Contractor issuing a notice to the Project Manager notifying the Project Manager of the failure to reply. The Project Manager then has a further two weeks to reply to the original notification and can do so in either of the ways set out above. Should the Project Manager fail to reply within this further two weeks, for which there is no provision to extend, then the Contractor's original notification is deemed to be accepted and the event notified is a compensation event. This automatic acceptance also constitutes an instruction to the Contractor to submit a quotation for the compensation event.

16.4 Other matters associated with notifying compensation events

16.4.1 Failure to give an early warning

Whenever the Project Manager notifies a compensation event, consideration must be given to whether the Contractor gave any early warning (see clause 15.1 and Section 6.4) that an experienced contractor could have given in relation to that event. The Project Manager must make this determination, as clause 61.5 requires the Project Manager to state this in the instruction for the submission of a quotation by the Contractor. This is the start of the procedure by which the Contractor is penalised for not giving an early warning. The wording of clause 61.5 is such that it appears that the Project Manager only gets one chance to raise this alleged failure by the Contractor.

Should a Contractor's notification be accepted by default under the provisions of clause 61.4 (see Section 16.3.2), the Project Manager will lose this opportunity to notify

---

[4] Certainly at any time up to the issue of the Defects Certificate (see clause 61.7) but arguably at any time in the future subject to any limitation in the relevant legal jurisdiction (e.g. as set by the Limitation Act 1980 in the UK).

such a failure to give an early warning and the Client will, as a result, potentially find itself in a situation where the Contractor has received more compensation than would otherwise be due.

In the event that the Project Manager makes a statement pursuant to clause 61.5 that the Contractor failed to give an early warning, then the provisions of clause 63.7 (see Section 17.5.1) will apply. In order to make it practically possible for the Contractor to be able to take these provisions into account when assessing the compensation event, the Project Manager will need to set out in the statement given under clause 61.5 what mitigating steps could have taken had the early warning been given in order to reduce the impact of the event to the advantage of the Client. Without such information, the Contractor will probably not be able to apply the provisions of clause 63.7. This will lead to the Project Manager rejecting the Contractor's quotation for this event, which may then escalate into a formal dispute requiring the services of the Adjudicator or the Dispute Avoidance Board (see Chapter 19), a situation that should be avoided whenever possible.

The drafting of clause 61.5 has been frequently criticised by commentators on the ECC. The best way to understand the clause is to consider the bigger picture about what the early warning mechanism is trying to achieve. One of the fundamental principles of the ECC is to use everyone's skills, in a spirit of mutual trust and cooperation, to create foresight which then allows the most cost- and time-effective methods to be adopted to the advantage of the project. One of the key tools in achieving this principle is the concept of early warnings (see clause 15 and Sections 6.4 and 6.5.1). By making it a mandatory requirement (see Section 1.4) to give early warnings such as an experienced contractor could give, a failure to give such a notice constitutes a breach of the contract. The right to damages arises from such a breach of the contract, a right which can be provided by express terms of the contract thereby avoiding the need to refer to common law for the remedy. In the ECC, clause 61.5 is the trigger that initiates that express contractual right to damages for the Client. These damages do not take the form of a payment from the Contractor to the Client. Instead, they take the form of a reduction in the amount paid to the Contractor by reducing the evaluation of the compensation event to that which the Contractor would have received, had the early warning been given. This step provides the Client with the advantage of any instructions the Project Manager could have issued as a result of the foresight gained from that early warning, which would have mitigated the impact of the event on the Client. It is possible that such mitigation could have reduced the impact to zero and this may well be the assessment that is determined.

16.4.2 Latest time for notifying an event

It is stated clearly in clause 61.7 that the latest date for notifying a compensation event is the issue of the Defects Certificate (see Section 9.6). The setting of this date, some weeks after Completion (see Section 12.5.2), envisages that situations which constitute compensation events could occur after Completion. A simple practical example of such an occurrence could be where the Scope has not required a piece of work to be completed by Completion: for example, some form of planting which should only be carried out at certain times of the year. When that work is then carried out after Completion but before the defects date and the issue of the Defects Certificate thereafter, it is possible that there could be a compensation event associated with that work. This provision therefore allows for the Contractor to be compensated for any such event.

The Project Manager must be aware of this limitation, as the most likely source of a compensation event at this time will be that there will be a need to issue an instruction changing the Scope in connection with any work being carried out after Completion.

Contractors should also note this condition, as they are required to notify compensation events within eight weeks of becoming aware of the event. This period for notification will start to reduce for any event that occurs in the eight weeks prior to the issue of the Defects Certificate. As the date of the issue of the Defects Certificate is flexible between the defects date and the end of the last defect correction period (see clause 44.3 and Section 9.5), Contractors need to be alert to the possibility from eight weeks before the defects date, being the earliest date that a Defects Certificate could be issued of the pending end of the period in which they can notify compensation events. In practice, this situation should not arise very often as the events the Contractor should have to notify are unlikely to occur after the Completion. Contractors must, however, remember that their obligation to notify compensation events extends to notifying matters, such as a change to the Scope, which the Project Manager should have notified but did not (see clause 61.3 and Section 16.3.1).

16.4.3 Giving assumptions

In addition to considering whether the Contractor has issued an early warning that could have been given in connection with a compensation event, the Project Manager must also consider whether or not to state any assumptions about the basis on which the Contractor should assess the compensation event when preparing the quotation. This provision is set out at clause 61.6 and allows the Project Manager to remove some of the uncertainty that can exist within the preparation of quotations for things that have not yet happened and which cannot be reasonably forecast. The power to state assumptions is clearly one for the Project Manager and the Project Manager alone. The Contractor, following the wording, does not have any part to play in the process.

The reasoning behind this provision and the way in which the Project Manager can use it to exert cost control are discussed in detail as part of the following chapter concerning the assessment of compensation events (in particular, Section 17.4).

16.5 Quotations: Substance

16.5.1 What is a quotation?

As with the term 'compensation event', the ECC does not set out a definition of what a quotation actually is. From clause 62.2 it is can be seen that a quotation is a proposal from the Contractor for changes to the Prices and any delay to the Completion Date and Key Dates as assessed by the Contractor.

Users should note that, in accordance with clause 62.2, the quotations are based on the Contractor's assessment. In carrying out this assessment, the Contractor will make forecasts which will be at the Contractor's risk. The Project Manager's only way of exerting control over these forecasts is to state assumptions under clause 61.6.

16.5.2 Status of quotations: If instructed

All quotations submitted by the Contractor to the Project Manager as a consequence of an instruction from the Project Manager must be considered by the Project Manager and replied to in accordance with the requirements of clause 62.3 and the general principles of clause 13.3. The ECC appears to contain an anomaly in that the Contractor can be instructed to provide more than quotation for a particular compensation event based on different ways of dealing with that event. The Project Manager is given three options for the reply to a quotation (see clause 62.3 and Section 16.6.2), none of which cater for the rejection of an alternative quotation on the grounds that another quotation for that compensation event has been accepted.

Quotations submitted in response to an instruction from the Project Manager have no formal status under the ECC until accepted by the Project Manager. In the event that such a quotation is not accepted, it is suggested that it will provide the basis for the Contractor's submission to the appropriate dispute resolution forum (see Chapter 19) should the Contractor be unsatisfied with a subsequent assessment for that compensation event by the Project Manager (pursuant to clause 64, see Sections 16.7.1–16.7.3).

16.5.3 Status of quotations: If not instructed

Under clause 62.1, the Contractor may submit quotations to the Project Manager other than those that the Project Manager has instructed for other methods of dealing with the compensation event. As such quotations are submitted in accordance with a provision of the ECC, there would appear to be no doubt that they must be responded to by the Project Manager as if the quotation had been instructed (see Section 16.5.2).

Where the Project Manager has decided that a compensation event notified by the Contractor under clause 61.3 is not a compensation event, and given a decision to that effect in accordance with clause 61.4, the Contractor may in any event (and without any right to do so under the Contract) submit a quotation for that notified event. As such quotations have not been instructed, it would appear that the Project Manager is not obliged to consider them under the provisions of clause 62.3. It is suggested that the only purpose such quotations would serve would be to act as a statement of the Contractor's contention should the dispute over the status of the notified event be referred to any form of dispute resolution.

16.5.4 Cost of preparing quotations

There is no provision in main options A and B that allows the Contractor to recover the cost of preparing quotations for compensation events. The Contractor is obliged to provide quotations by clause 62.1. As the Contract requires the Contractor to Provide the Works (see the definition at clause 11.2(15) and Section 7.2) and that term requires all incidental actions required by the Contract to be done then it can only be concluded that the cost of providing quotations is included in the Contractor's Prices for these two main options. Should the time on site be extended by a compensation event, then the resources that are preparing the quotations and who are based in the Working Areas will form part of the assessment for such a compensation event.

Under main options C and D, the cost falls under the definition of Defined Cost at clause 11.2(24), but only where the people preparing the quotations are doing so in the Working Areas. There would be no change to the Prices so there would be a detrimental effect on the Contractor's Share.

Under main option E, the cost falls under the definition of Defined Cost at clause 11.2(24) but only where the people preparing the quotations are doing so in the Working Areas.

Under main option F, the cost of preparing quotations is part of the Contractor's process for carrying out the work the Contractor is to do, much in the same way as main Options A and B (see above).

16.6 Quotations: Submission and reply

16.6.1 The requirement for quotations

The quotation procedure maintains the principle that everything should be out in the open at the earliest possible time and then resolved. The ECC therefore provides detailed provisions which are consistent with this principle and the spirit of mutual trust and cooperation. These are contained in clause 62 and are supplemented as appropriate by some of the other subclauses within core clause 6.

Whenever the Project Manager has instructed the provision of a quotation as a result of a compensation event recorded by the Project Manager notifying the event (pursuant to clause 61.1; see Section 16.2) or the confirmation of a Contractor's notification by the Project Manager (pursuant to clauses 61.3 and 61.4; see Sections 16.3.1 and 16.3.2), then clause 62.1 follows the spirit of mutual trust and cooperation. It requires the Project Manager and the Contractor to discuss alternative methods of dealing with a compensation event, with the objective of providing the best solution for the Client. The best solution would consider not only time and money, being the matters covered by a quotation, but also the technical solutions available to the Client, which could result in differing life cycles or maintenance issues. As a result of this consultation, the Contractor can be required to supply more than one quotation for each compensation event as instructed to do so by the Project Manager.

The only limit on the type and number of alternative quotations is an implied one that they must be practicable. Although the ECC doesn't expressly say so, it must be presumed that the Project Manager needs to specify in the instruction to submit quotations the different ways of dealing with the compensation event for which the Contractor must provide quotations.

The clause is worded so widely that the type of alternatives envisaged is not only limited to different construction methods, but also different resource levels and corresponding lengths of time that will produce varying costs for the same construction solution. The provision of quotations for different time periods is an alternative to either the Project Manager or the Contractor proposing an acceleration pursuant to clause 36.1 (see Section 12.6.4).

Clause 62.1 also allows the Contractor to submit alternative quotations, in addition to any alternatives that the Project Manager has instructed, for other methods of dealing with the compensation event which the Contractor considers practicable. As with alternative quotations instructed by the Project Manager, the wording is wide and the above comments apply equally. As the ECC allows the Contractor to submit such quotations, the Project Manager has no alternative other than to consider the submission and reply as required by clause 62.3 (see Section 16.6.2).

16.6.2 Procedure and timetable

As with all procedures in the ECC, a timetable is provided that covers the submission of a quotation by the Contractor and the reply to that quotation by the Project Manager. The requirements are set out at clauses 62.3 and 62.4. The procedure takes effect after a compensation event has been notified and the Project Manager has instructed that a quotation is to be provided for that compensation event.[5]

The procedure starts by requiring the Contractor to submit the required quotation or quotations (see Section 16.6.1) within three weeks of being instructed to do so. Nothing more is said in the procedure about the content of the quotation (see clause 62.2 and Section 16.5.1) or how these are put together (see clause 63 and Chapter 17); the clause simply sets out the time for submission. The next step in the procedure is that the Project Manager is required to reply to each quotation within two weeks of the date on which that submission was received (see clause 13.2 and Section 6.2). The Project Manager is required to reply in one of three stated ways, as set out in the bullet points in clause 62.3. The first bullet point, which is the one that represents the goal of the procedure, allows the Project Manager to reply by stating that the quotation is accepted. If that is the reply made, then that compensation event will be implemented (see clause 66.1 and Section 16.9.1) and the accepted quotation treated as final and binding on the parties subject to the Client's right to refer the assessment to dispute resolution (see Chapter 19). The Contractor would not be able to instigate dispute resolution proceedings, as the acceptance of its quotation creates a position whereby there is nothing for the Contractor to dispute.

Should the Project Manager not accept the quotation at this time, the clause provides two other options.

The option given to the Project Manager by the third bullet point is to inform the Contractor that the Project Manager does not accept the quotation submitted by the Contractor, and will instead carry out an assessment of the impact of the compensation event in question. If the Project Manager elects to use this option, the Project Manager

---

[5] As per the notification procedures under clauses 61.2 and 61.4; see Sections 16.2, 16.3.1 and 16.3.2.

is effectively saying to the Contractor that the quotation has not been prepared properly in accordance with the rules in the ECC or that the effects of the event forecast by the Contractor are unreasonable. Project managers are advised to use this power sparingly as otherwise it could easily be seen as provocative and contrary to the spirit of mutual trust and cooperation. The effect in practical terms is that the use of this third bullet point, while within the Project Manager's discretion, acts negatively on the relationship between those involved and has a disproportionate effect on the success of the project. In the event that the Project Manager is considering using this power, it is worth considering the option provided by the second bullet point (see Section 16.6.3) as the opening move is more effective, especially when coupled with good communications and a discussion with the Contractor to 'clear the air'.

16.6.3 Revised quotations

The second bullet point at clause 62.3 allows the Project Manager to reply to a quotation submitted by the Contractor by instructing it to submit a revised quotation but only after explaining the reasons for doing so. This requirement to explain the reasons for requiring a revised quotation is stated in clause 62.4. Unlike many other clauses in the Contract that list the reasons the Project Manager can give when rejecting a submission from the Contractor, no such list of reasons is given in connection with the instruction for a revised quotation. Presumably, the Project Manager can give any reason considered to be appropriate in the circumstances. That being the case, the Project Manager has a free hand to use whatever reasons considered to be appropriate.

In practice, these reasons are going to be limited to the content of the quotation to which this reply is being made. This will cover things such as the forecast resource levels being wrong or being applied over too long a period, that the rates used are not in accordance with whichever rules apply, that adjustments made to the Accepted Programme are incorrect or similar valuation type issues. More detailed comment concerning the assessment of compensation events, both in terms of the contractual provisions and the practical application, is given in Chapter 17.

A matter that is unclear in respect of revised quotations is whether an instruction for a revised quotation can include assumptions about the compensation event in accordance with clause 61.6 (see Section 16.4.3). There are no clear words in the ECC that either allow or prevent the Project Manager from stating such assumptions in an instruction for revised quotations. The User Guide, at Volume 4 (NEC Panel, 2017e), says that the Project Manager can add or alter any assumptions when giving this instruction. This comment is eminently sensible and reasonable, especially in view of the overriding requirement to work in a spirit of mutual trust and cooperation, that the Project Manager should state assumptions in an instruction for a revised quotation where considered to be appropriate (perhaps after discussion with the Contractor).

Clause 62.4 concludes by requiring the Contractor to submit any revised quotation that has been instructed by the Project Manager within three weeks of being instructed to provide the revision. This is of course the same period as allowed for the submission of the quotation in the first instance. The Contract is silent as to what the Project Manager does when the revised quotation is received from the Contractor. Common sense suggests that revised quotations should be treated in the same way as the original quotation and that, if necessary, the process could be repeated over and over again. In practice, the potential endless repetition of the process seldom happens as, after one or two revised quotations have been received, if project managers still feel that the quotation is unacceptable they will inevitably elect to make their own assessment.

16.6.4 The timetable simplified

While relatively straightforward to understand, the procedure and timetable reviewed above can be distilled into the following step-by-step list:

Step 1. Project Manager instructs Contractor to submit quotation(s).
Step 2. Within three weeks Contractor submits quotation(s).
Step 3. Within two weeks Project Manager replies with one of the three decisions allowed by Clause 62.3.
Step 4. Either
  (a) no further action is required as the Project Manager has decided (i) to accept the quotation or (ii) to make an own assessment; or
  (b) within three weeks the Contractor resubmits the quotation.

This simplified view shows that the timetable is not complicated. In practice, project managers and contractors should endeavour to make this overall process simpler. Ideas for how this can be achieved are set out in Section 16.10.4.

Another way of looking at the procedure is to consider the flow diagram at Figure 16.2, which as well as covering the four steps above also adds the procedure included in the final part of clause 61.4 to cover instances where the Project Manager has not replied to a quotation submitted by the Contractor.

16.6.5 Extended time for quotations and replies

Following the general principle at clause 13.5 that the period for replying to a communication can be extended, the periods for the submission of quotations by the Contractor and the reply by the Project Manager can, in accordance with clause 62.5, be extended by agreement. The agreement must be made before the period in question expires and must be confirmed to the Contractor by the Project Manager. This clause does not refer to revised quotations but logic suggests that the rules should apply equally.

16.6.6 Failure to reply to quotations

The publication of the third edition of the ECC saw the introduction, maintained in the fourth edition, of a new provision relating to the failure by the Project Manager to reply

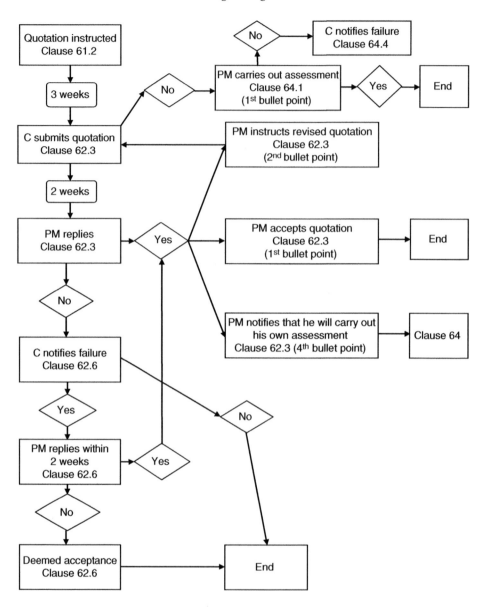

Figure 16.2 Clause 62 – Quotation Procedure.

to a quotation. This provision was introduced as a reaction to a much-stated complaint by User Group members. One of the common causes of the compensation event mechanism breaking down was the failure of project managers to reply to quotations that had been submitted, not just in the required period but at all. The new provision that has been provided is at clause 62.6 and works in an identical way to that at clause 61.4 (see Section 16.3.2).

In the event that the Project Manager fails to reply to the submission of a quotation within the required timescale, then this provision can be invoked at any time thereafter[6] by the Contractor. This provision is initiated by the Contractor issuing a notice to the Project Manager notifying him that the Project Manager has failed to reply to the quotation. The Project Manager then has a further two weeks to reply to the quotation and can do so in any of the ways provided for at clause 62.3 (see Section 16.6.2). Should the Project Manager fail to reply within this further two weeks, for which there is no provision to extend, then the Contractor's original quotation is deemed to be accepted. This automatic acceptance would then be binding on the Client unless it chose to initiate dispute resolution proceedings (see Chapter 19).

If the Contractor had submitted more than one quotation for the compensation event in question, then when it issues its notice to the Project Manager reminding the Project manager of the failure to reply, the Contractor must state which of these multiple quotations it proposes should be accepted. The Project Manager is not bound by this proposal but in the event that the Project Manager doesn't reply after receiving such a reminding notice, then the quotation proposed by the Contractor from however many alternatives there are will be accepted by default.

The purpose of this new clause is to prevent the Project Manager from delaying the assessment and implementation of compensation events. Some Project Managers have been criticised in the past for delaying the process of finalising compensation events by, for example, waiting for the work to be done in order to test the accuracy or otherwise of the Contractor's forecasts. This clause can prevent such delays from happening if invoked promptly by contractors.

The clause does not provide an automatic default for failure to reply to the quotation submission. It is the failure to reply after the Contractor has issued a notification recording the initial failure to reply that brings about the deemed acceptance. As each such notification should be on a separate communication (see clause 13.7 and Section 6.2), the Project Manager can have no excuse for not being aware of the obligation.

16.7 Assessments by the Project Manager

16.7.1 The trigger events

Under the ECC the Project Manager becomes obliged to carry out the assessment of a compensation event should one of four trigger events occur. These four trigger events are set out as four bullet points in clause 64.1. Under the first, third and fourth bullet points, the obligation arises if the Contractor has failed in some way. Under the second bullet point, it is a decision of the Project Manager that triggers the obligation to act.

The first bullet point details that, should the Contractor fail to submit a quotation for a compensation event together with the details required by the ECC within the timescale

---

[6] Certainly at any time up to the issue of the Defects Certificate (see clause 61.7), but arguably at any time in the future subject to any limitation in the relevant legal jurisdiction (e.g. as set by the Limitation Act 1980 in the UK).

stated,[7] then one of the trigger events has occurred. The third and fourth bullet points are related to the submission and rejection for a stated reason of a programme required by the ECC.[8] The third relates to a simple failure of the Contractor to fail to submit a required programme by the time required, while the fourth results from the Contractor not having submitted a programme which is sufficiently detailed or complete to allow the Project Manager to accept that programme.

The second bullet point is the counterpart of the third bullet point at clause 62.3 (see Section 16.6.2), which allows the reply to state that the Project Manager will be carrying out the assessment in response to a quotation submitted by the Contractor. In this circumstance, the decision to carry out the assessment is one that is at the discretion of the Project Manager. As an alternative the Project Manager could have elected to reply to the Contractor's quotation as provided for by the other two bullet points in that clause.

The wording of clause 64.1 is such that the Project Manager does not have any option once a trigger event has happened. The obligation on the Project Manager is so clear that, should the Project Manager fail to carry out the assessment, then a breach will occur for which the Client will be liable. It is suggested that the remedy for such a breach could be that set out at clause 51.3, i.e. interest on a correcting amount when in a later certificate the amount the Project Manager should have assessed is corrected, even if the correction is including the amount for the first time. The intent of the clause is to ensure that the compensation event mechanism is completed by having the assessment done by the Project Manager when the Contractor fails to do what is required of it. The Project Manager could find that clause 64 creates a heavy burden in this respect.

16.7.2 Assessing the programme

When carrying out the assessment of a compensation event, the Project Manager is allowed by clause 64.2 to assess the programme where there is either no Accepted Programme or the Contractor has not submitted a revised programme as required by the Contract. The Project Manager then uses this assessment of the programme in the assessment of the compensation event. These two provisions match the third and fourth bullet points in clause 64.1 (see Section 16.7.1), thereby providing the Project Manager with the ability to overcome the failures by the Contractor that resulted in the Project Manager becoming obliged to carry out the assessment in the first instance.

Practically, while this clause provides the Project Manager with the necessary tools to carry out the assessment, it does place the extra burden on the Project Manager of effectively having to prepare an up-to-date revision of the programme in order to overcome the failure of the Contractor to do so. There is no mechanism in the ECC, which allows the Client to recover any additional cost from the Contractor, which may be incurred in having this exercise carried out.

---

[7] Three weeks at clause 62.3 or a longer time agreed under clause 62.5; see Sections 16.6.2 and 16.6.5.
[8] See clauses 31.2, 31.3, 32.1 and 32.2 and Sections 12.2 and 12.3.

16.7.3 Timing of the Project Manager's assessment

Under clause 64.3, the period in which the Project Manager is allowed to carry out the assessment of a compensation event starts as soon as the need for the performance of this duty becomes apparent, i.e. as soon as one of the trigger events at clause 64.1 has occurred. The Project Manager is then allowed to take the same time as the Contractor had to prepare its quotation in which to prepare the assessment and send details of it to the Contractor. This period would be the three weeks allowed in clause 62.3, unless there had previously been an agreement under clause 62.5 to extend that period in which case the extended period would apply. It must be noted that clause 62.5 does not provide for any extension to the time allowed to the Project Manager to carry out any required assessment.

16.7.4 If the Project Manager fails to do the assessment

Clause 64.4 includes a provision, similar in structure to that contained at clauses 61.4 and 62.6, to cater for those situations where the Project Manager fails to carry out the assessment of a compensation event within the time period allowed by clause 64.3. In this version of the 'deemed acceptance' procedure, it is first of all necessary for the Contractor to have submitted a quotation within the timescale required by the ECC. If the Contractor has not submitted a quotation in the required timescale and in the required form, then the provisions of this clause cannot operate under any circumstances.

Providing that the Contractor has submitted a compliant quotation then, should the Project Manager not provide an assessment within the permitted period, the Contractor may issue a notice to the Project Manager recording that failure. Should the Project Manager then fail to reply with an assessment in the next two weeks, then the quotation that the Contractor originally submitted in respect of the compensation event in question is deemed to have been accepted.

16.8 Proposed instructions

The ECC provides at clause 65.1 that the Project Manager can instruct the Contractor to provide quotations for a proposed instruction. The wording of this subclause relates to the giving of quotations for a proposed instruction rather than for the compensation event that would arise should the proposal become firm. This is slightly odd in that the rest of the ECC requires quotations for compensation events. Although there is a clear disparity in language and terminology, this should not be treated as significant as the intent is clear and the principle not unfamiliar to users of other types of construction contract.

This provision makes it clear that the Contractor does not put the proposed change into effect, i.e. the Contractor does not carry out the work for which it has been requested to supply a quotation. The instruction given by the Project Manager simply requires the Contractor to provide a quotation, not to do any work.

It is suggested that if the Project Manager uses this power in practice, which many do, then the Project Manager should make it abundantly clear on the face of the instruction that the Contractor does not carry out the work unless instructed at a later date to do so. In the event that it is not clear and the Contractor does carry out the work then, in practice, it is likely that should the matter find its way to one of the dispute resolution forums (see Chapter 19), then the result would be in favour of the Contractor.

Clause 65.1 also requires that the instruction from the Project Manager states the date by which the proposed change may be given. This is a very sensible proviso as to not knowing when the instruction could be given could make it difficult for the Contractor to forecast numerous elements of the quotation, in terms of both time and money.

The Contractor is required by clause 65.2 to submit the instructed quotation within three weeks of the instruction (the same period as in clause 62.3 for the submission of a quotation (see Section 16.6.2). The Project Manager then replies by no later than the date stated in the original instruction given under clause 65.1 by which the proposed instruction might be given. The Project Manager is allowed three types of reply, as set out in the three bullet points in clause 65.2. The second of these three bullet points is the issue of an instruction to carry out the previously proposed instruction together with confirmation of the compensation event and acceptance of the quotation. The first bullet point allows the Project Manager, after giving reasons, to instruct a revised quotation (see clause 62.4 and Section 16.6.3).

The third option is to state that the quotation is not accepted. The position created by this third point is also created if the Project Manager does not reply to the quotation within the time allowed by clause 65.2. Should either cause result in the quotation not being accepted, then clause 65.3 allows the Project Manager to issue an instruction for the carrying out of the change, accompanied by confirmation that it is a compensation event and an instruction to submit a quotation. Should the power at clause 65.3 be used, then effectively the instruction is treated as if it were given under clause 27.3 (see Section 7.5.2) and that the procedures at clauses 61.1 and clause 62 (see Sections 16.2 and 16.6) applied.

16.9 Implementing compensation events

16.9.1 Procedure and timing

The ECC requires that each compensation event is 'implemented', which presumably means 'to put into action'. A compensation event is implemented, as set out in clause 66.1, when one of three events occurs in relation to that compensation event. The first and second of these occur when the Project Manager either notifies the Contractor that a quotation submitted by the Contractor is accepted or notifies the Contractor of the assessment that the Project Manager has carried out under clause 64.1. The third occurrence comes about in the event that one of the Contractor's quotations is deemed to be accepted by the operation of either clause 62.6 (see Section 16.6.6) or clause 64.4 (see Section 16.7.4).

It follows that, as each of the three triggers for implementation are when a quotation becomes 'accepted' in accordance with the contract, compensation events will not be implemented in the event that the quotation procedures are not followed.

This step of implementation will, in practice, often occur after the work has been done. The implementation of a compensation event is not a condition precedent to the Contractor's obligation to obey an instruction given by the Project Manager or to put an instruction into effect. Clause 27.3 clearly obliges the Contractor to obey an instruction issued by the Project Manager, providing that the instruction has been issued under the Contract. Further, there is nothing in clause 66.1 that says that the act of implementing a compensation event precedes the Contractor's obligation to comply with this other clause.

16.9.2 Effect of implementation

By reading clause 66.1 in conjunction with clause 66.2 it appears that implementation is a two-stage process. The first stage is the notification, by the Project Manager, of the proposed changes to the Prices, the Completion Date and the Key Dates. These changes will all be contained in the compensation event assessment that becomes implemented under whichever of the three occurrences set out in clause 66.1 occurs (see Section 16.9.1).

The second stage is the application of these proposed changes. For Main Options A and C this will be a change to the activity schedule (see clause 63.14 and Section 16.9.2) and for Main Options B and D it is a change to the bills of quantities (see clause 63.15 and Section 16.9.2), both of which are then used for subsequent assessments of the Price for Work Done to Date in accordance with core clause 5. For Main Option E the change to the Prices is included in the next forecast produced by the Contractor pursuant to clause 20.4 (see Section 7.2). For Main Option F, clause 63.17 (see Section 16.9.2) the change is made to the prices for the Contractor's work.

16.10 Practical issues

16.10.1 Project Manager's instructions and other notifications

As set out in Section 16.2, the strict interpretation of clauses 61.1 and 13.7 together requires the Project Manager to issue three separate communications every time that the Contractor is notified of a compensation event arising from an instruction, a changed decision by either the Supervisor or the Project Manager or the issue of a certificate required by the Contract. The examples given in Volume 4 of the User Guide (see Section 6.3.4) shows the issue of two separate communications in respect of this particular procedure.

Practically, and acting as required by clause 10.1, it is suggested that the parties agree that this administrative burden can be reduced (to the mutual benefit of all concerned).

For example, it could be agreed that an instruction to change the Scope can include a statement that the instruction constitutes a compensation event and a complimentary instruction to submit a quotation or confirmation that a quotation has already been submitted. Indeed, many parties draft up a standard Project Manager's Instruction form which includes the two additional statements on it, hence negating the need for two separate communications.

16.10.2 Events that do not constitute compensation events

The fourth of the five bullet points in clause 61.4 setting out reasons why an event would not be a compensation event has the effect of meaning that an event which fits the description of one of the compensation events listed in the ECC (see Section 15.3), but which does not have any impact of any of the Defined Cost, Completion or any Key Date, is not registered as a compensation event. Many users, including the author, find this both alien to their professional instincts and to good practice. These users consider that if an event has happened which qualifies as a compensation event, then that event should be logged as a compensation event. When it is subsequently agreed that there is no effect on any of the three factors identified in the bullet point, the log should then record this agreement. The practical benefit of this approach is that it provides the Project Manager, the Contractor and the Client with a permanent record and an audit trail in the event that someone resurrects an event which had previously been determined to have no effect.

A simple practical example could be that an instruction has been issued by the Project Manager instructing a change to the colour of the paint in a particular area. The instruction is given well before the work is started and the paint has not been procured. The change to the Scope is a compensation event (see clause 60.1(1) and Section 15.3.1). Assuming that there is nothing unusual, the labour content to apply the paint would not change and the paint cost would be the same. The effect of the event would therefore be nil in respect of all three things that could be affected. By recording the event and the agreement that there is no impact, should the personnel on the project change there is a permanent record of the agreed outcome of the instructed change.

It doesn't take much effort from the team to achieve this position and to make the record. This minor effort has the added advantage that, if the Project Manager discards the use of this fourth bullet point, then tension and a potential dispute cannot be created by denying the existence of a compensation event when an event clearly qualifies under the list. By saying that such an event is a compensation event the Project Manager can then obtain agreement that there is no impact from the event, the parties can record this agreement and the trust and cooperation is maintained or even enhanced.

16.10.3 Not identifying compensation events when instructing change

The vast majority of compensation events on most projects will be the result of instructions issued by the Project Manager to change the Scope. A common complaint by contractors in practice is that many such instructions arrive in the form of revised drawings which the contractor is instructed to comply with. The instruction received is generally a simple covering note to the effect that the contractor is to comply with a list of drawings. The instruction does not identify the changes that have been made and are contained on the drawings. Neither does it confirm that a compensation event or events have occurred or instruct the provision of a quotation. Such instructions do not comply with clause 61.1 but, under clause 27.3, the contractor is obliged to obey the instruction.

In the better instances, the drawings highlight what revisions have been made or at least show the aspects of the drawings that have been changed by the use of 'cloud' indicators around the altered parts. Sadly, in many instances there is no indication of what has been changed. Without diverging into questions of bad practice in the preparation of the revised drawings, whatever has or has not been provided in such circumstances puts the Contractor in the position whereby it has to review all such drawings and identify the changes. Each such change has then to be notified as a compensation event by the Contractor as required by the second bullet point of clause 61.3. This creates a considerable amount of extra work for the Contractor and has a negative impact on the spirit of mutual trust and cooperation. Contractors understandably find it hard to understand why a designer, when making a change, cannot simply list that change; this would enable the Project Manager to have a readymade list identifying such matters, which could be used as the basis of the Project Manager's own notification of the compensation events. With that more detailed information included in the instruction to the Contractor, the carrying out of the assessment would also become an easier task, both in the preparation by the Contractor and the checking by the Project Manager.

Users should recognise that the creation of such circumstances only serves to delay the efficient management of the compensation event procedures and results in resources, which could otherwise be used to contribute to the success of the project, being diverted into such exercises. Further, as this common cause of compensation events (i.e. changes to the Scope) is one that should be notified by the Project Manager, the eight-week time period for notification by the Contractor at clause 61.3 does not apply as it is ruled out by the exception at the end of that clause.

16.10.4 Simplifying the quotation procedure

The procedure for submitting and replying to quotations, as set out in the ECC, is very rigid and formalised. It can be carried out at arm's length without any face-to-face communication between the Project Manager and the Contractor. If followed rigidly, this formal approach can (and indeed does) waste a lot of time and resources. This is a common thread within comments made by users of the ECC and from many commentators. In order to overcome this, users need to stand back and appreciate what the purpose of the compensation event procedure actually is in practical terms.

The compensation event procedure recognises that the construction process involved in all the industry sectors which employ construction is one that will inevitably require change from the concept that was included in the original contract. Ours is an industry which, on the whole, builds prototypes in a variable environment. The primary two undertakings given by a contractor when pricing a scheme is that it will build the project as defined at that time for a stated sum of money and in a stated period of time. The contracts used to establish this relationship spread the risk of various events between the two parties. Recognising that the original project definition will change and that risk events that are at the employer's risk may occur, the contracts allow for and provide mechanisms to cater for those changes. Under the ECC that is known as the compensation event procedure.

Under the ECC, compensation events are the mechanism for changing the Prices, the Completion Date and Key Dates, these being the three things linked to the Contractor's primary undertakings. Users should also recognise that, while under Main Options A, B and F the amount of money attached to a compensation event directly effects the amount that the Client will pay the Contractor, under Main Options C and D the money will change the target and any inaccuracies will be shared between the parties as a function of the Contractor's share mechanism (see clause 54 and Section 13.7). As with so many things, the most efficient way to achieve an agreed adjustment to these three things is by discussion and negotiation. On projects where the compensation event mechanism has been worked effectively and efficiently, those involved, instead of formally submitting quotations for every event, sit down round the table at regular intervals to discuss and agree about the events that have arisen. The interval between these discussions is rarely more than a week. At each discussion, those involved consider two classes of compensation events: those which are new and have not been discussed before and those which have been discussed before.

In respect of the new events, the focus is on reaching the swift agreement (there and then) of all events which can be assessed and agreed. In construction, many changes are relatively minor in effect, being a few tens or hundreds of the relevant currency unit, and the assessment of these items is straightforward for the professionals involved. The discussion identifies these events and carries out an assessment which is acceptable to all in an efficient manner. This is then recorded on a log and becomes the accepted quote. A simple but meaningful build-up of the amount is kept for audit purposes and the permanent record. If those present at the discussion consider that the factors involved in a new event are too complicated to be resolved there and then, the actions required to progress that event are identified and allocated to the person best placed to progress them. These events are then added to the ongoing list of outstanding events to agree.

A simple example of this would be a change of material from type A to type B[9] where the labour to fit both types is the same. Knowing the number and the difference in cost of the two material types, the team can quickly and accurately establish the change in Prices. There is no need to follow the formal procedure.

The second part of the discussion considers those events that have previously been identified as being too complicated to be resolved the first time they came to the discussion. If the actions that were allocated have been carried out and the assessment of the event can now be concluded, then the discussion takes that path. If not, then the further actions required are agreed and the matter is taken away again. Some parties use this

---

[9] This would be a change to the Scope and therefore a compensation event under clause 60.1(1); see Section 15.3.1.

process of discussion to identify that certain events require the submission of a formal quotation and set the timetable for that submission among themselves.[10]

An example of this might be the encountering of a physical condition,[11] the solution to which and the impact on cost and time is still being considered by the team. The team can set a timetable in which to gather information and start to determine reasonable forecasts of the impact. The team can even consider whether it would be prudent for the Project Manager to state an assumption (pursuant to clause 61.6; see Section 16.4.3) about the impact. By progressing in this way and reporting back to the next discussion, the team can progress the assessment of complicated compensation events without becoming restricted by the rigid timetable and in a manner that is more likely to bring about the early resolution of the assessment of that event.

The key is that effective discussion can lead to swift agreement which saves time and effort by both the Contractor and the Project Manager. It also creates transparency and trust. The effect is that the process is simplified and the teams that adopt this approach regularly achieve the completion and agreement of all compensation events by the time that the Project Manager certifies Completion (under clause 30.2, see Section 12.5.2) or not long after.

16.10.5 The time periods are not long enough!

A commonly heard complaint about the procedure for submitting quotations and replying to them is that the time periods provided in the ECC are not long enough. The author does not share this point of view. Users who consider that the time periods of three weeks and two weeks (see clause 62.3 and Sections 16.6.2 and 16.6.3) are not long enough for dealing with their part of the process should remember that a contractor, when pricing the whole of a project at tender stage, is generally allowed in the region of four to six weeks. It is suggested that three weeks to forecast the effect of a compensation event, which on the whole only affects part of a project, is more than sufficient when compared to the time in which contractors price complete schemes from scratch. In the rare instance that the circumstances are such that three weeks is genuinely insufficient, users should always remember that clause 62.5 allows for periods to be extended by agreement.

16.10.6 Compensation events and Main Option E

In practice, compliance with the compensation event procedure under Main Option E does not seem to have any tangible benefit to either the Client or the Contractor. If anything, the operation of the mechanism could result in the Client's cost being higher as it is likely that the people involved will be paid for as part of the Defined Cost under this cost reimbursable method of payment.

---

[10] Using the timetable at clause 62.3 and the power to extend those times contained in clause 62.5; see Sections 16.6.2, 16.6.3 and 16.6.5.
[11] A compensation event under clause 60.1(12), see Section 15.3.12.

This is recognised by the ECC to the extent that the changes resulting from the assessment of compensation events under this Main Option (see clause 66.2 and Section 16.9.2) are applied to the forecast of the outturn cost prepared at monthly intervals by the Contractor (as required by clause 20.4; see Section 7.2). In practice, it would seem to be more cost effective (i.e. cheaper for the Client) to show the effect of matters that would qualify as compensation events on revised programmes (as submitted under clause 32; see Section 12.3) and in the forecast under clause 20.4.

16.10.7 What do we do when the procedures are not followed?

One of the most common questions raised concerns what should or can be done if any step in the procedures covered by this chapter is not followed. Most of these positions are covered in the procedures themselves. Nonetheless, in order to assist the reader who has not identified the steps that can be taken, they are all gathered together here.

As there are several possible failings in this respect, the sensible place to start is where the Project Manager fails to act as required. This is because the Contract provides for most of them.

Following the procedures in clause order the first failing by the Project Manager is where the Project Manager fails to notify a compensation event, as required by clause 61.1 (see Section 16.2), having issued an instruction or one of the other matters covered by that clause. The remedy to this position is contained in clause 61.3 (see Section 16.3.1), which gives the Contractor the obligation to notify any compensation event that has not been notified by the Project Manager. The caveat contained in the last few lines of clause 61.3 means that such a notification is not required within the eight-week period stated. What the Contractor does need to remember is that clause 61.7 (see Section 16.4.2) prevents the notification of a compensation event after the issue of the Defects Certificate (see Section 9.6). In summary, where the Project Manager fails to notify under clause 61.1 it is within the hands of the Contractor to rectify that situation.

Where the Project Manager fails to reply to the notification of a compensation event by the Contractor under clause 61.3, as is required from the Project Manager by clause 61.4, then clause 61.4 (see Section 16.3.2) provides that the Contractor can notify this failure to the Project Manager. If the Project Manager then fails to reply to the original notification within a further two weeks the Contract determines that the matter notified is a compensation event and automatically requires the Contractor to submit a quotation.

The same procedure also applies via clause 62.6 (see Section 16.6.6) where the Project Manager has failed to reply to a quotation submitted by the Contractor in accordance with clause 62.3 (see Section 16.6.2). It must be noted that if the Contractor has not issued the quotation on time or to the required level of detail then the Contractor may find that this remedy will not stand up if argued in adjudication.

This procedure referred to in the previous two paragraphs is also found at clause 64.4 where the Project Manager has failed to provide an assessment as required by clause 64.3 (see Section 16.7.3) subsequent to one of the triggers at clause 64.1 (see Section 16.7.1) having arisen. The problem for the Contractor here is that if it did not submit a quotation in the first place, then this version of the procedure cannot apply.

What the above paragraphs tell the user is that if the Project Manager does not do something that is required by the Contract then, providing that the Contractor has been doing what it is required to do, the Contract will determine that a compensation event will be deemed to have occurred and subsequently a quotation will be deemed to have been accepted. If this happens, the Project Manager will be powerless and all that the Client can do is either accept the position or refer the matter to the adjudicator.

Turning to failures of the Contractor to follow the procedures the first potential failure is that of failing to notify a compensation event as required by clause 61.3. If the event is one that the Project Manager should have notified under clause 61.1, then providing that the Contractor issues a notice before the issue of the Defects Certificate, it has not lost anything. If the required notice is not issued by the date of issue of the Defects Certificate, then all is lost. For any of the events that should normally be notified by the Contractor (see Section 16.3.1), if the Contractor has failed to notify within the eight-week period stated in clause 61.3 the intent of the Contract is that the Contractor no longer has any rights in respect of that event. Unless a court were to find that the wording intended to create the condition precedent is not sufficient to ensure that position there is no remedy.

Where the Project Manager has failed to issue a reply or to assess a compensation event that the Project Manager is required to assess (see clauses 61.4, 62.6 and 64.4) and the Contractor does not exercise its right to issue a notification of this failure the contract is silent as to what happens next. It does not appear that the period for reply (clause 13.3, see Section 6.2) applies as these are not positions where the Contractor is required to reply. Indeed, the option the Contractor is given arises as the result of a failure of the Project Manager. That being the case, if the Contractor does not issue the notification that it is within its power to issue, it would appear that the Contractor could still seek the resolution of the particular matter up to the last time which it could invoke the dispute resolution procedures in Option W1, Option W2 or Option W3 (see Chapter 19).

Another area where the Contractor can fail is to fail to submit a quotation either within the three-week period required by clause 62.3 (see Section 16.6.2) or containing the detail required by clause 62.2 (see Section 16.5.1). What happens here is quite simple; in accordance with the first bullet point of clause 64.1 (see Section 16.7.1), the Project Manager becomes obliged to carry out the assessment. If the Project Manager then fails to carry out that assessment, the Contractor is not at fault in that respect and should be able to refer the event in question to the Adjudicator up to the last time that it could invoke that procedure.

Having considered all the potential failures referred to above, unless the Contractor leaves its option to rectify the position until too late, the only type of failure that will be fatal will be the failure to notify any compensation event which it (the Contractor) is required to notify under clause 61.3 and which did not arise from some action of the Project Manager covered by clause 61.1.

Chapter 17
Compensation Events: Assessment

17.1 Introduction

The rules laid down by the ECC for the assessment of compensation events result in a procedure that is not only fundamentally different to other standard forms of contract used in the construction industry but requires considerable thought as to how the rules should be applied. From reading the User Guide, the background to the principles and procedures appears to be based on the concept that the Contractor should recover its actual costs resulting from a compensation event, rather than be at risk that contract rates used as the basis of valuation will cause it to incur a loss or, from the Client's point of view, give the Contractor a windfall gain.

The intent of the ECC is to expedite the procedure and to arrive at an accepted quotation for the effect of each compensation event as soon as possible. Waiting for the varied work to be done or the additional cost to be incurred does not figure in the scheme.

This approach appears to be the cause of many of the problems that have arisen in practice. Further, it also results in the Contractor taking considerable risk in connection with Client's defaults which, under other widely used standard contractual arrangement or at common law, it would not be exposed to. This is particularly relevant in relation to any 'claim' element that is included within the assessment of a compensation event. In effect the Contractor is required to take risk as a result of a default by the Client which, if remedied at common law by damages, would ensure that the Contractor was reimbursed its actual costs and was thereby returned to the position it would have been in but for the default by the Client.

Unfortunately, the undoubted good intentions of the drafting body have in all probability not achieved the desired result. Instead, they have created a system which, without understanding and a certain amount of practical application, can be difficult to administer, can put the Contractor at considerable risk, and contains the potential for disputes. In respect of the assessment of the impact of compensation events, these matters and concerns will be addressed in this chapter.

How the impact of a compensation event should be assessed will be considered separately for the two main areas of cost and time. While cost will be dealt with first, it should become apparent that when carrying out an assessment in practice it will be time that users consider first. The determination of the effects of time, whether on a single activity

---

A Practical Guide to the NEC4 Engineering and Construction Contract, First Edition. Michael Rowlinson.
© 2019 John Wiley & Sons Ltd. Published 2019 by John Wiley & Sons Ltd.

or on the programme as a whole where delay to planned Completion occurs, will be the factor that then determines the amount of resource to be considered in the assessment of the cost.

17.2 Changes to the Prices

17.2.1 Basic principle

The basis for the assessment of changes to the Prices is set out at clause 63.1 and is formed on the principle of assessing the difference between the Defined Cost of doing the work before the event occurred and the Defined Cost of doing the work after the event occurred. The Fee (see clause 11.2(10) and Section 13.4.7) is then added to this difference. Put another way, this translates into the cost of the change in the resources needed to carry out the work. The value of the original work included in the tender price plays no part in the valuation. This principle is confirmed by reference to two paragraphs from page 61 of Volume 4 of the User Guide (NEC Panel, 2017e), which state:

> Where the work to be done is changed, it is important that the assessment is based upon the change in forecast or recorded Defined Cost. The clause gives no authority for the price for the originally specified work to be deleted or for the forecast Defined Cost of all work now required to be used as the basis for a new price.
>
> If the Scope originally included a piece of work (a) which is now to be replaced by a piece of work (b), the compensation event is assessed as the difference between the forecast Defined Cost of (b) and the forecast Defined Cost of (a). The Fee is then added to this difference and the resulting total amount is used to change the Prices, by either adding them to, or subtracting them from, the original price in [the] pricing document.

In theory, this should provide a neutral effect on the profit or loss that the Contractor makes on the project. In practice, however, because the effect is often forecast before the work is done, an element of risk is introduced which creates the potential for an increase or decrease in the Contractor's profit.

Returning to the content of clause 63.1, this contains three bullet points, the first of which covers the cost of work done in connection with the compensation event before the dividing date established by the last two paragraphs of the clause. The said paragraphs determine that where the compensation event arises due to the Project Manager giving an instruction or notification, changing a previous decision or issuing a certificate, then the dividing date is the date of the communication setting out that change. For all other compensation events the dividing date is the date that the compensation event in question was notified by either the Project Manager or the Contractor to the other. It must be remembered that the Contractor is obliged by clause 27.3 to obey and put into effect any instruction issued by the Project Manager or Supervisor.

Contrary to this provision, it is not difficult to envisage situations where the changed work required by such instructions is commenced and even completed before the three-week period for the submission of the quotation has expired. Similarly, many events that are at the Client's risk may require emergency or immediate action (for example, to safeguard the works or the public). The User Guide suggests, and no practical reason presents itself to the contrary, that where the work has been done prior to the submission of the quotation a retrospective valuation should not be used in such circumstances. The User Guide states that this is because the intent of the ECC is that assessments will usually be forecasts of the effect of the event in which the Contractor takes some risk. By adopting this approach, the ECC is attempting to prevent the assessment of compensation events becoming retrospective and effectively becoming cost reimbursable.

Further, if retrospective assessments based on a record of resources used became the norm, it is feared that this would lead to gamesmanship being played in the timetable for the submission and agreement of quotations where one or other of the Contractor or the Project Manager would attempt to manoeuvre to gain what they perceive to be an advantageous position. The development of this approach would lead to the taking of adverse positions and create conflict within the team that would potentially result in formal disputes arising. It is, sadly, common to see these types of positions being created in practice by individuals who believe that, by taking their time in either the submission or acceptance of a quotation, they can gain an advantage or simply create a position where they are comfortable that they cannot be criticised. Those that adopt this approach do so contrary to the intent of the ECC.

The second bullet point clearly refers to the forecast of Defined Cost yet to be incurred but necessary to complete the work required by the compensation event in question. Again, this forecast is calculated using Defined Cost rather than true actual cost. No rules are set in relation to the establishment of forecasts and this can become an area ripe for differences of opinion between the Contractor and the Project Manager.

The final bullet point relates to the Fee that results from the assessment under the first two bullet points. The Fee is defined at clause 11.2(10) and has previously been covered in Section 13.4.7. The Fee is added to the Defined Cost assessed, regardless of whether the Defined Cost is a positive or a negative. The User Guide is clear that, if the effect of a compensation event is only to delete future work, the assessment is the forecast decrease in the Defined Cost plus the Fee.

17.2.2 Cost risk allowances

When assessing a compensation event, clause 63.8 expressly allows the assessment to include for cost risk allowances (also time risk allowances; see Section 17.3.1). The test for whether such allowances should be included is that they should have a significant chance of occurring, be at the Contractor's risk and that the risk will not be a compensation event in its own right. The principle of recognised factors such as allowances for material wastage and downtime due to inclement weather should be easy to agree, leaving just the actual level of allowance made to be discussed. However, it is likely that risk allowances for delay, disruption and matters such as winter working will be more controversial and cause some disagreement. In such circumstances, Project Managers will have to decide whether to either instruct revised quotations, instruct assumptions (clause 61.6; see Sections 16.4.3 and 17.4) or make their own assessments.

When carrying out their own assessments (as required by clause 64.1; see Section 16.7.1), Project Managers must take account of clause 63.8 and ensure that they include suitable allowances for matters which are at the Contractor's risk and would qualify for inclusion under this clause, in respect of cost. To fail to do so would give the Contractor grounds for initiating the dispute resolution procedures (see Chapter 19) under which, in such circumstances, the Contractor would probably have some success.

17.2.3 Assumption that the Contractor reacts competently and promptly

Clause 63.9 states that assessments are made on the basis that the Contractor reacts competently and promptly to the compensation event. In respect of matters of cost, this requirement is not difficult to apply and acts to protect the Client against any inefficiency of the Contractor. However, this clause (as discussed in Section 17.3.2) is more difficult to operate in terms of time. Competence would include varied aspects of performance such as buying the right materials, employing people with the right skills, complying with health and safety requirements and operating in the most commercially appropriate way.

17.2.4 Defined Cost – Which Schedule?

In all the Main Options except Main Option F, the definition of Defined Cost takes the user, by one path or another, to either the Schedule of Cost Components (see Section 14.2) or the Short Schedule of Cost Components (see Section 14.3). As it is the change in the Defined Cost that determines the assessment of the impact of each compensation event upon the Prices then, for these five Main Options, users must recognise which version of the Schedule applies to the Main Option that they are using.

For Main Options A and B, the definition of Defined Cost at clause 11.2(23) determines that it is the Shorter Schedule of Cost Components that is used.

For Main Options C, D and E, the definition of Defined Cost at clause 11.2(24) determines that it is the Schedule of Cost Components that is used.

17.2.5 Use of rates and lump sums

Clause 63.2 allows the Project Manager and Contractor to agree that compensation events can be assessed using lump sums and rates. The clause does not prescribe what the source of such rates and lump sums should be. For Main Options B and D, the source of these rates and lump sums could be the Bill of Quantities (see clause 11.2(22) and Section 13.5.2) to which a pro rata type basis can be applied to calculate and agree suitable rates where there is no corresponding item in the document. For Main Options A and C where the only source of reference is the Activity Schedule (see clause 11.2(21) and Section 13.5.1), the basis for establishing any such rates is not as easy to define. It appears that what the ECC is trying to do is to provide a sensible alternative where the use of either version of the Schedule is impractical or uneconomic.

17.2.6 Reduction in Prices

The ECC includes a somewhat puzzling provision at clause 63.3 that restricts the reduction of the Prices to compensation events where it is stated that the effect can be that the Prices are reduced. The clause does not contain any statement as to which compensation events this strange provision applies to. Instead, the user must refer to other clauses to find where the necessary statement is made. At clause 63.4 a general provision is made which, which being a core clause, applies to all Main Options. This confirms that the Prices will be reduced if the event is a change in the Scope (clause 60.1(1); see Section 15.3.1) or is the correction of an assumption stated by the Project Manager (clause 60.1(17); see Sections 15.3.17, 16.4.3 and 17.4). Within Main Options A, B, C and D and the Secondary Options further provisions connected to the potential to reduce the Prices are included. The easiest way to identify them is by a simple list:

- *Clause 60.4*: Options B and D;
- *Clause 60.6*: Options B and D;
- *Clause 63.12*: Options A and B;
- *Clause 63.13*: Options C and D;
- *Clause X2.1*: Secondary Option X2;
- *Clause X12.3(6)*: Secondary Option X12; and
- *Clause X12.3(7)*: Secondary Option X12.

Clauses 60.4 and 60.6 relate to the use of a Bill of Quantities and were considered in the list of events at Sections 15.3.22 and 15.3.24.

Clauses 63.12 and 63.13 refer to any change to the Scope proposed by the Contractor being exempted from this rule (see clause 16 and Section 6.6.3) and covers how any such saving is dealt with. For Main Options A and B, which are subject to clause 63.12, the Prices are reduced by the amount that results from multiplying the assessed reduction by the value engineering percentage. This term is in *italics* (see clause 11.1 and Section 6.6.1), indicating that is to be found in Contract Data Part One (see Section 21.3). By this mechanism, the Contractor is rewarded for its suggestion as to how to reduce the cost of the project to the Client.

For Main Options C and D, clause 63.13 determines that there is no reduction in the Prices. This does not mean that the Contractor gets to keep all the saving. The result is that when the Contractor's Share mechanism (see clause 54 and Section 13.7) the saving will be shared between the Client and the Contractor thereby creating a similar position to clause 63.12 under main Options A and B.

Clause X2.1 (see Section 20.2) provides that if the effect of a change in the law is to reduce the Prices, then such reduction is applied. By doing so, when the Client has accepted the risk of the effects, if the law is changed, then it will also gain the advantage should such a change reduce the Prices. This is most likely to occur in respect of tax changes in favourable economic times.

Clauses X12.3(6) and X12.3(7) (see Section 20.4) refer to changes to the Partnering Information and partnering timetable, respectively. Given that these two documents potentially have a very wide coverage, it is reasonable that reductions as well as increases in the Prices should be a potential result.

The effect of these provisions is not as onerous for the Client as it may first appear. Changes to the Scope, as stated previously, are likely to constitute the majority of compensation events. With the exception of the correction of assumptions and the other matters covered, these changes are the only practical way in which the Prices could be reduced.

17.3 Changes to the Completion Date and Any Key Dates

The way that the ECC assesses a delay to the Completion Date or a Key Date is noticeably different than under the more traditional standard forms of construction contracts used across the world. The significant and telling difference is that, under the ECC, the delay to completion is measured as the delay to planned Completion. Under the more traditional forms of contract, delay is measured by the effect on the delay event upon the completion date as defined under those contracts. In other words, under those forms, any project float[1] would have to have been taken up by delays of either the Contractor's or the Client's responsibility before the completion date is changed. In UK law, this has been confirmed by the Courts in cases such as *Balfour Beatty Building Ltd v Chestermount Properties Ltd* (1993) 62 BLR 1. This case and others like it have confirmed that the project owns the float and whoever delays the project first gets the benefit of any project float.

Under the provisions of the ECC, there is no doubt that the Contractor owns any project float. This is determined by the wording of clause 63.5, which states that the delay to the Completion Date or to a Key Date is assessed by establishing the delay caused by the event to planned Completion,[2] in respect of the Completion Date, and to the planned date for meeting the Condition associated with the relevant Key Date.[3] The duration of this delay is then used to extend the Completion Date or Key Date. The wording is particular in that it refers to the planned date for either factor being later than the equivalent date before the event occurred.

The use of planned Completion as the yardstick by which delay is assessed makes it safe, from the Contractor's point of view, to plan to complete the works early and to show such early completion on the Accepted Programme, as the Client will not be able to take advantage of any such project float. On the other hand, any activity float[4] that is available in the Accepted Programme[5] will still be available to absorb the impact of delaying events, whether a Client or a Contractor delay.

---

[1] Defined here as float between the end of the critical path of the project and the contractual date by which the contractor is obliged to complete the works.
[2] See the second bullet point at clause 31.2, which requires this to be shown on every programme submitted for acceptance; see Section 12.2.
[3] See the fifth bullet point at clause 31.2, which requires this to be shown on every programme submitted for acceptance; see Section 12.2.
[4] Defined here as being the float attached to an activity that is not on the critical path, the duration of which is equal to the amount of delay that can be suffered by that activity before there will be any impact on the critical path and therefore on planned Completion.
[5] See the first sub-bullet point to the sixth bullet point at clause 31.2, which requires (activity) float to be shown on every programme submitted for acceptance; see Section 12.2.

In order to assess the impact of a particular compensation event, the effect of that event needs to be fed in to the Accepted Programme either as a change to the timing or duration of an existing activity or as a new activity or activities with appropriate links into the existing bars. If this change results in planned Completion being delayed, then the Completion Date is delayed by the same period. Similarly, if there is an effect on the planned date associated with a Key Date, then the Key Date is delayed by that period. Clause 63.5 makes it clear that the Accepted Programme to be used in making this assessment is the one that is current at the dividing date.[6] It also states that in carrying out the assessment of the change to planned Completion, the assessor can only change activities on the programme which are not completed or which are impacted on by the compensation event. This restriction prevents other parts of the works, which are not affected by the compensation event, from being re-sequenced or otherwise changed as part of the assessment. Any such changes will have to be left to the next revision of the programme under clause 32 (see Section 12.3).

There is no provision in the ECC that provides for the reduction of the time for carrying out the works. It is not possible to set an earlier Completion Date. This restriction is a cause for concern for project managers when considering whether to make assumptions with regard to the effect of a compensation event pursuant to clause 61.6 (see Sections 16.4.3 and 17.4). As any correction of such assumptions is in itself a compensation event (see clause 60.1(17) and Section 15.3.17), this second compensation event cannot reduce the time for completion should the assumptions made prove to have given rise to a longer than necessary delay to planned Completion. Accordingly, project managers should be conservative in the assumptions they give but not in the giving of assumptions where appropriate.

17.3.1 Time risk allowances

When assessing a compensation event, clause 63.8 expressly allows the assessment to include for time risk allowances (also cost risk allowances; see Section 17.2.2). The test for whether such allowances should be included is that they should have a significant chance of occurring and be at the Contractor's risk and that the risk will not be a compensation event in its own right. The principle of recognised factors such as allowances for downtime due to inclement weather should be easy to agree, leaving just the actual level of allowance made to be discussed. However, it is likely that risk allowances for delay, disruption and matters such as winter working will be more controversial and cause some disagreement. As time risk allowances are required to be included on every programme submitted for acceptance,[7] it should be possible to determine what allowances have already been included by reference to the Accepted Programme and make appropriate adjustments or additional allowances. In the event that this does not prove to be possible, then in such circumstances, project managers will have to decide whether to instruct revised quotations, instruct assumptions (clause 61.6; see Sections 16.4.3 and 17.4) or make their own assessments.

---

[6] For a discussion about the 'dividing date' see Section 17.2.1.
[7] See the second sub-bullet point to the sixth main bullet at clause 31.2 and Section 12.2.

When carrying out their own assessments (as required by clause 64.1; see Section 16.7.1), Project Managers must take account of clause 63.8 and ensure that they include suitable allowances for matters which are at the Contractor's risk and would qualify for inclusion under this clause, in respect of time. To fail to do so would give the Contractor grounds for initiating the dispute resolution procedures (see Chapter 19) under which, in such circumstances, the Contractor would probably have some success.

17.3.2 Assumption that the Contractor reacts competently and promptly

At clause 63.9 it is stated that assessments are made on the basis that the Contractor reacts competently and promptly to the compensation event and that the Accepted Programme can be changed. Given that the Contractor needs to mitigate the impact of any delay, then this requirement can be interpreted as saying that the Contractor is obliged to change its arrangements if it can, and presumably whenever and however required, in order to accommodate the effects of a particular compensation event.

What the ECC does not provide for is what happens if the Accepted Programme cannot be changed, such as in the case of a track possession on a railway that cannot be delayed or rearranged. On consideration of this type of situation, two possibilities become apparent. Either the Project Manager proposes an acceleration (see clause 36.1 and Section 12.6.4), which in itself could be difficult as the Contractor can decline to consider the proposal if the acceleration required cannot be achieved or the effect on time is assessed as if the Accepted Programme could be changed and the Prices are adjusted to reflect the notional delay. In the second instance, the Contractor would receive payment for delay, which presumably it would have to spend in effectively accelerating, but not an extension to the time. Neither of these options can be said to be satisfactory if the Accepted Programme cannot be changed because of external constraints. The Contractor cannot fairly be disadvantaged as the result of a risk that it does not carry.

The procedures in the ECC seem to run into a dead end on this particular point. In the rare event that it is encountered, the parties will need to work together to overcome the problem in a sensible and practical manner. If they do not, it would appear that the only way forward would be expensive dispute resolution (see Chapter 19), which would almost inevitably prove to be to the disadvantage of them both.

17.4 Project Manager's assumptions

The power given to the Project Manager at clause 61.6 to state assumptions about the effect of a compensation event appears to be vastly under used in practice, probably because it is not understood by those who have the power to use it. The power exists in order to remove some of the uncertainty that can exist within the preparation of quotations for things that have not yet happened and that cannot be reasonably forecast. By stating assumptions and instructing the Contractor to prepare the quotation on the basis of those assumptions, the Project Manager can remove the uncertainty. In doing so, the

Project Manager can exert a certain amount of cost control over the assessment of the compensation event so that the Client ends up paying for what happens, as opposed to the Contractor including risk factors for things that might not happen.

Under clause 61.6, the power to state assumptions is clearly one for the Project Manager and the Project Manager alone. The Contractor, following the wording of the clause, does not have any part to play in the process, except that the application of the spirit of mutual trust and cooperation could result in the Contractor suggesting to the Project Manager that the instructing of assumptions in certain instances could be of advantage to the parties.

In order to appreciate the purpose of clause 61.6, it is important to understand the difference between 'assumptions' and 'forecasts' in the preparation of quotations for compensation events. If the Contractor bases its quotation on forecasts of its own, then it must be aware that, as established in clause 66.3, the assessment of a compensation event is only revised if the Contract provides for such a revision. As there is no provision providing for the correction of a Contractor's forecast if recorded information later shows those forecasts to have been wrong, then the risk is entirely the Contractor's. On the other hand, if an assumption given by the Project Manager is later found to have been wrong, then the Project Manager notifies a correction, and this correction becomes a compensation event in its own right under clause 60.1(17). The distinction that must be appreciated is that forecasts are at the risk of the Contractor and assumptions are at the risk of the Client. If the Contractor submits a quotation and states within that quotation that it has based the assessment 'on the following assumptions' (or words to that effect), but the assumptions used have not been stated by the Project Manager, these so-called assumptions are (in fact and contractually) forecasts and subject to the provisions of clause 66.3.

Two questions commonly arise from the discussion above:

- How can the Contractor protect itself where the Project Manager does not state assumptions?
- How can the Project Manager exert a reasonable level of cost control on the Client's behalf?

In respect of the first of these questions, it is inevitable that, where a Contractor feels at risk from the uncertainty affecting any aspect of a compensation event, it will load its quotation in order to reduce that risk. Indeed, the inclusion of risk allowances in quotations is envisaged by the contract (see clause 63.8 and Section 17.2.2). The Project Manager will possibly see such allowances in the quotations received as being unacceptable and leaving the Project Manager open to criticism by the Client if the Contractor's quotation is accepted. The Project Manager then ends up making an assessment under clause 64, which, if the allowances later prove to be inadequate, the Contractor will be able to challenge through the dispute resolution procedures (see Chapter 19). This of course is not a desirable situation.

On the other hand, and in respect of the second question, the Project Manager may feel that the Client can be protected by the stating of assumptions about the effect of all compensation events. By this precautionary method, the Project Manager will create a situation whereby all assumptions, excepting those that prove to be right, will be corrected with the result that the Client pays for what happened and the Contractor's risk contingencies will be negated.

A word of warning is however needed here: clause 63.2 states that a correction of an assumption is one of the situations where the Prices are reduced if the effect of the compensation event is to reduce total Defined Cost. However, what the ECC does not allow for is any time previously added to the Completion Date to be reduced if assumptions on which that time was assessed subsequently prove wrong. Accordingly, if project managers do state assumptions about the effect of compensation events, they need to apply caution. Project managers should (if anything) understate the assumption in order that corrections will only ever result in additions to time, thereby avoiding creating a position that will disadvantage their client.

In practical terms, by working in a spirit of mutual trust and cooperation, the Project Manager and Contractor can communicate about the effect of compensation events as soon as either of them is notified (under either clause 61.1 or 61.3; see Sections 16.2 and 16.3.1) by the other. They can agree on the forecasts to be used (albeit at the Contractor's risk) or the assumptions that need to be stated by the Project Manager, and the level such assumptions should be pitched at. The ECC does not require this step, but it seems a sensible and practical approach that will prevent disputes arising.

17.5 Other related matters

17.5.1 Failure to give early warning

It is in the assessment of a compensation event, as set out in core clause 63, that one of the consequences of not complying with the early warning procedure (see clause 15 and Section 6.4) becomes apparent. The effects of non-compliance impinge on the Contractor; no penalties are apparent for any failure of the Project Manager to notify an early warning matter. Should the Project Manager fail to make such a notification, it would appear that the Contractor will be compensated by means of the compensation event mechanisms should the Project Manager's failure result in one of the events occurring.

In accordance with clause 61.5, if the Project Manager decides that the Contractor failed to give an early warning of an event that an experienced contractor could have given, then the Project Manager is required to notify this decision to the Contractor when instructing the Contractor to submit quotations (under either clause 61.2 or 61.4; see Sections 16.2.1 and 16.3.2). The wording of the clause suggests that the Project Manager only gets this one opportunity to make such a decision and to then give the required notice. When instructing a revised quotation (under clause 62.3; see Section 16.6.3), it is not clear if the Project Manager could notify the Contractor that it had failed to give an early warning (having failed to do so when instructing the Contractor to submit a quotation in the first instance).

When giving notice of this decision, there is no express requirement in clause 61.5, which requires the Project Manager to state what would or could have done to mitigate the effects of the event had the early warning have been given. In practical terms, this must be considered as an oversight in the drafting. Where the Project Manager has given such a notice then (in accordance with clause 63.7), the assessment of the compensation event to which the notice applies must be made, in terms of both cost and time, as if the Contractor had given early warning of the event. In such circumstances, unless the Project Manager sets out what would have been instructed to be done in mitigation had the Contractor given early warning, the Contractor must somehow predict what could have been instructed by the Project Manager when preparing its quotation. Should the Contractor be left to make such predictions, it is not difficult to anticipate that the views of the Contractor and Project Manager will differ, thereby leading to a disagreement.

The best way for this to be avoided is for the Project Manager to set out, when giving the notice required by clause 61.5, exactly what would have been instructed at the time. By doing so, the Project Manager will give the Contractor a clear indication of what the Project Manager considers the basis of the assessment should be and will also allow the Contractor to understand the thinking behind the decision. By having this understanding, the Contractor is put in a position whereby it can question whether the mitigation steps stated by the Project Manager would have been feasible. Although this exchange may result in a disagreement, it will be one that is open and clear. Such disagreements are more likely to be resolved, under the spirit of mutual trust and cooperation, by those involved sharing their views; formal dispute resolution based on poor communications when both sides are entrenched in their positions can therefore be avoided.

17.5.2 Assessing the effects of ambiguities or inconsistencies

Should any compensation event have arisen solely because of an ambiguity or inconsistency between any of the other documents (see clause 17.1 and Section 6.6.5) or simply contain an element of such a matter, then the assessment of any such event is based on the rules set out in clause 63.10. The ambiguity or inconsistency will have been resolved, as required by clause 17.1, by the Project Manager stating how it is to be resolved. This might be an instruction to change the Scope and therefore a compensation event under clause 60.1(1). The clause follows the *contra proferentem* principle of contract construction, by interpreting the ambiguity or inconsistency against the party who wrote the Scope concerned. This provision recognises that within the ECC there can be both Scope written by or on behalf of the Client, and Scope written by the Contractor in relation to any design that the Contractor is responsible for.

If the ambiguity or inconsistency was contained within Scope provided by the Client, the assessment would be based on the factor or factors most favourable to the Contractor. On the other hand, if the ambiguity or inconsistency was contained within Scope provided by the Contractor, the assessment would be based on the factor or factors most favourable to the Client. Such an approach does not appear to cause problems in practice, as users can see the sense behind this provision.

17.5.3 Change to the Condition for a Key Date

Under the ECC, Key Dates and Conditions (see definition at clause 11.2(11) and Section 12.5.3), if used, come as a pair: one cannot exist without the other. While there is a provision to change a Key Date as a result of the impact of a compensation event (see clause 63.5 and Section 17.3), there is no provision to change a related Condition. The ECC, however, recognises at clause 63.11 that a change to the Scope may result in a Condition, as described in Contract Data Part One, becoming incorrect. In such circumstances this clause requires the Project Manager to correct the description of the Condition. Any impact on either the Prices or the Completion Date resulting from this correction of such a description is incorporated into the assessment of the offending compensation event.

17.6 Practical issues

17.6.1 The Accepted Programme contains all the clues

Whenever the assessment of a compensation event is carried out, then those carrying out this function should recognise that many of the clues to the information that they need to assemble in order to complete this exercise are contained within the Accepted Programme (see clause 32.1 and Section 12.3).

In respect of time, it is the relevant Accepted Programme (being the Accepted Programme current at the dividing date (see clause 63.1 and Section 17.2.1)) that provides the baseline against which the impact of the event is assessed. The impact of the event on the Accepted Programme will determine whether there is any delay to either planned Completion or the planned date for achieving a Key Date. This, of course, determines whether the assessment needs to include for any delay and consequent extension of time in connection with any of the dates forming part of the Contractor's obligations. It is not the place of this guide to explore the mechanics of how such delay analysis is carried out. That said, users must recognise that several of the compensation events relate directly to the Accepted Programme (see clauses 60.1(2), 60.1(3) and 60.1(5) and Sections 15.3.2, 15.3.3 and 15.3.5) and that clause 63.5 makes direct reference to the assessment of delay being related to the dates shown on the Accepted Programme. Without an Accepted Programme, the Contractor cannot assess the effect of a compensation event on time.

When considering the effect of a compensation event on the Prices, the relevant Accepted Programme is again the source for the baseline information against which to forecast any change. The method of assessing the change to the Prices, as set out at clause 63.1, takes the user to the definition of Defined Cost.[8] This, in turn, for all but Main Option F, takes the user to either the Schedule of Cost Components (see Section 14.2) or the Short Schedule of Cost Components (see Section 14.3). The mechanism for valuing work under the Schedules is entirely based on the resources to which the components of cost apply. As the Accepted Programme includes the principal resources that the Contractor plans to use on each activity,[9] this provides the baseline against which to forecast any change caused by the compensation event.

In practice, the Accepted Programme is the baseline that should be used to measure the changes to both time and the Prices caused by compensation events. This document

---

[8] See clauses 11.2(23) (for Main Options A and B), 11.2(24) (for Main Options C, D and E) and 11.2(25) (for Main Option F) and Sections 17.2.4, 13.4.5 and 13.4.6.
[9] See the penultimate bullet point at clause 31.2 and Section 12.2.

provides a record of the Contractor's proposed method of carrying out the works, including the resources to be used. It is therefore a vital reservoir of information and clues that can be used to forecast the effect of each compensation event. As the Accepted Programme is itself a forecast, then its use in this way is consistent with its preparation. This of course introduces risk for the Contractor, but there is no doubt that this approach is the intent of the ECC and, with the publication of the third edition, was recognised as being the case. Nothing has changed in the fourth edition in this respect.

17.6.2 The time effect of a number of small compensation events

The concept of the compensation event mechanism is that the effects in respect of both time and money are considered and included in each assessment on a stand-alone basis. In practice, this concept causes immense difficulties, as often it is impossible to determine that a single compensation event has any effect on planned Completion, a position that could be said to be 'delay by a thousand variations'.[10] When numerous examples of such events occur, the cumulative effect is to cause a measureable delay. The ECC does not provide a mechanism to compensate the Contractor in such circumstances, however. The Contractor therefore has to include a small amount of time in each assessment, which is unrealistic and will probably not be accepted by the Project Manager, or forgo any adjustment to the Completion Date, which can result in the Contractor incurring delay damages (as provided for at Secondary Option X7; see Section 12.7.3).

In practice, many users get around this situation by agreeing to ignore the effects of smaller compensation events on the Completion Date within the assessment of the effect on the Prices arising from such events. Instead, they agree to consider the compound effect of the smaller compensation events at pre-agreed intervals. This is often the same interval as the regular revision of the programme (see clause 32.2 and Section 12.3). If this agreement is made, the Contractor and the Project Manager review the progress over the latest interval. Where there has been any delay, they consider whether the smaller compensation events when compounded together have caused any of that delay. In effect, this is a 'Windows Analysis' type exercise on the progress between each revision of the programme.

For example, if the last period between revisions of the programme showed a delay of, say, one week (five working days), they would analyse why this delay had occurred. This might result in an analysis which determined that: two working days' delay was due to a large compensation event where the delay had been included in that assessment; one working day was due to a breakdown of vital Equipment (which is the Contractor's problem); and the remaining two working days were down to the cumulative effect of several other small compensation events. The Project Manager and the Contractor would then create another compensation event in which the time (two working days) and related costs would be assessed in isolation from the direct cost effects of the actual events which contributed to that delay.

---

[10] Thanks to a former colleague of mine, Andrew Duncan, who coined this phrase while we were working on a project. There were more than 4000 small value variations which cumulatively had caused delay, but it was impossible to demonstrate the effects of any single one of them (this project was not under the ECC).

It must be recognised that there is no strict contractual basis for this approach under the ECC. However, using the spirit of mutual trust and cooperation and by adopting a pragmatic view, many users have implemented this approach. Those that have used this method have found that it provides a fair and reasonable outcome which both the Client and the Contractor are satisfied with. That being the case, it must be recommended as a practical approach that is worth considering.

17.6.3 Use of estimators

For some decades now, contractors in the United Kingdom have routinely employed quantity surveyors as part of their site management teams. One of the primary tasks assigned to these quantity surveyors has been the management and pricing of variations on the project, so when the contract for the project is the ECC, it is they who have primarily been responsible for the assessment of compensation events. The way that quantity surveyors have traditionally been trained means that many of them have little or no experience of estimating the impact of a compensation event prior to the work being carried out. Instead, they have been trained to monitor, record and measure before retrospectively valuing the work that has been done. As the assessment of compensation events under the ECC is based on forecasts, many contractors have found that the process can be more effectively managed by estimators rather than quantity surveyors. This has resulted in the employment of estimators as part of the contractor's onsite commercial team (mainly on larger projects), where their skills have been used to complement those of the quantity surveyors, planners and construction specialists who input into the assessments.

17.6.4 Assessing compensation events when the procedures have not been adhered to

A question that often arises concerns the assessment of compensation events where the procedural mechanism within the ECC has not been followed as required in core clauses 61 and 62 (see Sections 16.2 to 16.6). Providing that the compensation event is not barred by the provisions of clause 61.3 (see Section 16.3.1), which would end the matter there and then, we need to consider what happens where other parts of the mechanism have failed. For example, a common occurrence is that the Contractor does not submit a quotation within the three weeks required by clause 62.3 (see Section 16.6.3) and the Project Manager, having become obliged to carry out the assessment as required by clause 64.1 (first bullet point) (see Section 16.7.1) fails to deliver that assessment within the three weeks allowed by clause 64.3 (see Section 16.7.3). So what happens then?

What is quite clear is that the Contractor is still entitled to have the Prices changed; there is no loss of rights caused by these failures. The Courts in Northern Ireland addressed what happens when the mechanism breakdowns in the judgement in *Northern Ireland Housing Executive v Healthy Buildings (Ireland) Ltd* [2017] NIQB 43. This case considered a contract which was using the NEC3 Professional Services Contract as the conditions and considered a number of issues, including the assessment of compensation events where neither party had complied with the contractual requirements. As the processes in the core clause 6 of the NEC3 Professional Services Contract are the same as the processes in core clause 6 of the NEC4 Engineering and Construction Contract, then this case is directly applicable to the subject of this book.

In this judgement the Court considered preliminary issues, the pertinent one in relation to this Section being:

> [6] The questions for determination by the court have been agreed as follows:
>
> (1) On the true construction of the contract, and in particular Clauses 60 to 65 of the contract, is the assessment of the effect of the compensation event calculated by reference to the forecast Time Charge or the actual cost incurred by the consultant?
> (2) Are actual costs relevant to the assessment process in Clauses 60 to 65 of the contract?

The Judge concluded that the answer to both of the above was yes. This answer does not sit easily with the question at [6](1) above, but by reading the judgement the reasoning has the cumulative effect of determining that when assessing a compensation event that was carried out before the assessment was done, then that assessment should be on the basis of actual time and cost records. The Judge said at paragraph [54] of the judgement *'Why should I shut my eyes and grope in the dark when the material is available to show what work they actually did and how much it cost them?'*

So to summarise the position, if the mechanism is not followed and by the time the assessment is carried out some or all of the work has been carried out, then the retrospective records should be used and not forecasts as required by clause 63.1. As the case dealt with a scenario where the contractual mechanism had broken down, it is suggested that the assessment would still be based on the forecast for work done after the dividing date where for the event in question the contractual mechanisms were still in play.

Chapter 18
Termination

18.1 Introduction

This chapter concerns a mechanism that will generally only be used when one party or the other has breached the contract in such a serious way that the relationship engendered by the ECC has broken down and the injured party wishes to bring the relationship to an end. That said, there will also be circumstances that arise where the Client in particular may have reasons other than a breach by the Contractor to bring the contract to an end. Such circumstances are also dealt with by core clause 9.

The only purpose of core clause 9 is to provide a mechanism whereby the performance of all the main obligations by the parties, other than those which expressly survive and are maintained by the core clause, are brought to an end. Unless such circumstances arise, users will find no need to consult this clause during a project. Nevertheless, users should be aware of the contents of this core clause as it contains the trigger to what are very serious consequences should one of the parties breach an obligation set out elsewhere. As the ECC does not provide cross-references, it is not unknown for users to find that a failure to comply with some other provision acts to give the other party the right to trigger the termination mechanism.

18.2 Reasons for termination

A total of 22 reasons for termination, grouped into eight categories, are provided in clause 91. Some of these categories are triggered by a breach of the Client, others by a breach of the Contractor, some by a breach of either party or even by an event triggered by a third party. As there is no distinct separation into different clauses, any party deciding to initiate the mechanism must be sure that the reason they are going to rely on is one which gives them the right to terminate. If a party tries to terminate using a reason which does not give it a right to do so, the Project Manager has a role to play which acts as a check in such circumstances (see clause 90.1 and Section 18.4).

The reasons themselves are identified by the inclusion of a unique 'R' number attached at the end of each such item. These numbers play a part in determining the other procedures and accounting factors to be taken via the termination table (see clauses 90.2 and 90.3 and Sections 18.2 and 18.5.1).

A Practical Guide to the NEC4 Engineering and Construction Contract, First Edition. Michael Rowlinson.
© 2019 John Wiley & Sons Ltd. Published 2019 by John Wiley & Sons Ltd.

18.2.1 Client and Contractor defaults

Of the 22 reasons for termination, the first 10 listed (see clause 91.1), numbered Reasons 1–10 inclusive, refer to the default of either party should they become insolvent. Separate reasons are given to cater for various acts by either an individual or a company/partnership and are drafted based on English law. The introductory sentence to this category makes reference to an equivalent fault that is intended to create a satisfactory position in relation to the law in other jurisdictions. In practice, users should confirm that the terms used can be satisfactorily matched with the terminology used in the jurisdiction which will apply to the contract (see clause 12.2 and Section 6.6.8). If they cannot, then Secondary Option Z needs to be used to amend the reasons to keep the reasons in line with the relevant law.

18.2.2 Client-only defaults

There are two reasons, which are Client defaults but which cannot also be a Contractor fault. The first of these is Reason 16 (see clause 91.4), which concerns a breach by the Client whereby it fails to pay the Contractor an amount due under the Contract (see clause 51.1 and Section 13.2.2) within 13 weeks of the date by which the payment should have been paid. The wording clearly links the 13-week period to the date that the payment was required to be made by[1], rather than the date that is was certified (clause 51.1) or became due (clause Y2.2). This is a crucial matter to note in practice as the difference will be at least three weeks and possibly more if the Contract Data has been used to extend the payment period, as is often the case.

This right to terminate arises automatically on expiry of the 13-week period; it does not require any notice from either the Contractor or the Project Manager prior to the expiry of the period in order to make the Client aware of the position. However, if a payment was not being made it is difficult to conceive that the parties would not already be discussing the issue or that a deep-seated dispute had not already arisen.

Although there are no express words to this effect, it is also suggested that the amount certified by the Project Manager would include any award made by the Adjudicator that either opened up and revised a previous certificate issued by the Project Manager or directed that a payment be made to the Contractor. Further, for contracts in the United Kingdom that fall under the provisions of the Housing Grants, Construction and Regeneration Act 1996 (regardless of whether Secondary Option Y(UK)2 forms part of that contract), this reason for termination is separate from the right to suspended performance under section 112 of that Act.

The second is Reason 19 (see second bullet point at clause 91.6), where the Contractor is given the right to terminate should the Project Manager have given an instruction for him to stop or not start any work (see clause 34.1 and Section 12.6.2), and then not rescinded that instruction within 13 weeks of having issued the first instruction in

---

[1] See clause 51.2 and Section 13.2 or, for contracts including Secondary Option Y(UK)2, see clause Y2.2 (final date for payment) and Section 13.9.

circumstances where the reason for the Project Manager issuing the instruction was due to a default of the Client. The ECC does not give any clues as to what such a default would be, but in practice, it would cover matters such as not providing access to the Site (see clause 33.1 and Section 12.6.1) or not providing something that the Client is to provide as stated in the Scope (see clause 25.2 and Section 7.5.2). The wording is wide enough that any breach by the Client that results in the Project Manager instructing the Contractor to stop or not to start will create the right to terminate. A breach by the Client without a consequent instruction from the Project Manager would not create the right for the Contractor to terminate. The need for both a breach by the Client and an instruction to stop or not start work from the Project Manager are expressly set out in the wording.

18.2.3 Contractor-only defaults

There are a total of seven reasons linked to a default by the Contractor (contained in three different clauses) that give rise to the ability for the Client to terminate. (It could be said that one of these seven reasons actually covers two different potential defaults, meaning that there are eight such reasons in total.) The first four are contained in the three bullet points at clause 91.2 and are numbered Reasons 11–13 inclusive. All of these reasons give the right to the Client to terminate only if the Project Manager informs the Client that the Contractor has defaulted in one of the ways, and that the Contractor has not stopped defaulting within four weeks of the Project Manager's notification to the Client. The Project Manager will need to notify this failure to stop defaulting to the Client at the end of the four-week period. In order to make this mechanism work, the Project Manager will need to copy the original notification to the Client that the default has occurred to the Contractor, in order to put the Contractor on notice that it has four weeks in which to stop the default.

It must be noted that it is the Client who must initiate the termination and that the right to do so is discretionary; the Client can choose not to take this step after the four weeks have expired even if the Contractor has not stopped defaulting. The Project Manager's role is simply to inform the Client that the default has occurred in the first instance, and then to inform the Client that the default still exists after the four weeks. The purpose of the four-week period is to give the Contractor the chance to rectify the defect that has been notified to the Client by the Project Manager, hence the reason that the Project Manager needs to simultaneously inform the Contractor that the Client has been notified of a default.

The default given as Reason 11 concerns a substantial breach of the Contractor's obligations to the Client. In practice, the wording of this Reason presents difficulties because the use of the word 'substantial', while employed to prevent termination occurring for minor infringements, introduces an element of subjectivity. Where such subjectivity exists, what one person considers to be substantial may in the eyes of another person seem insubstantial. Accordingly, for the Project Manager to issue a notice to this effect and for the Client to act upon it creates the risk that, if challenged through the dispute resolution procedures (see Chapter 19), the Client can be found to be the one in breach of the contract for terminating for a reason not provided by the Contract. The ECC offers a safeguard against this, but the financial consequences of the safeguard are more onerous than the correct application of the mechanism (see Section 18.3).

The second bullet point, which is Reason 12, actually covers two potential defaults by the Contractor, both of which can be considered objectively. These two breaches refer to a failure to complete the obligation to provide a guarantee (see Secondary Option X4 and Section 20.3) or a bond (see Secondary Option X13 and Section 20.5) if required by the Contract. Either of these defaults can only occur if one or other of the Secondary Options concerned is included in the Contract.

The final part, Reason 13, refers to a breach by the Contractor of its obligation not to appoint a Subcontractor until the appointment of that Subcontractor has been accepted by the Project Manager as set out at clause 26.2 (see Section 8.3). Again, this Reason uses the word 'substantial', which creates the same problems in practice as discussed above.

Clause 91.3 contains the next two reasons linked to a default by the Contractor and is worded in almost identical terms to clause 91.2. The difference here is that, instead of referring to the Contractor not having put the default right, this clause refers to the Contractor not having stopped the default. The mechanism of the Project Manager notifying the default and then notifying the failure, in this case, to stop is the same as for clause 91.2 and in the same period of four weeks. The reasons themselves both start with the word 'substantially', which brings the same practical problem of interpretation as discussed above for Reason 11. The matters to which the 'substantially' requirement is prefixed are, for Reason 14, the hindering by the Contractor of the Client or Others (see definition at clause 11.2(12) and Section 5.11). This would be a breach of any of clauses 20.1 (Section 7.2), 25.1 and 27.1 (both Section 7.5.2) by the Contractor. For Reason 15, the matter is the breaking of a health or safety regulation in breach of clause 27.4 (Section 7.5.2).

The next reason in this category is Reason 18 (see clause 91.6), where the Client is given the right to terminate should the Contractor have committed a default that has resulted in the Project Manager giving the Contractor an instruction to stop work or not to start any work and that instruction remains active for 13 weeks after it was issued. In order for such an instruction to remain in place for such a period, the default is likely to have to be for a very serious reason. The more common causes of such instructions are a failure in respect of health and safety obligations, but such incidents do not usually result in projects being stopped for more than 13 weeks unless it becomes apparent that the Contractor is simply not competent to perform in safe manner. This tends to demonstrate that there was an error in the selection procedure. Accordingly, such events are likely to be rare.

The final reason in this category is Reason 22 (see clause 91.8), which arises if the Contractor has committed a Corrupt Act (see definition at clause 11.2(5) and clause 18 and Section 6.6.6). If such an act is done by the Contractor, there is no need for any notice from the Project Manager; the Client can trigger the termination immediately. However, before doing so the Client needs to consider whether one of the two exceptions included in clause 91.8 applies. These both refer to the Corrupt Act being by a Subcontractor or Supplier. The first exception would apply if the Contractor was not and should not have been aware that the Corrupt Act was being perpetrated by the Subcontractor or Supplier. The second one is that the Contractor informed the Project Manager that the Corrupt Act had occurred as soon as the Contractor became aware of it and the Contractor took immediate action to stop the Act. In order to determine whether either of these exceptions apply the Client will need to consult with both the Project Manager and the Contractor in order to investigate the circumstances that prevail in order to decide whether the termination can be carried out or whether the Contractor has acted as required and can be excused as a result of its actions.

18.2.4 For other reasons

A total of three of the reasons arise as a result of something occurring over which neither party has any control. These are set out at Reasons 17, 20 and 21. Should Reason 17 or 20 arise, then either party may elect to initiate the termination provisions; in the case of Reason 21, it is only the Client who can elect to use this as the trigger event to the procedure.

Reason 17 (clause 91.5) arises when the parties have been released from their contractual obligations by the law. The clause does not say that the law is that of the country in which the Site is located. As the contract is governed by the law of the country as stated in Contract Data Part One (see clause 12.2 and Section 6.6.8), it must be this law that is applicable when determining if this reason has arisen. If this were not the case, then, in practice, the parties would find themselves in an impossible position.

Reason 20 (clause 91.6) gives either party the right to terminate, should the reason for the Project Manager's instruction under clause 34.1 to stop or not start any work be due to something other than a default of either party[2] and that such an instruction had not been rescinded in the stated period of 13 weeks.

The sole right of the Client to terminate at Reason 21 (clause 91.7) arises where a 'prevention' (see clause 19.1 and Section 6.6.7) event has occurred. Unlike the compensation event for such an event (see clause 60.1(19) and Section 15.3.19), there is no caveat attached to this Reason that would render it inoperable if one of the other Reasons also applied.

18.3 *Secondary Option X11*

Sections 18.2.1–18.2.4 consider the 22 reasons set out in clause 91 that create a reason for one or both parties to terminate the contract. Secondary Option X11, if used, creates in clause X11.1, a separate right for the Client to terminate for any reason. This creates a right that is as wide ranging as it suggests. The Client can terminate the contract for any reason whatsoever. To do so will not constitute a breach of the contract, as the ECC gives him this right. Clause X11.2 then creates the position whereby the Contractor is recompensed for what it has done and for the theoretical fee that it would have earned on the remaining work (see Section 18.6.3), but it does not constitute a breach.

In earlier editions, this right was included in core clause 90. That it wasn't listed in clause 91 seemed to have caused some confusion as to whether the Client could terminate for any reason (or convenience as it is often referred to). By separating this right from core clause 9 and making it a separate secondary Option it will now be abundantly clear

---

[2] A default by either party would trigger either Reason 18 or Reason 19.

that the Client has the right by simply looking at the secondary Options included in the Contract. As these are part of the first entry in Contract Data Part One, whether it is included or not will be very easy to spot.

18.4 Implementing termination

Under clause 90.1 in the event that a reason arises that gives either party the right to terminate, and they choose to exercise the discretion given by the ECC to terminate, then the party which decides to act in this way initiates the process by notifying the Project Manager and the other party of its intent. In that notice, the Client or the Contractor gives details of the reason that it relies on as being the trigger that allows it to exercise this right.

On receipt of such a notice, the Project Manager is required to check on the reason stated. If this check shows that the reason notified is one that is set out in the contract and one that gives the party who has issued the notice the right to terminate, then the Project Manager is required to promptly issue a certificate confirming the termination. The issue of the termination certificate by the Project Manager brings to an end the obligations of the parties set out in the other core clauses and instead triggers a fresh set of procedures and rights as set out in core clause 9. Clause 90.4 expressly states that, after the issue of a termination certificate, the Contractor does nothing further to Provide the Works (see clause 20.1 and Section 7.2).

18.5 Procedures after termination

18.5.1 Under the core clause

Immediately after the Project Manager has issued a termination certificate, the actions and obligations of the parties are determined by the procedures to be followed identified against the relevant Reason for termination as set out in the Termination Table (see clause 90.2). Clause 90.3 confirms that the parties implement these procedures immediately at this juncture in time. The procedures to be implemented vary depending on the Reason that the termination was invoked. In simple terms, if the termination was the result of a default by the Contractor, the procedures are more favourable for the Client than if the termination was the result of a default of the Client.

The Termination Table, which is located immediately after clause 90.2, clearly sets out by reference to the terminating party and the 'R' number they have cited when terminating the procedures, each identified by a 'P' number, which apply. The 'P' numbers are expanded on at clauses 92.1 and 92.2.

Users will note that Procedure P1 applies in all cases. This procedure confirms that the Client always has the right to complete the works that comprise the project. The way in which the Client achieves this is not stated and no limitations are placed on the Client in this respect. This procedure also gives the Client the right to use any Plant and Materials (see definition at clause 11.2(14) and Section 14.2.3) providing that it has title (see clauses

70.1 and 70.2 and Section 10.2) to those goods. This wording acts to capture any Plant and Materials that are not in the Working Areas but have been marked by the Supervisor as being for the contract (see clause 70.1).

Where the Client has terminated for Reasons R1–R15, R18, or R22, i.e. for a Contractor default, then Procedure P2 will also apply. Under this procedure, the Client is given the discretion to instruct the Contractor to leave the Site, but not the Working Areas (see definition at clause 11.2(20) and Section 6.6.4), and when doing so to remove its Equipment and Plant and Materials. There appears to be a contradiction here between Procedure P1 and the Client's title to Plant and Materials arising from clause 70.2 and this procedure. As the Client has title to all the Contractor's Plant and Materials when they enter the Working Areas, then Procedure P1 allows the Client to make use of all such items when completing the works. However, Procedure P2 gives the Client the right to instruct the Contractor to remove such items from the Site. Clause 70.2, however, refers to the Project Manager giving permission for the Plant and Materials to be removed, at which time the title would pass back to the Contractor. As Procedure P2 gives the Client the right to instruct the removal of unwanted Plant and Materials, this power would supersede that of the Project Manager at clause 70.2. The result would still be the same, i.e. the title would pass back to the Contractor.

Procedure P2 also provides for the Client to require the Contractor to pass the benefit of any subcontract or other contract related to the project to the Client. The procedure uses the word *assign* while what would be required legally in the majority of instances would be the novation of existing contracts from the Contractor to the Client.

Procedure P3 applies where the Client has terminated for one of the Contractor default reasons, i.e. Reasons R1–R15, R18 or R22 and also for Reasons R17 and R20 where either party has been given the right to terminate and it has been the Client who exercised that option. This procedure allows the Client to use the Contractor's Equipment to which the Contractor has title in order to finish the project. A duty for the Project Manager is included here, as the Project Manager is required to notify the Contractor when any such Equipment is no longer required. At this time, the Contractor returns to Site and removes the items covered by any such notice. It should be emphasised that this procedure does not give the Client title to the Contractor's Equipment. The Client does not have any right to sell or otherwise dispose of the Contractor's Equipment; it only has the right to use it for as long as needed in relation to the contract works only.

The final procedure, Procedure P4, applies when either the Client has terminated as a result of a prevention event (see clause 19.1 and Section 6.6.7, and clause 91.7 and Section 18.2.4), or when the Contractor has initiated the termination for any of the reasons that permit him to take this step. This procedure simply obliges the Contractor to leave the Working Areas and to remove all Equipment while doing so. Procedure P4 never applies when either P2 or P3 apply.

18.5.2 Under Secondary Option X11

Where the Option applies, clause X11.2 sets out the Procedures to be used if the Client terminates using the right created by clause X11.1. The Procedures that are applied under such circumstances are P1 and P2. They apply exactly as described above in relation to the core clause.

18.6 Assessing the amount due after termination

18.6.1 Under the core clause

The Project Manager is required to assess and certify the amount due within 13 weeks of the Project Manager issuing the termination certificate (see clause 53.1 and Section 13.6). The amount certified is the difference between the amount due on termination and that which has been paid previously by the Client to the Contractor. Whatever payment is due from one party to the other is then paid within three weeks of that certificate. This requirement assumes that the party that is due to make the payment is solvent and in a position to pay the amount due. This is not always the case, as one of the more common reasons for termination is insolvency, i.e. one of Reasons R1–R10, in which case such a transaction is unlikely to happen.

Once a termination certificate has been issued, the second paragraph of clause 90.3 creates a position whereby any payment for which a certificate has been issued (under either clause 51.1 (see Section 13.2.2) or clause Y2.2 (see Section 13.9)) but which has not been paid does not have to be made. Any monies due at such a time would effectively be withheld until such time as the amount due after termination had been calculated and was due for payment (i.e. 13 weeks after the date of the termination certificate). This provision is only active where it is the Client who has initiated the termination as a result of a Contractor default (see Sections 18.2.1 and 18.2.3). This position is very similar to that allowed by the Housing Grants, Construction and Regeneration Act 1996, as amended by the Local Democracy, Economic Development and Construction Act 2009, in respect of insolvency. Secondary Option Y(UK)2, at clause Y2.4 (see Section 13.9), includes the Act compliant provision, which is restricted to when the Client has terminated for one of Reasons 1–10 (i.e. an insolvency event).

As with the procedures to be followed (see Section 18.5), the factors to be considered in assessing the amount due are identified against the relevant Reason for termination as set out in the Termination Table. The Termination Table, which is located immediately after clause 90.2, clearly sets out by reference to the terminating party and the 'R' number they have cited when terminating the 'amount due' factors, each identified by an 'A' number, which apply. The 'A' numbers are expanded on at clauses 93.1 and 93.2.

Amount Due A1 applies in all cases no matter what Reason has been used to initiate the termination and regardless of which party initiated the process. The requirement here is for the Project Manager to carry out an assessment of the value of the work carried out up to the time of the termination by the Contractor. This assessment is made up of five principle elements, the first of which is an assessment of the amount due as a normal payment i.e. the amount due determined in accordance with clause 50.3 (see Section 13.4).

To that amount is added the Defined Cost[3] of Plant and Materials, which are either in the Working Areas or to which the Client has title (see clauses 70.1 and 70.2 and Section 10.2), and the Defined Cost that has been reasonably incurred by the Contractor in the expectation that it would complete the whole of the Works. Also included in the assessment are any amounts that have been temporarily retained by the Client, such as retention (see Secondary Option X16 and Section 13.10.3). In addition to any amounts deducted under clause 50.3, any outstanding balance of any advanced payment (see Secondary Option X14 and Section 13.10.2) is also deducted from the assessment.

In all circumstances, where the termination has been initiated by the Contractor for one of the Reasons it is allowed to use or by the Client for any Reason other than those set out at clauses 91.1–91.2, then Amount Due A2 will form part of the assessment. Under this element, the Contractor will be paid the Defined Cost incurred after termination in removing Equipment from the Working Areas.

Where the Client has initiated the termination on the grounds of a default by the Contractor[4] then Amount Due A3 provides for a deduction from the assessment of the forecast additional cost of completing the works that will be incurred by the Client. The concept of this deduction is that the breach by the Contractor has resulted in the Client incurring a loss that it is entitled to recover as damages at law. The difference in practice is that at law damages generally have to be proved by the claimant. Under this clause the Project Manager is required to forecast that loss, a process which by its very nature and timing will mean that there is a considerable amount of estimating included in the assessment. Often it will be many months if not years before a project is completed and the resulting true loss to the Client can be established based on actual expenditure. As the ECC requires the Project Manager to complete the assessment within 13 weeks of termination (see clause 53.1 and Section 13.6), the final cost data is unlikely to be available to him. The assessment must therefore be based on forecasts as to the costs that will be incurred in completing the works, and these costs will, in all likelihood, include contingencies and other provisional allowances. As this type of assessment is the basis of the agreement between the parties, then that approach is one that the Project Manager is obliged to carry out by the contract. There is no mechanism within the ECC for this assessment to be opened up and re-visited at any subsequent time other than by reference to dispute resolution (see Chapter 19).

Amount Due A4 applies where the Contractor has terminated because of default of the Client. This provides that the assessment will include an allowance for the direct fee percentage applied to the balance of the work not carried out by the Contractor. For Main Options A–D, this balance is established by calculating the difference between the Price for Work Done to Date (see Section 13.4) and the total of the Prices at the Contract Date, i.e. the tender sum. For Main Options E and F, the first forecast of the total Defined Cost

---

[3] See the definition of Defined Cost at clause 11.2(23) (Main Options A and B), 11.2(24) (Main Options C, D and E or 11.2(25) (Main Option F), and Section 13.4.
[4] See Reasons R1–R15, R18 and R22, clauses 91.1–91.3 and 91.6–91.7 and Sections 18.2.1–18.2.4.

for the project is substituted for the total of the Prices. In effect, this gives the Contractor its loss of profit on the balance of the contract sum.

18.6.2 Under the main options

Under three of the main options, the amount due as calculated under A1 and A2 is refined by the application of additional rules. Under main option A, when calculating the amount due for the normal payment at the time of the termination, clause 93.3 provides that the first bullet point of the definition of Price of Work Done to Date (PWDD) at clause 11.2(29) concerning the completion of groups of activities is disregarded. Despite this relaxation, it still means that the Price for any activity that has been commenced but not completed at the date of termination is not included in the calculation.

For main options C and D, clauses 93.4 (main option C) and 93.5 (main option D) determine how the Contractor's share[5] is calculated before being applied to the amount due under A1 and A2. Clause 93.6, which applies to both these main options, then confirms that the share of any saving is added to the amount due and that the share of any overspend is deducted from the same amount.

18.6.3 Under Secondary Option X11

Where the Option applies clause X11.2 sets out the 'Amount Due' components to be applied if the Client terminates using the right created by clause X11.1. The Amount Due components that are applied under such circumstances are A1, A2 and A4. They apply exactly as described above in relation to the core clause.

18.7 Practical issues

18.7.1 Maintaining title after termination

When termination takes place as a result of the Contractor becoming insolvent (see Reasons R1–R10 at clause 91.1 and Section 18.2.1), the Client may encounter a problem in maintaining title to goods which (under the contract) it technically owns. Under clause 70.2 (see Section 10.2), whatever title the Contractor has to Plant and Materials passes to the Client when the goods are brought into the Working Areas. From the checks the Project Manager has carried out on the conditions of subcontract under clause 26.3 (see Section 8.3), the Project Manager may have confirmed that title of goods brought to the Working Areas passes to the Contractor and hence to the Client upon delivery or, at worst, upon payment. However, while this is the undertaking the subcontractors have given, they may not have secured title to the goods in the same way from their supply chain. As a result, in the event that the Contractor becomes insolvent, it is likely that

---

[5] See clauses 54.1–54.4 (main option C) and clauses 54.5–54.8 (main option D), and Section 13.7.

the subcontractors' supply chain will claim title to goods that the Project Manager and Client believed had passed to the Client. In practice, such claims are often expressed by the physical removal of any unfixed goods (see Section 10.1) from the Working Areas.

In the United Kingdom, this is a difficult position to protect in practical terms. It is a civil dispute under contract in which the law enforcement agencies will not want to become involved. The subcontractors and their supply chain react quickly by physically removing goods. All the Client can do is tighten up the payment procedure so that no unfixed Plant and Materials are paid for unless evidence is provided that the Contractor has legal title to the goods concerned; even this is not guaranteed.

18.7.2 Avoiding duplication when assessing the amount due

When assessing the amount due after termination, the Project Manager should recognise that there is potential for the duplication of costs that could be included in more than one of the elements that may apply or even within different parts of the same element. In Amount Due A1, for example, depending on the Main Option in use, the amount due under the first bullet point could include sums that are also covered by the second bullet point. This is a situation of which the Project Manager should be aware; the Project Manager must take steps to ensure that costs are not inadvertently included in more than one element to the detriment of the Client.

Chapter 19
Resolving and Avoiding Disputes

19.1 Introduction

The Schedule of Options at the start of the ECC states that one of the three dispute resolution Options must be selected. The way this is worded makes the inclusion mandatory (see Section 1.4). The choices presented to users are an option for use where the UK's Housing Grants, Construction and Regeneration Act 1996 (HGCRA96), as amended by the Local Democracy, Economic Development and Construction Act 2009[1] (the Act) applies, which is referred to as Option W2 (see Section 19.3), or two alternatives that are designed to be used everywhere in the world, including the United Kingdom, when the Act does not apply. The first of the alternatives, known as Option W1 (see Section 19.2) creates a three layer system starting with a review by senior representatives of both parties, followed by contractual adjudication and then an optional final review by a tribunal if required. The second alternative, known as Option W3 (see Section 19.4), employs a Dispute Avoidance Board (DAB) followed by the option of a final review by a tribunal. The choice is therefore relatively straightforward.

Users in the United Kingdom will have to confirm that the work they intend to be carried out under the contract constitutes a 'construction operation' as defined at section 105(1) of the Act and are not one of the excluded operations at section 105(2). The result of this determination will identify their option or options.

Users elsewhere in the world will need to consider whether legislation exists in the jurisdiction within which they are operating, which would make Option W1 noncompliant with the law. This comment is made as some countries, for example Australia, New Zealand, Singapore, Malaysia and Ireland, have followed the success of the UK Act by introducing their own similar legislation. It is not within the ambit of this guide to consider whether either of the dispute resolution options is compatible with that legislation. If changes are needed, then Option Z (see Section 20.12) will need to be employed to facilitate the necessary amendments. Otherwise they, and UK users when the Act does not apply, can choose between Option W1 and Option W2.

---

[1] The Local Democracy, Economic Development and Construction Act 2009 applies to all new contracts entered into from 1 October 2011. It amends rather than replaces the 1996 Act. The 1996 Act applied to all qualifying contracts entered into from 1 May 1998.

A Practical Guide to the NEC4 Engineering and Construction Contract, First Edition. Michael Rowlinson.
© 2019 John Wiley & Sons Ltd. Published 2019 by John Wiley & Sons Ltd.

This chapter presents a conflict with the rest of this guide. This guide is primarily about how to practically use the ECC to make all the project management procedures contained within it work to effectively manage a contract. By seeking to work in this way, users should develop good inter-party communications (see clause 13 and Sections 6.2 and 6.3) and create and enhance a spirit of mutual trust and cooperation. The result of this approach is that while the parties may occasionally have differences of opinion, the working environment they have created will enable them to openly discuss any such issues, achieve a resolution that is acceptable to all and maintain the joint goal of bringing the project to a successful conclusion. That being the case, the dispute resolution options should not be needed.

Further, this guide was never intended to consider dispute resolution in any depth. There are numerous other texts available which consider the finite points of such procedures and the problem of case law serving to subtly change the rules as new issues are considered by the courts. As a result, Sections 19.2–19.4 contain only a general review and outline of the procedures contained within the three options available to users. Should it be necessary to use these procedures, then users are advised to seek specialist advice in these areas.

19.2 Option W1

This dispute resolution Option follows the model included in previous editions of the ECC and which, prior to 1 May 1998 when the provisions of the HGCRA96 became effective for new construction contracts that fell within its ambit, was also used in the UK. The publication of the fourth edition introduced a new first step in the process, being a review and negotiation of disputed matters by Senior Representatives (see Section 5.6) from both parties, following which any remaining disputes can be referred to an Adjudicator (see Section 5.7). Only after the Adjudicator has notified a decision to the parties can either of them elect to refer the dispute to the tribunal for a final and binding determination. The reference to the Senior Representatives, then the Adjudicator, and finally to the tribunal must be made in that order and within a set timetable.

The Option itself considers the considerations and actions of the Senior Representatives, the appointment of the Adjudicator, the referral of the dispute to the Adjudicator, the powers of the Adjudicator and a final review by the tribunal.

19.2.1 Identity of the Senior Representatives

In Contract Data Parts One and Two, users will find under the heading of 'Resolving and avoiding disputes' two spaces each in which to identify their respective senior representatives. Whilst two spaces are provided, more or less could be named. Volume 4 of the User Guide states that the numbers should be equally matched between the parties and that the individuals named should not have day to day involvement with the project so that they can bring a fresh view to any dispute.

The ECC does not make any statement allowing the Senior Representatives to be changed, but from a practical basis this must be permissible where necessary.

Practically there is nothing to stop the Parties from creating an escalation process within the senior representatives allowing the matter being considered being handed up levels of management until a solution is found.

19.2.2 Referring a Dispute to the Senior Representatives

Clause W1.1(1) requires that any dispute that arises under the contract is referred to the Senior Representatives (see Section 5.6). The reference must be made in accordance with the Dispute Reference Table, which is found after clause W1.1(4). The Dispute Reference Table establishes four categories of dispute. The first three of these are specific to disputes related to particular circumstances, these being:

- an action or inaction of the Project Manager or Supervisor;
- a programme or compensation event matter being treated as being accepted (i.e. under clauses 31.3 (final paragraph), 61.4 (final paragraph), 62.6 or 64.4); and
- an assessment of Defined Cost which is treated as being correct.

The fourth category covers any other dispute not caught by the three categories above. The Table also provides that either Party can refer any of the dispute categories except for the second one, which can only be referred by the Client. The first three categories must be referred to the Senior Representatives within four weeks of a stated occurrence (different for each category). The fourth category must be referred when the dispute arises. The timing requirement in relation to the fourth category, which is not as definite as the other three categories, appears to create grounds for disputes about when it was referred instead of the about the substantial matter concerned. Parties will need to act quickly if such a dispute arises if they are not to face an argument that they are time barred from referring the dispute concerned. The only good thing is that this fourth category is likely to be the smallest category, in terms of numbers, to be referred as the first three categories cover the common areas of disagreement.

A Party refers a dispute by notifying it to the Senior Representatives, the Project Manager and the other Party. The notification sets out the nature of the dispute that it wishes to resolve. Clause W1.1(2) then requires each party to submit to the other Party a statement of case in respect of the matter referred to in the notice. The clause limits the length of the statement to ten sides of A4 paper. The statement can be supported by an unlimited amount of evidence. By agreement, the Parties can vary this restriction on the length of the statements. What is slightly strange about this clause is that it does not require the statements to be sent to the Senior Representatives. As the Senior Representatives are part of the Parties organisations, it is easy to see that it is expected that the people involved at project level will pass a copy of what they send and of what they receive to the Senior Representatives in their own organisation.

19.2.3 The Senior Representatives actions

What the Senior Representatives do once a matter is referred to them is set out in clause W1.1(3). This allows the Senior Representatives to use whatever process they agree to use and to hold as many meetings as they determine are required in order to try to resolve the dispute. The clause restricts the process to a period of three weeks and makes no provision for that period to be extended.

No later than the end of the three-week period, the Senior Representatives produce a list of the issues that they have considered and report to the Project Manager and each Party as to what is agreed and what is not agreed. The Project Manager then puts those issues that have been agreed to into effect.

As allowed by clause W1.1(1), any issue that has not been resolved by the Senior Representatives can be referred to Adjudication (see the following sub-sections).

Clause W1.1(4) makes it clear that none of the documentation submitted to the Senior Representatives nor any record of any of the discussions between the Senior Representatives is either disclosed or used or referred to in subsequent proceedings before the Adjudicator or the tribunal. This is an important clause as it effectively puts the process followed by the Senior Representatives onto a without prejudice basis. By doing so, it permits the Senior Representatives to be open with each other, safe in the knowledge that anything that they say or suggest cannot be used outside of those negotiations.

19.2.4 The notice of adjudication

In respect of any of the issues that the Senior Representatives report as not having been resolved, or if the Senior Representatives fail to report on within the period allowed, either Party can issue a notice of adjudication to the other Party and the Project Manager within two weeks of the list of not agreed issues or when it should have been given; see clause W1.3(1). In accordance with clause W1.3(2), the two week period can be extended by agreement between the Contractor and the Project Manager before the end of the period in which the notice is due.

The same clause also states that any matter that is not so referred cannot be referred by either Party to the Adjudicator or the tribunal at any future time. This makes compliance with this timetable a must for a party that wishes to continue to pursue resolution of the disputed matter.

19.2.5 Appointment of the Adjudicator

The appointment of the Adjudicator or any replacement is covered by clauses W1.2(1) to W1.2(4). This is facilitated by entries in Contract Data Part One, where a pre-selected Adjudicator or panel of adjudicators can be provided. If this person or persons are not available or not named, then an application for an appointment is made to a nominating body, which is identified as a fallback. Whoever is appointed by whatever means is engaged using the current version of the NEC4 Dispute Resolution Service Contract at the starting date for the works (see Section 12.5.1). Once appointed, clause W1.2(5) establishes that neither the Adjudicator nor the Adjudicator's colleagues are liable for any of their actions in the adjudication unless such an action is taken in bad faith.

19.2.6 Referral to the Adjudicator

Any dispute that has not been referred to but not resolved by the Senior Representatives can be referred to the Adjudicator.

The party making the referral submits its referral to the Adjudicator within one week (clause W1.3(1)) and, by clause W1.3(3), includes all the information upon which it intends to rely and which the Adjudicator is required to consider. In accordance with clause W1.3(6), a copy of this communication to the Adjudicator (as well as all other communications between either party and the Adjudicator) is copied to the other party at the same time. Clause W1.3(3) then allows for any further information to be provided by a party, which is easily interpreted as either party, being submitted to the Adjudicator within four weeks or some longer period that is agreed between the Adjudicator and the parties. Whether such submissions are made in turn or simultaneously is not specified, thereby leaving it up to the parties and the adjudicator to decide and agree between them.

A further issue for the Contractor to consider when a referral under the ECC is being made is whether the matter is also disputed under a subcontract to which it is a party. In the event that it is, and the subcontract permits such a dispute to be joined with a main contract dispute about the same matter, then under clause W1.3(4), the Contractor can elect to have the subcontract dispute considered by the Adjudicator at the same time as the main contract dispute. In effect, the disputes are considered by the same adjudicator who issues a decision which is temporarily binding under both contracts. The advantage of this approach is consistency of decision and a reduction in the overall costs of the procedures.

19.2.7 Powers of the Adjudicator

The powers of the Adjudicator are set out at clause W1.3(5). These allow the Adjudicator to review and revise any action or inaction of either of the Client's agents, i.e. the Project Manager and the Supervisor, which is related to the dispute. The Adjudicator can also alter any quotation that has been deemed to be accepted by operation of one of the mechanisms in core clause 6 (see clauses 61.4, 62.6 and 64.4 and Sections 16.3.2, 16.6.6 and 16.7.4). When the wording of this provision is considered carefully, it reveals that essentially the Adjudicator will have the necessary power to decide any dispute between the parties.

As an example, consider a dispute over termination of the contract by the Client, a more complicated position than a simple dispute over the assessment of a compensation event. In this example, it is the Client who has instigated the termination. The Adjudicator is not, however, given the power to consider an action of the Client, so some may conclude that such a dispute is outside the jurisdiction of the Adjudicator. Here we have to remember that, on receipt of a termination notice, it is an action by the Project Manager that results in the termination certificate being issued (see clause 90.1 and Section 18.4). As the Adjudicator has the power to review and revise an action of the Project Manager, this can be applied to the decision of the Project Manager to issue the termination certificate, thereby effectively considering whether the Client had a valid reason for notifying the termination in the first instance.

In exercising this power under clause W1.3(5), the Adjudicator can:

- take and use the initiative in ascertaining the facts;
- take and use the initiative in ascertaining the law;
- obtain information from either party by issuing directions and specifying the time for the provision of such information; and
- instruct either party to take any other action, such as opening up work or carrying out tests, within a specified time.

Using these powers, the Adjudicator considers the information submitted by the parties in connection with the dispute and, in accordance with clause W1.3(8), reaches a decision within four weeks of the end of the period for submitting information to the Adjudicator or such longer period which the parties agree. If this decision involves assessing time or money, then clause W1.3(7) requires that any such assessment is carried out by the Adjudicator in accordance with the rules for assessing a compensation event (see clause 63 and Chapter 17).

Once made then, in accordance with clause W1.3(10), the Adjudicator's decision is binding on the parties and could be given contractual and legal effect through the courts if necessary. Should the decision contain any clerical errors, then clause W1.3(11) allows for it to be corrected within two weeks of the decision. Otherwise, unless either party notifies the other within four weeks of the decision (see clause W1.4(2)) that it intends to refer the matter to the tribunal, then the decision becomes final and binding.

Once the Adjudicator has notified the decision the Adjudicator has no further powers, except for the power to correct clerical errors as referred to above, in connection with the dispute. Should the dispute be subsequently referred to the tribunal (see Section 19.2.8), neither party can call the Adjudicator as a witness in those proceedings (see clause W1.4(6)).

While this process is taking place, the Client, Contractor, Project Manager and Supervisor are required by clause W1.3(9) to continue as if nothing has changed. The contract continues to operate, despite actions having been taken which the Adjudicator's decision may revise.

19.2.8 Review by the tribunal

The tribunal is the body identified in Contract Data Part One by that name and which will have the power to make a final and legally binding decision in connection with any dispute which is referred to it. In the United Kingdom and most other jurisdictions, users will be restricted to a choice of either referring the matter to litigation, where the body concerned will be the courts, or referring the matter to arbitration, where the body concerned will be either a single arbitrator or a panel of arbitrators, depending on the rules in operation. Where arbitration is selected, the rules will either be those selected by the parties or imposed by the law of the country concerned. These details, as required by clause W1.4(5), are set out in Contract Data Part One.

Clause W1.4(1) prevents the reference of any dispute to the tribunal until it has firstly been referred to the Adjudicator. Should either party then wish to refer the matter to the tribunal, that party must notify the other of that intent within either four weeks of being informed of the Adjudicator's decision (clause W1.4(2)) or, in the event that the Adjudicator does not notify the decision within the timescale provided by the contract, within four weeks of when the Adjudicator should have notified the decision (clause W1.4(3)). Both the clauses refer to the dissatisfied party simply notifying the other party of its intent to refer the dispute to the tribunal. Neither clause requires the actual referral to the tribunal to be made within four weeks or any other specified period. The dissatisfied party can then make the actual referral to the tribunal at any subsequent time subject to any limitations imposed by the relevant law[2].

Once an actual referral is made to the tribunal, that body determines and settles the dispute in accordance with whatever rules or procedures apply to it. The power of the tribunal will be determined by the law, which in many cases will imply far more detailed rules into the contract than are set out in clause W1.4. When such a referral is made, either party is at liberty to introduce and present any submissions or evidence it wishes to make; it is not restricted to just those matters presented to the Adjudicator (see clause W1.4(4).

19.3 Option W2

While sharing many of the principles and approaches used in Option W1 (see Section 19.2), Option W2 has been drafted specifically to satisfy the provisions of the HGCRA96, which became effective for new construction contracts that fell within its ambit entered into after 1 May 1998. The amendments made to the HGCRA96 by the Local Democracy, Economic Development and Construction Act 2009, which apply to all new construction contracts entered into from 1 October 2011 are incorporated into the current provisions. As Option W2 is similar to Option W1 in many ways, the rest of this section is in the same format as the preceding section.

This model provides for an optional 'Senior Representatives' process but requires that all disputes that the parties cannot resolve by agreement through that process to be referred to the Adjudicator (see Section 5.6). Only after the Adjudicator has issued a decision to the parties can either of them elect to refer the dispute to the tribunal for a final and binding determination. The onward reference to the tribunal must be made within a set timetable.

The Option itself considers the Senior Representatives process, the appointment of the Adjudicator, the referral of the dispute to the Adjudicator, the powers of the Adjudicator and a final review by the tribunal.

---

[2] In the UK, the Limitation Act 1980 would apply requiring the referral within 6 years for contracts signed under hand or 12 years for contracts signed under seal or as a deed, both periods running from Completion of the works.

19.3.1 Identity of the Senior Representatives

In Contract Data Parts One and Two, Users will find under the heading of 'Resolving and avoiding disputes' two spaces each in which to identify their respective senior representatives. Whilst two spaces are provided, more or less could be named. Volume 4 of the User Guide states that the numbers should be equally matched between the parties and that the individuals named should not have day to day involvement with the project so that they can bring a fresh view to any dispute.

The ECC does not make any statement allowing the Senior Representatives to be changed, but from a practical basis this must be permissible where necessary.

Practically there is nothing to stop the Parties from creating an escalation process within the senior representatives allowing the matter being considered being handed up levels of management until a solution is found.

19.3.2 Referring a Dispute to the Senior Representatives

Clause W2.1(1) allows a dispute arising under the contract to be referred to the Senior Representatives if the parties agree. Such an agreement will be required for each dispute that arise and is to be referred to the Senior Representatives. This mechanism is not mandatory, as the HGCRA96 requires that any contract which is subject to the legislation must be capable of being referred to adjudication 'at any time'. If the Senior Representatives process was mandatory, as it is in Option W1, then this statutory requirement would not be complied with. The consequence would be that the contractual mechanism would be replaced with the statutory Scheme.

Subject to the required agreement having been made, a Party refers a dispute by notifying it to the Senior Representatives, the Project Manager and the other Party. The notification sets out the nature of the dispute that it wishes to resolve. Clause W1.1(2) then requires each party to submit to the other Party a statement of case in respect of the matter referred to in the notice. The clause limits the length of the statement to ten sides of A4 paper. The statement can be supported by an unlimited amount of evidence. By agreement the Parties can vary this restriction on the length of the statements. What is slightly strange about this clause is that it does not require the statements to be sent to the Senior Representatives. As the Senior Representatives are part of the Parties organisations it is easy to see that it is expected that the people involved at project level will pass a copy of what they send and of what they receive to the Senior Representatives in their own organisation.

19.3.3 The Senior Representatives actions

What the Senior Representatives do once a matter is referred to them is set out in clause W2.1(3). This allows the Senior Representatives to use whatever process they agree to use and to hold as many meetings as they determine are required in order to try to resolve the dispute. The clause restricts the process to a period of three weeks and makes no provision for that period to be extended.

No later than the end of the three-week period, the Senior Representatives produce a list of the issues that they have considered and report to the Project manager and each Party as to what is agreed and what is not agreed. The Project Manager then puts those issues which have been agreed into effect.

As allowed by clause W2.1(1), any issue that has not been resolved by the Senior Representatives can be referred to Adjudication (see the following sub-sections).

Clause W2.1(4) makes it clear that none of the documentation submitted to the Senior representatives nor any record of any of the discussions between the Senior Representatives is either disclosed or used or referred to in subsequent proceedings before the Adjudicator or the tribunal. This is an important clause as it effectively puts the process followed by the Senior Representatives onto a without-prejudice basis. By doing so, it permits the Senior Representatives to be open with each other, safe in the knowledge that anything that they say or suggest cannot be used outside of those negotiations.

19.3.4 The notice of adjudication

The trigger for the appointment procedure to commence, as set out at clause W2.3(1), is the issuing by the party who wishes to commence the adjudication of a notice of adjudication to the other party. In this notice, the notifying party must briefly describe the dispute and the decision which it seeks. In order to be consistent with the HGCRA96, it is also advisable in practice to include reference to the contract and to give the details of the parties to that contract. If the Adjudicator is named in Contract Data Part One, then a copy of this notice of adjudication should also be sent to the Adjudicator. It is then incumbent on the Adjudicator to confirm to the parties within three days whether or not the Adjudicator can act in the matter.

19.3.5 Appointment of the Adjudicator

The appointment of the Adjudicator or any replacement is covered by clauses W2.2(3) to W2.2(6). This is facilitated by entries in Contract Data Part One, where a pre-selected Adjudicator or panel of adjudicators can be provided. If this person or persons are not available or not named, then an application for an appointment is made to a nominating body which is identified as a fall back. Whoever is appointed and by whatever means is engaged using the current version of the NEC4 Dispute Resolution Service Contract at the starting date for the works (see Section 12.5.1). Once appointed, clause W2.2(8) establishes that the Adjudicator or the Adjudicator's colleagues are not liable for any of their actions in the adjudication unless such an action is taken in bad faith.

Under clause W2.2(1), any dispute can be referred to the Adjudicator at any time. This is a fundamental requirement of the HGCRA96. In order to be consistent with the requirements of the HGCRA96, all time periods in Option W2 concerning the adjudication stage are stated in days. To further follow these requirements clause W2.2(2) states that, when calculating such periods in days, then those days that constitute public holidays are not counted (see section 116 of HGCRA96).

In the event that the named Adjudicator (if there is one) either cannot act or fails to reply, then the party who issued the notice of adjudication can apply to the adjudicator nominating body for the appointment of an adjudicator who can act in the matter. The adjudicator nominating body is required to make the required appointment within four days of the request to appoint.

19.3.6 Referral to the Adjudicator

After triggering the process for appointing an adjudicator, clause W2.3(2) then requires that the party who issued the notice of adjudication must issue its referral to the Adjudicator within seven days of the said notice and include all the information upon which it intends to rely and which the Adjudicator is required to consider. In accordance with this clause, a copy of all this information must be provided to the other party at the same time. That strict compliance with this seven-day period, from issue of the notice of adjudication to referral of the dispute to the Adjudicator, is required has been confirmed by case law[3] which refers to all adjudications carried out under the ambit of the HGCRA96 and not just Option W2 of the ECC.

Clause W2.3(2) then allows for any further information to be provided by a party (which is easily interpreted as either party) being submitted to the Adjudicator within 14 days, or some longer period which is agreed between the Adjudicator and the parties. Whether such submissions are made in turn or simultaneously is not specified, thereby leaving up to the parties and the adjudicator to decide and agree between them. In accordance with clause W2.3(6), any communication between either party and the Adjudicator must be copied to the other party at the same time it is issued. This requirement is worded such that the Adjudicator must copy any communication to one party to the other party.

A further issue for the Contractor to consider when a referral under the ECC is being made is whether the matter is also disputed under a subcontract to which it is a party. In the event that it is, and the subcontractor consents to such a dispute being joined with a main contract dispute about the same matter, then under clause W2.3(3) the Contractor can elect to have the subcontract dispute considered by the Adjudicator at the same time as the main contract dispute. In effect, the disputes are considered by the same adjudicator who issues a decision which is temporarily binding under both contracts. The advantage of this approach is consistency of decision and a reduction in the overall costs of the procedures.

19.3.7 Powers of the Adjudicator

The powers of the Adjudicator are set out at clause W2.3(4). These allow the adjudicator to review and revise any action or inaction of either of the Client's agents, i.e. the Project Manager and the Supervisor which is related to the dispute. The Adjudicator can

---

[3] See *Hart Investments v Fidler & Ors* [2007] BLR 30.

also alter any quotation that has been deemed to be accepted by operation of one of the mechanisms in core clause 6 (see clauses 61.4, 62.6 and 64.4 and Sections 16.3.2, 16.6.6 and 16.7.4). When the wording of this provision is considered carefully, it reveals that essentially the Adjudicator will have the necessary power to decide any dispute between the parties.

As an example consider a dispute over termination of the contract by the Client, being a more complicated position than a simple dispute over the assessment of a compensation event. In this example, it is the Client who has instigated the termination. However, the Adjudicator is not given the power to consider an action of the Client so some may conclude that such a dispute is outside the jurisdiction of the Adjudicator. Here we have to remember that, on receipt of a termination notice, it is an action by the Project Manager that results in the termination certificate being issued (see clause 90.1 and Section 18.4). As the Adjudicator has the power to review and revise an action of the Project Manager this can be applied to the decision of the Project Manager to issue the termination certificate, thereby effectively considering whether the Client had a valid reason for notifying the termination in the first instance.

In exercising this power under clause W2.3(4), the Adjudicator can:

- take and use initiative in ascertaining the facts;
- take and use initiative in ascertaining the law;
- obtain information from either party by issuing directions and specifying the time for the provision of such information; and
- instruct either party to take any other action, such as opening up work or carrying out tests, within a specified time.

Using these powers, the Adjudicator considers the information submitted by the Parties in connection with the dispute and, in accordance with clause W2.3(8), reaches the decision within 28 days of the dispute having been referred. The time for making the decision may be unilaterally extended by the referring party by up to 14 days or longer at the agreement of both parties. If this decision involves assessing time or money, then clause W2.3(7) requires that any such assessment is carried out by the Adjudicator in accordance with the rules for assessing a compensation event (see clause 63 and Chapter 17).

Clause W2.3(7) also provides that should the Adjudicator's decision change any amount which had previously been notified as being due then any payment decided by the Adjudicator is due no later than seven days after the date of the decision. Clause W2.3(8) also allows the Adjudicator to decide how the Adjudicator's fees and expenses shall be allocated between the parties within the decision.

In the event that one or both of the parties do not provide information to the Adjudicator, or comply with any instruction which the Adjudicator is empowered to give within the time specified, then the Adjudicator is given the power by clause W2.3(5) to continue and make a decision based solely on the information that has been received. This clause does not comment on whether the Adjudicator should or should not consider any information which is received late before making the decision. In such circumstances the Adjudicator would have to consider matters of natural justice, a subject that has been considered numerous times by the UK courts in relation to adjudications carried out under the HGCRA96 (the study of which is outside the scope of this guide).

Clause W2.3(10) makes provision for circumstances where the Adjudicator does not make a decision within the time allowed. This provision allows the parties and the Adjudicator to agree to extend the period further. Practically, parties should be cautious of this clause, as it does not say when such an agreement can be made. It is suggested that, in order to comply with common law, then such an agreement would need to be made before the original time for making the decision had expired. Once the original time period expires, the Adjudicator automatically becomes *functus officio* and thereby loses all power to act in any way whatsoever, including the power to agree an extension of the time for making a decision. Without the power to act, any agreement by the Adjudicator would be worthless.

In accordance with clause W2.3(11), once made then the Adjudicator's decision is binding on the parties and could be given contractual and legal effect through the courts if necessary. Should the decision contain any typographical or clerical errors that have arisen either by accident or by omission, then clause W2.3(12) allows for the decision to be corrected within five days of the decision being given. Otherwise, unless either party notifies the other within four weeks of the decision (see clause W2.4(2)) that it intends to refer the matter to the tribunal, then the decision becomes final and binding.

Once the Adjudicator has notified the decision the Adjudicator has no further powers, except for the power to correct any typographical or clerical errors as referred to above, in connection with the dispute. Should the dispute be subsequently referred to the tribunal (see Section 19.3.8), neither party can call the Adjudicator as a witness in those proceedings (see clause W2.4(5)).

While this process is taking place, the Client, Contractor, Project Manager and Supervisor are required by clause W2.3(9) to continue as if nothing has changed. The contract continues to operate, despite actions having been taken that the Adjudicator's decision may revise.

19.3.8 Review by the tribunal

The tribunal is the body identified in Contract Data Part One by that name and which will have the power to make a final and legally binding decision in connection with any dispute that is referred to it. In the United Kingdom and most other jurisdictions, users will be restricted to a choice of either referring the matter to litigation, where the body concerned will be the courts, or referring the matter to arbitration, where the body concerned will be either a single arbitrator or a panel of arbitrators, depending on the rules in operation. Where arbitration is selected, the rules will either be those selected by the parties or imposed by the law of the country concerned. These details, as required by clause W2.4(4), are set out in Contract Data Part One.

Clause W2.4(1) prevents the reference of any dispute to the tribunal until it has firstly been referred to the Adjudicator. Should either party then wish to refer the matter to the tribunal, that party must notify the other of that intent within four weeks of the notification of the Adjudicator's decision (clause W2.4(2)). The clause refers to the dissatisfied party simply notifying the other party of its intent to refer the dispute to the tribunal. The clause does not require the actual referral to the tribunal to be made within four weeks or any other specified period. The dissatisfied party can then make the actual referral to the tribunal at any subsequent time, subject to any limitations imposed by the relevant law[4].

Once an actual referral is made to the tribunal, that body determines and settles the dispute in accordance with whatever rules or procedures apply to it. The power of the tribunal will be determined by the law, which in many cases will imply far more detailed rules into the contract than are set out in clause W2.4. When such a referral is made, either party is at liberty to introduce and present any submissions or evidence it wishes to make; it is not restricted to just those matters presented to the Adjudicator (see clause W2.4(3)).

19.4 Option W3

Option W3 is a new addition introduced as part of the fourth edition. Its primary dispute resolution methodology is the DAB, which reviews any potential disputes and assists the Parties in resolving the matter before it becomes a full-blown dispute. In the event that the dispute is not avoided by the use of the DAB, then any matter that has been referred to the DAB but not resolved can be referred to the tribunal.

The Option provides a mechanism that includes the formation of the DAB and then sets out what the parties and the DAB must do in respect of each potential dispute. The provisions regarding the use of the tribunal are essentially the same as for Options W1 and W2. The following sub-sections consider this option in the same way as used for Options W1 and W2 above.

19.4.1 Setting up the Dispute Avoidance Board

The first thing to be decided, by the Client, after deciding to use Option W3 is whether the DAB will consist of one or three members. The selection is recorded by a deletion in Contract Data Part One. If the decision is that it will be a three-member DAB, then each party, by an entry in the Contract Data, nominates one member of the DAB. In accordance with clause W3.1(1) the Parties then agree on the identity of the third member. If the Parties cannot agree on the identity of the third member, then either Party may apply to the DAB nominating authority, named in Contract Data Part One, to choose a member for them. In accordance with clause W3.1(4), this nomination is made within seven days of the application. The same process is used to nominate a replacement, should the previously chosen member no longer be available.

Both the ECC and the User Guide are silent about who nominates the single member on a one-member DAB. It would seem to be appropriate for the nominating authority

---

[4] In the UK the Limitation Act 1980 would apply requiring the referral within 6 years for contracts signed under hand or 12 years for contracts signed under seal or as a deed, both periods running from Completion of the works.

to make the selection in such circumstances. Should either party select the single DAB member, then there would always be a suspicion that the DAB would favour that Party, hence the suggestion that it would be appropriate for the nominating authority to make the choice.

As required by clause W3.1(2), the DAB member(s) are appointed using the NEC4 Dispute Resolution Service Contract current at the starting date (see Section 12.5.1).

When appointed, the DAB is required to act impartially (see clause W3.1(3)), hence the need for it to be independent. Where a three-member DAB is in place, the third member should chair the DAB (as confirmed in the User Guide) in order to act as the balancing factor between the Parties' nominated members. Clause W3.1(7) confirms in carrying out their actions the members of the DAB and any of their employees or agents have no liability to the Parties unless they do so in bad faith. This is a standard type provision for such individuals; the same position is found in Options W1 and W2 in relation to adjudicators.

19.4.2 Actions of the Dispute Avoidance Board

The role of the DAB is to advise and assist the Parties in connection with potential disputes in order to enable the Parties to resolve those matters before they become actual disputes (see clause W3.2(1)). There is a clear distinction between potential disputes and actual disputes in this requirement; the DAB does not review actual disputes. To prevent this situation from arising, the terms create the position whereby no dispute can be considered as anything other than being 'potential' until after the DAB has made a recommendation.

Clause W3.1(5) sets the scene for the DAB's actions by establishing that the DAB will make regular visits to the Site for the purpose of inspecting the progress and to become familiar with any potential disputes. The interval of the visits is set out in Contract Data Part One. The way that this entry is laid out suggests that the minimum interval would be monthly, but it could well be a multiple of months between visits. These visits should occur at the prescribed interval from the starting date (see Section 12.5.1) to the defects date (see Section 9.5), unless the Parties (not the DAB) agree that a scheduled visit is not required. To counterbalance the potential for cancellation, the clause also allows for additional visits by the DAB at the request of the Parties.

It is required by clause W3.1(6) that prior to every Site visit by the DAB, the parties propose an agenda for the visit. The Parties will need to send the agenda to the DAB before the visit and in sufficient time to allow the DAB to consider the proposal and decide what the final agenda should cover.

As to dealing with potential disputes, clause W3.2(3) requires that such a matter is notified to the other party and the Project Manager by the Party initiating the potential dispute. The dispute is then referred to the DAB between two and four weeks after the notification. That it is referred to the DAB is required by clause W3.2(2) and clause W3.2(1) requires the DAB to assist the Parties in resolving any such potential dispute before it becomes a dispute.

In order to assist the DAB the Parties are required by clause W3.2(4) to make copies of the contract, progress reports and any other material they consider to be relevant to the potential dispute to the DAB in advance of the Site visit. The DAB considers such information before they attend Site. Clause W3.2(5) requires the DAB to visit the Site and inspect the works. The length of the visit is not prescribed; it could be a day or longer, depending on what the Parties and the DAB consider to be necessary. Whilst at the Site, the DAB is required to review all potential disputes with the Parties with the aim of helping the Parties to settle those potential disputes before they become disputes. There is no prescribed manner in which the DAB should assist the Parties. The wording in the ECC suggests that this could be a form of assisted negotiation or mediation, or in the form of the Conciliation process that was included in some of the old Institution of Civil Engineers Conditions of Contract (albeit that the conciliation process was more formal). In addition, clause W3.2(6) allows the DAB to take the initiative when reviewing potential disputes. What the DAB can do is not restricted; the clause states that the DAB can ask for further information but this is only one step that might be employed.

Clause W3.2(5) also includes a requirement for the DAB to produce a note of the visit to the Site. It does not say who this note is distributed to. Given the purpose of the DAB and the advisory nature of the service it provides, it seems sensible that this note of each visit is issued to both Parties and the Project Manager.

More importantly, if the Parties have been unable to resolve any potential dispute by the end of the Site visit, then the DAB is also required (by clause W3.2(5)) to provide a recommendation for resolving any such unresolved potential dispute. Such a recommendation might be a suggestion of how the Parties seek and obtain a satisfactory settlement of the issues, or simply a statement of what the DAB considers the answer to be to that particular issue. A combination of these two approaches would satisfy the requirement of the process.

In the event that the Parties either resolve a potential dispute with the assistance of the DAB or whether the Parties agree to adopt the recommendation made by the DAB, then the matter is closed. If no such resolution of a potential dispute is achieved, then the next stage is to refer the matter to the tribunal.

19.4.3 Review by the tribunal

The tribunal is the body identified in Contract Data Part One by that name and which will have the power to make a final and legally binding decision in connection with any dispute that is referred to it. In the United Kingdom and most other jurisdictions, users will be restricted to a choice of either referring the matter to litigation, where the body concerned will be the courts, or referring the matter to arbitration, where the body concerned will be either a single arbitrator or a panel of arbitrators, depending on the rules in operation. Where arbitration is selected, the rules will either be those selected by the parties or imposed by the law of the country concerned. These details, as required by clause W3.3(4), are set out in Contract Data Part One.

Clause W3.3(1) prevents the reference of any dispute to the tribunal until it has firstly been referred to the DAB. Should either party then wish to refer the matter to the tribunal, that party must notify the other of that intent within four weeks of the notification of the DAB's recommendation (clause W3.3(2)). The clause refers to the dissatisfied party simply notifying the other party of its intent to refer the dispute to the tribunal. The clause does not require the actual referral to the tribunal to be made within four weeks or any other specified period. The dissatisfied party can then make the actual referral to the tribunal at any subsequent time, subject to any limitations imposed by the relevant law[5].

Once an actual referral is made to the tribunal, that body determines and settles the dispute in accordance with whatever rules or procedures apply to it. The power of the tribunal will be determined by the law, which in many cases will imply far more detailed rules into the contract than are set out in clause W3.3. When such a referral is made, either party is at liberty to introduce and present any submissions or evidence it wishes to make; it is not restricted to just those matters presented to the Adjudicator (see clause W3.3(3)).

19.5 Practical issues

19.5.1 A dispute resolution option is not selected

In the Schedule of Options at page 1 of the ECC, it states that one of the dispute resolution options must be chosen. The use of the word *must* makes the selection of one of Option W1, Option W2 or Option W3 mandatory (see Section 1.4). Similarly, in the first entry in Contract Data Part One where the Client is required to list the Options that will form the Contract, a space is provided for the chosen dispute resolution option to be entered. It might, therefore, be considered that, unless one of these three options is chosen and incorporated into the Contract, then any Contract formed without such an inclusion would be incomplete, inoperable and potentially unenforceable.

In practice, all the jurisdictions known to the author would act to provide a dispute resolution system should such provisions not be included in a contract. In the United Kingdom, for example, any contract that qualified as a 'construction contract' under the HGCRA96 would automatically include adjudication, as this Act implies such a system into any qualifying contract that does not include compliant provisions. In addition, all such contracts and any other contract which is outside the provisions of that Act would always be under the jurisdiction of the courts. Similar arrangements exist in many other countries that have introduced adjudication requirements; where such provisions do not apply, the courts are always available to disputing parties.

19.5.2 Continuing as if nothing is wrong

Prior to the issue of a decision by the Adjudicator, both clause W1.3(9) of Option W1 and clause W2.3(9) of Option W2 require the Parties and the Project Manager and Supervisor to continue with the performance of the contract and their duties and obligations under it as if the matters referred to the Adjudicator are not disputed. In the event that

---

[5] In the United Kingdom the Limitation Act 1980 would apply, requiring the referral within 6 years for contracts signed under hand or 12 years for contracts signed under seal or as a deed, both periods running from Completion of the works.

an adjudication is taking place before Completion (see definition at clause 11.2(2) and Section 12.5.2), those involved are, in practice, going to find it difficult to comply with this requirement.

The nature of adjudication is such that, in order to get the necessary points over in the short timescale allowed, the submissions made by each party to the Adjudicator are often likely to be of a direct and forthright nature. The language, tone and presentation of these submissions will not be in a style that would in any way act to promote the spirit of mutual trust and cooperation. Indeed, there is a real danger that the language used in submissions to an adjudicator could destroy even the best of working relationships that have been established prior to any adjudication being commenced. This being the case, there is good reason for the Client and Contractor to explore other ways of achieving the resolution of any difference before it escalates into a dispute.

This issue is of more concern to those operating under Option W1, as the timescale for notifying disputes and then referring the matter to the Adjudicator is only four weeks from the Senior Representatives discussions. Under Option W2, the ability to refer a dispute to the Adjudicator at any time allows the parties to take much longer over exploring ways to resolve their differences before the need to refer any matter to an adjudicator arises.

Sections 19.5.3 and 19.5.4 consider practical ways of preventing this escalation.

19.5.3 Identifying differences before they become disputes

A practical approach that parties can use to prevent differences escalating into disputes, while at the same time building the required spirit of mutual trust and cooperation, is to create an environment of openness in relation to such matters. Readers of this guide will no doubt recognise that the vast majority of differences between parties to construction contracts relate to time and/or money. An environment is created through the payment and compensation procedures within the ECC where the Contractor, by submitting payment applications at each assessment date and quotations for compensation events as required, should be regularly updating the Project Manager with what it considers to be the current value of work carried out (see Section 13.4) and the adjustments it considers should be made to the Prices and Completion Date (see Sections 16.5, 17.2 and 17.3). Depending on the Main Option in use, other similar mechanisms can also be in use. With respect to the payment procedure, the Project Manager is obliged to give details of the value certified to the Contractor for each assessment (see Section 13.2) and to reply to each quotation received for a compensation event (see Section 16.6.2).

By comparing the views of the Contractor and the Project Manager in respect of the payment assessments and the assessment of compensation events, the Contractor and the Client have a readymade indicator showing them what (if any) differences exist between them in respect of issues of time and/or money. By regularly reviewing any differences between the respective positions of the Project Manager and the Contractor, the parties will create an environment where they can discuss any such issues with the Project Manager and work towards resolving such matters without the need for recourse to the Adjudicator.

These discussions can lead to the process in any one of Options W1, W2 or W3 being triggered. With both W1 and W2 the Parties can refrain from instigating the processes by discussing the matter between themselves, whereas in Option W3 all such potential disputes should be notified to the DAB, who will then assist the Parties in reaching a settlement.

19.5.4 Resolving differences before they become formal disputes

Having identified differences as suggested above, the parties can then set up a forum that reviews such matters. The first step in any such process is normally to step the matters that require consideration up to a level of management just above those directly involved in the project. If the organisations concerned are large enough, it is possible to put in place more than one level of review above the project personnel. By stepping matters up in this way, the parties allow a less involved view of the issues that are creating differences. They often find that a sensible resolution can be found, which, when it is explained to the project personnel, prevents any damage to the spirit of mutual trust and cooperation.

It is also possible to introduce other resolution techniques such as dispute review boards or mediation, which can be used before invoking a formal procedure such as adjudication. If such methods are considered, it is often necessary to include additional provisions in Option Z to cater for the mechanisms that it is agreed shall be used.

The addition of the Senior Representatives to Options W1 and W2 in the fourth edition have made this practical approach into a formal part of the process. Notwithstanding those processes, given the timetables that in Option W1 in particular that have to be followed can force the parties down a much more formal route than they want to take in the early days of an issue. There is nothing wrong with being informal knowing that the formal processes can always be initiated by either Party at any time.

Where Option W3 is being used then early engagement with the DAB is key to providing the environment where their assistance can be beneficial.

19.5.5 Evolving common law

In the United Kingdom and other countries, which have introduced a statutory requirement to include adjudication procedures within construction contracts, the common or case law applicable to such procedures is having the effect of continually ironing out technical issues related to these statutory requirements. Parties who are operating in such jurisdictions should be aware of any changes required to the conditions of the ECC created by the evolving nature of such cases, and ensure that at the Contract Date (see clause 11.2(4) and Section 12.3) any such changes are set out in Option Z. In the event that the common law evolves further after the Contract Date and makes the provisions in a contract noncompliant with the law at the time a dispute is referred to the Adjudicator, the parties will have to accept that the default statutory provisions may apply in replacement of whatever they have included in the contract. If this situation arises, it cannot be avoided and should not in itself become a further area of dispute.

Chapter 20
Secondary Options

20.1 Introduction

This chapter collects together all of the Secondary Options that have not been included in a previous chapter where it was relevant to make reference to Secondary Options directly relevant to the core clause being considered. Hence, the contents of this chapter are not necessarily related to each other. That being the case, the Secondary Options considered here are simply listed in number order. The amount of comment against each Secondary Option is what is considered necessary and is in no way commensurate with the importance of that Secondary Option.

In order for any Secondary Option to be applicable to a particular project, it must be listed in the first entry in Contract Data Part One as one of the operative Options. Any Secondary Option not included in this list would not form part of the Contract.

20.2 X2: Changes in the law

The inclusion of this Secondary Option acts to allocate the risk of a change in the law of the country where the Site is located to the Client. If this Secondary Option is not used, then the risk would sit with the Contractor as there would no mechanism under which it could seek compensation for a change in the law.

The primary purpose of this Secondary Option is to create a compensation event as the mechanism by which the Contractor is compensated for any change in the relevant law. The measure by which the right is triggered is the occurrence of a change in the law after the Contract Date (see definition at clause 11.2(4) and Section 12.3). The way this clause is worded does not act to exclude any change in the law that was known of at the time of tender but which did not occur before the Contract Date. As it is unclear about who should notify this compensation event then it would seem essential that the Contractor should do so under clause 61.3 (see Section 16.3.1).

In the event that such a compensation event were to occur, then it is made clear that it is one of the compensation events that could result in a reduction of the Prices (see clause 63.3 and Section 17.2.6).

A Practical Guide to the NEC4 Engineering and Construction Contract, First Edition. Michael Rowlinson.
© 2019 John Wiley & Sons Ltd. Published 2019 by John Wiley & Sons Ltd.

20.3 X4: Ultimate holding company guarantee

This Secondary Option is only of use where the Contractor is a subsidiary within a larger group of companies. In such circumstances, this Secondary Option can be used by the Client to secure a guarantee of the Contractor's performance from its ultimate holding company. Users should be clear that this is the company at the top of the corporate structure and not the immediate or intermediate owner within that same corporate structure. Should this Secondary Option be used, then it is necessary for the Client to include the form of the ultimate holding company guarantee within the Scope (see definition at clause 11.2(16) and Section 22.2.28). The Client should note that there is no entry in Contract Data Part One that refers to this Secondary Option and therefore there is no reminder to include the reference.

In the event that an ultimate holding company guarantee is required, then this could be provided by the Contractor before the Contract Date (see definition at clause 11.2(4) and Section 12.3). If it is required but was not provided before the Contract Date, then the clause requires it to be provided within four weeks of the Contract Date. Should the Contractor fail to provide the guarantee within the required period, then the Client is given the discretionary right to terminate the Contract.[1]

Where this Secondary Option is in use, contractors should be aware that the wording of the clause is such that once the Contract is formed, there is no right to negotiate the wording of the guarantee set out in the Scope. The Contractor must therefore consider the wording before it enters into the Contract, and ensure that any changes that are required to that wording are made before the Contract is formed.

20.4 X12: Multiparty Collaboration

20.4.1 Introduction

The Secondary Option for Multiparty Collaboration is the longest of this type of additional clause that can be added to the core contract. It is strictly for use where the Promotor requires several Partners who are involved in the same overall project, who are not necessarily in contract with each other or even all in contract with the Promotor, to be brought together in an arrangement that provides for coordination and cooperation between the activities under the various contracts between the several Partners concerned. Such an option is not required where the intent is simply to enable the two parties to a contract to partner, as this is achieved by the use of a standard ECC without the need for any additional clauses. The Partners' rights under the contract between them and another party, each of which is covered by the general rule of privity of contract, are protected and all such contracts are referred to as an Own Contract (see clause X12.1(3)).

It is made abundantly clear by clause X12.2(6) that the use of this Secondary Option does nothing to establish any legal relationship between any of the Partners who are not party to the Own Contract. The way this clause is worded does not suggest that

---

[1] See Reason 12 at clause 91.2 (second bullet point) and Section 18.2.3.

any Partner outside of any of the Own Contracts brought under this overarching arrangement would, in the UK, gain any third party right to enforce any term of an Own Contract to which it was not a party. The use of this Secondary Option is compatible with the use of Secondary Option Y(UK)3 (see Section 20.11).

In order to make this Secondary Option work, four entries in Contract Data Part One of each Own Contract must be completed and a Core Group must be established. This Core Group will then decide on how they will work to achieve the Promotor's objectives.

20.4.2 Entries in Contract Data Part One

The Secondary Option requires four entries in Contract Data Part One. The first is the name and address of the Promotor. This may or may not be the Client under the ECC between the two parties. As several separate contracts related to the same overall project will all have this Secondary Option active, it is possible that more than one of the employers under those contracts will not be the Promotor. Even if the Client is the Promotor, the Promotor's identity must be entered in the appropriate place in Contract Data Part One.

The second entry requires information as to the location of the Schedule of Partners (see clause X12.1(2) and Section 20.4.4 below). This Schedule of Partners identifies all of the Partners at the time the contract is entered into and can be added to later when new partners are agreed upon.

The third entry requires the Promotor's objective for the project to be clearly set out. This does not require voluminous detail. It simply requires a concise and explicit statement as to what the Promotor wishes to achieve, preferably containing quantifiable objectives against which the performance of the Partners can be gauged.

The fourth entry requires the location of the Partnering Information, a document that is referred to further below. The meaning of this term is defined at clause X12.1(6), which in simple terms gives it the same type of standing as the Scope under the Own Contract, including the ability for it to be changed by an instruction given under the Own Contract.

20.4.3 Role of the Partnering Information

As the Partnering Information has to be identified in each of the Own Contracts before they come into existence, it is something that the Promotor would need to prepare and include in each of the Own Contracts to which it is a party. By including a requirement about which other Own Contracts that those it contracts with must include the Partnering provision, the Promotor can ensure that the Partnering Information is passed on to other future Partners and becomes embodied in their Own Contracts. For example, a Promotor may enter into, say, three separate contracts with contractors for different parts of an overall project. Each one of those would be an ECC with the Promotor as Client and the three contractors as the Contractor, one per contract. The Promotor may require that the subcontractors appointed by the contractors for a defined list of major trades are also Partners and includes this requirement in the documentation for each of the three contracts it enters into. The Contractor under each such contract must then include Secondary Option X12 in each relevant subcontract and, in this way, pass the Partnering Information to those subcontractors.

The role of the Partnering Information is to specify how the partners should collaborate with each other. It also sets out various requirements for shared facilities and ways of working. Examples of such requirements are included in the guidance given at Volume 2 of the User Guide (page 24). In basic terms, the types of things to be included would be requirements that would enhance the outcome of the project by facilitating more efficient communications and by removing duplication and inefficiency from all the operations on the project.

20.4.4 Identifying the Partners

All the Partners are identified by being named in a document created for the purpose and called the Schedule of Partners. This requirement is set out at clause X12.1(2) and its location is identified in the second entry relating to Option X12 in Contract Data Part Two. Initially, it would be created by the Promotor and shared with every party who is named as a Partner. However, what the Promotor can complete will not satisfy all of the requirements for the contents of this Schedule. The definition of this document at clause X12.1(2) includes a statement that the Schedule of Partners includes an entry as to the objectives of the Partners. Presumably, these objectives could be the business objectives of each Partner related to this project and its Own Contract. Guidance as to the contents of the Schedule of Partners is given at page 23 of Volume 2 of the User Guide and an example of the layout can be found at page 92 of Volume 4.

Clause X12.1(1) also states that the Promotor will always be a Partner. If the Promotor and the Client are not the same person, it would be normal to find that the Client is already a Partner as a result of the provisions of another contract.

20.4.5 Actions of the Partners

In order to facilitate its dealings with the other Partners, each of the partners is required to nominate its representative (see clause X12.2(2)) through which all communications will take place and who, in practice, would attend any meetings or other events required by the Partnering Information. In these dealings, each Partner is required by clause X12.2(1) to collaborate with the other Partners in order to achieve the objectives of the Promotor and those of every other Partner, as stated in the Schedule of Partners.

The Partners select a Core Group (see clause X12.2(4)) and the members of this Core Group are identified in another schedule, which in this instance is called the Schedule of Core Group Members (see clause X12.1(5)). The Promotor is not automatically a member of the Core Group, but it would be rare for the Promotor not to be a Core Group Member. The Promotor's representative chairs this Group.

When working together, clause X12.3(1) requires the Partners to collaborate with each other as set out in the Partnering Information. Again, the Partnering Information acts to protect the individual Partners by limiting the way every Partner can act. The Partners are also expressly required by this clause to work in the spirit of mutual trust and cooperation; this requirement is in addition to the same requirement under the Own Contract. Part of this working together requires each Partner, when asked by another Partner, to provide information that the asking Partner needs to carry out work in its Own Contract (see clause X12.3(2)). This requirement is, of course, consistent with the similar provision at clauses 25.1 and 27.1 (see Section 7.5.2) and seems superfluous as, providing that each Partner is subject to an NEC contract, then each Partner should already be obliged to share information in this way. Similarly, clause X12.3(3) brings into play an early warning requirement (see clause 15.1 and Section 6.4), which is included automatically in the Own Contracts where they are based on any of the NEC contracts.

A further extension of this requirement to cooperate is included at clause X12.3(8) where Partners are required to give advice and information to the Core Group and other Partners when asked to do so in relation to work that another Partner is to carry out. This obligation extends to openly sharing information about contingency, and risk allowances for both time and cost. The danger here is that commercially confidential information might be required to be shared which some Partners might refuse to do on the basis that it could adversely affect their market position. This Secondary Option does not consider this issue.

A useful addition to the way each Partner should be working under each of the Own Contracts is that the Partners are required to use common systems, i.e. computer systems, to share information between them (see clause X12.3(4)). It would be better to include this requirement in the Partnering Information, including the identity of the software to be used, so that all involved know what will be used when preparing their tenders.

In the event that the Core Group has made a decision that it is entitled to make in accordance with the Partnering Information (see clause X12.3(6)), then each Partner is required, by clause X12.3(5), to issue the necessary instructions under all Own Contracts in which it is involved. Not all Partners will be required to issue instructions; only those that have the power under an Own Contract will be able to initiate this action. Some Partners will be required to issue instructions under many Own Contracts, e.g. a contractor to each of its subcontractors where those subcontracts are classified as Own Contracts.

In respect of subcontractors, each Partner is also required to notify the Core Group of its intentions before actually subcontracting any work. This provision is set out at clause X12.3(9) and, except for imposing the obligation to notify the Core Group, it does not actually impose any further obligation or requirement. What users should be aware of is whether there is anything referring to subcontracting set out in the Partnering Information. Assuming that there is no such requirement, a Partner working under the ECC who intends to subcontract will be bound by the provisions of clause 26 (see Chapter 8) of its Own Contract in that respect.

20.4.6 Actions of the Core Group

Once selected by the Partners and established, the Core Group becomes the body that is responsible for not only deciding when Partners join or leave that Core Group but also how the Core Group itself works in achieving the objectives set (see clause X12.2(4)).

Any such changes are recorded in the Schedule of Core Group Members and Schedule of Partners, revised copies of both being issued to all Partners whenever there is a revision of any type (see clause X12.2(5)).

Clause X12.2(3) acts to restrict the power of the Core Group to those matters that are set out in the Partnering Information. This restriction gives comfort to the Partners, in particular those who are not members of the Core Group. These Partners will be aware of the matters that are subject to this influence at the time they prepare their individual tenders for the work that will be carried out under the particular Own Contract providing, of course, that the tender is successful and a contract is formed between them and the party who invited the tender. It also follows that in practice, when tendering, potential Partners must take account of not only the management time they will need in order to perform their part of the arrangement (including possible membership of the Core Group) but also of any matter that might arise from this relationship which would not constitute a compensation event.

The Core Group has two primary powers that it can exercise in order to allow it to work towards achieving the Promotor's objectives. The effect of these primary powers is that the Partners so affected can be required to issue instructions (see clause X12.3(5)) under their Own Contracts, the effect of which may be to create compensation events under those Own Contracts. In a complicated web of contracts all linked together by the use of this Secondary Option, this can create a situation, in practice, where inappropriate use of these powers can result in change; this change may in itself cause complications under individual contracts. These powers therefore need to be applied with caution and thought. That the Promotor's representative leads the Core Group at least creates a position whereby the Promotor, who ultimately is likely to be the party who pays for the effect of any such change, could veto the change if it so desired and could not deny knowledge of the change, hence reducing the potential impacts of such powers.

These primary powers are set out at clause X12.3(6), where the Core Group is given the discretion to change the Partnering Information, and at clause X12.3(7), which obliges the Core Group to produce a timetable showing the timing of the contributions of the Partners and then gives them the power to change that timetable. This timetable is issued to each Partner every time it is revised. Should a change to the timetable require the programme prepared under an Own Contract to be changed, then such a change also constitutes a compensation event. One danger of this mechanism in practice is that the change to the timetable might actually require acceleration (see clause 36 and Section 12.6.4) under an Own Contract. This requirement could create a clash between the requirements of the Core Group and the Own Contract, as the Contractor under the ECC could, as allowed by clause 36.1, refuse to consider the acceleration required. Should a Contractor make such an election, then the change to the partnering timetable would not be possible. The ECC does not set an order of precedence between these two conflicting terms.

20.4.7 Incentives

Secondary Option X12 includes its own incentive mechanism. For this reason, Secondary Option X20 must not be used in conjunction with X12, as it would cause duplication of like conditions. The Key Performance Indicators (KPIs) against which a Partner's performance is measured are set out in the Schedule of Partners rather than a separate schedule (see clause X12.4(1)). Accompanying each KPI is a stated sum of money which is paid to the Partner if the target set is achieved or improved upon by that Partner. The payment is to be made as a payment under the Partner's Own Contract, which will mean that it becomes one of the other amounts to be paid to the Contractor that form part of the amount due (see second bullet point of clause 50.3 and Section 13.4.2). In practice, unless the Promotor is also the Client, there will need to be an associated mechanism between the Promotor and the Client in order that the KPI payment earned by the Contractor can be paid to the Client.

A facility is included (see clause X12.4(2)) that allows the Promotor to add further KPIs and incentive payments to the Schedule of Partners should it so desire. This mechanism only works one way, as the Promotor is prevented from either reducing or removing KPIs that are already in the Schedule. In practice, this mechanism can be used to provide extra incentives for certain Partners to improve their performance if that will be for the good of the whole project and to the Promotor's advantage.

20.5 X13: Performance bond

The incorporation of this Secondary Option allows the creation of an insurance or bank-backed bond to be put in place that will provide the Client with some protection should specified circumstances occur. These specified circumstances are not set out in the clause but will form part of the wording of the required bond, which itself must be included in the Scope. Contractors, when pricing for the provision of such a requirement, must therefore pay attention to the wording of the bond to determine under what circumstances the bond may be called by the Client. How such circumstances are set out will, of course, determine the cost of the providing the bond. Contractors should note that there is no provision in the clause for any alternative wording to be negotiated; if a contractor requires alternative wording to that included in the Scope, it needs to complete such negotiations before entering into the Contract otherwise the opportunity will have been lost.

The sum that must be covered by the bond is set out in Contract Data Part One. The form of the this statement, i.e. a fixed sum or a percentage of say the tendered total of the Prices, is not prescribed, leaving it up to the Client to select when preparing the invitation to tender. The bond is to be provided by the Contractor either before the Contract Date (see definition at clause 11.2(4) and Section 12.3) or no later than four weeks after that date. The suitability of the bank or insurer is to be verified by the Project Manager rather than the Client and, to this end, the Project Manager is given a reason for rejecting the proposed provider of the bond.

It is common to see this Secondary Option used at the same time as Secondary Option X4.[2] Many contracting organisations see the requirement to provide both as a 'belt and braces' approach, and increasing numbers of them are refusing to enter contracts that

---

[2] Ultimate Holding Company Guarantee; see Section 20.3.

require both to be provided. There are sound commercial and accounting reasons for contracting organisations not to want both of these potential liabilities on their books in relation to the same contract. Clients should consider carefully what merit (if any) there is in having both forms of protection in place at the same time. Clients should realise that there is a financial implication that is routinely included in contracting organisations' pricing structure to cover the cost of providing and administering these types of guarantees when considering their requirements.

20.6 X17: Low performance damages

The use of this Secondary Option allows the Client to recover damages caused by work carried out by the Contractor not achieving a specified performance level as liquidated damages, rather than the assessment required from the Project Manager under clauses 45.2 or 46. The option can only be used where the Scope sets a performance requirement for part of the works. The Contractor is given the same opportunity to correct any Defect relating to performance within its defect correction period (see clauses 44.1 and 44.2 and Section 9.5). Only if the Defect has not been corrected in the required time, and as a result is recorded on the Defects Certificate (see clauses 11.2(7) and 44.3 and Section 9.6), will the Client's right to deduct the low performance damages arise. The amount of damages to be paid is then calculated by applying the scale set out in Contract Data Part One. This scale needs to be inserted prior to issuing the invitation to tender and must represent a reasonable pre-estimate of the damages that will be suffered if the performance levels are not achieved. In the UK, it is not permissible to make this provision a penalty, as such clauses are not allowed by the law.

In practice, this provision is sometimes used as part of projects where the Contractor is providing some type of process, and usually where the Contractor is designing either the process or key parts of the installation which have a direct bearing on the success of the process in use. Examples of market sectors where damages can be related to performance include:

- the amount of chemical required to treat water in a water purification plant or system related to the quality of the input and/or output;
- the amount of power required to operate a particular process; or
- the output from a power generation facility.

20.7 X18: Limitation of liability

The relevance and use of this Secondary Option will vary from one jurisdiction to another around the world. Its use allows the parties to cap the Contractor's liability to the Client in respect of a number of matters which are set out within the clause. Each of the liabilities that can be capped requires an entry in Contract Data Part One that sets the level of the cap. As with most entries in the Contract Data, there is no default mechanism provided thereby creating flexibility. The three basic options will always be 'nil', a 'stated sum' or 'unlimited'. Between these three, the Client can vary the level of the Contractor's liability across the factors to be considered. There is no limitation on how these basic options can be mixed together. Should any of the entries be left blank then clause X18.1 determines that there will be no limit applicable to that entry.

As it is generally the Client that selects the Secondary Options to be used contractors must be aware that, unless the option has been included and where the entries in Contract Data Part One do not act to limit the Contractor's liability, then it will be either unlimited or nil in respect of all matters. In the UK, the limitation of liability by contractors is generally sought where they are carrying the responsibility for some aspect of design, especially where that design is related to a process of some kind. Elsewhere in the world, the need will vary depending on the legal framework that will be applicable to the project. Users should consider taking appropriate legal advice to establish what their liabilities might be and what limits may need to be applied if they are unfamiliar with the relevant law.

As this guide is not meant to be a guide to the law, the various liabilities covered by this secondary option are simply listed below. No attempt has been made to explain the legal need or implication of any of these factors. Further information of this type is contained in Volume 2 of the User Guide (NEC Panel 2017c), but they also recommend taking appropriate legal advice specific to the project. The aspects covered by the clause which are all stated as being the Contractor's liability to the Client are:

- for the Client's indirect or consequential loss (clause X18.2);
- for loss or damage to the Client's property (clause X18.3);
- for Defects resulting from the Contractor's design and which the Supervisor has not listed on the Defects Certificate (clause X18.4); and
- for all matters, except for the exclusions listed, in relation to liabilities in contract, tort and delict, or the equivalent or allowed under the relevant law (clause X18.5).

The period for which the Contractor remains liable to the Client can also be limited by an entry in Contract Data Part One as referred to by clause X18.6.

20.8 X20: Key Performance Indicators

The adoption of this Secondary Option provides a mechanism which the Client can use to incentivise the Contractor to improve its performance in specified areas. In order to do this under clause X20.1 the Client creates an Incentive Schedule, the location of which is identified in Contract Data Part One. Within the Incentive Schedule, the Client lists a target of performance for each such specified area and states a sum to be paid to the Contractor should it achieve or better the target stated (see clause X20.4). The Client has the power under clause X20.5 to be able to add to the Incentive Schedule any further KPIs that it chooses to add, together with a related incentive payment. This power does not allow the Client to deduct an existing KPI from the Incentive Schedule.

This option requires the Contractor to report both its performance to date and forecast final performance against each of the KPIs from the starting date (see Section 12.5.1) until the Defects Certificate has been issued (see clause 44.3 and Section 9.5). These reports are to be submitted at the interval stated in Contract Data Part One (see clause X20.2). In the event that one of these reports shows that the Contractor will not achieve the target for any KPI, then the Contractor is required by clause X20.3 to inform the Project Manager of how it intends to achieve an improvement in its performance. This requirement does not impose an obligation on the Contractor to improve its performance to a level where the target will be achieved. In addition, it does not require the Project Manager to accept the proposal from the Contractor about how it intends to achieve the required performance or give the Project Manager any reasons why the proposal can be rejected. The effect of this mechanism is therefore that there is no breach by the Contractor should it fail to achieve any of the targets set in the Incentive Schedule. All that happens is that the Contractor does not receive the incentive payment.

20.9 X21: Whole Life Cost

The opening sub-clause (X21.1) to this new Option identifies what it is all about, which is the Contractor making proposals to change the Scope so that the cost of operating and maintaining an asset will be reduced over its life. This could, of course, involve an increased initial capital cost. The clause allows the Contractor to make such a proposal to the Project Manager. This is a different type of proposal to that encouraged under clause 16.1 (see Section 6.6.3).

Only if the Project Manager is prepared to consider the change is the Contractor required to provide a quotation. It is suggested that the Project Manager will probably need to consult with the Client before replying to the Contractor. Clause X21.2 sets out that the quotation should not only consider the proposed change to the Prices (fourth bullet point) and Programme (fifth bullet point) under the ECC between the Contractor and the Client, but also a detailed description of the change (first bullet point), a forecast of the reduced cost to the Client over the lifetime of the asset (second bullet point) and an analysis of the risks to the Client that would result from this change (third bullet point).

Clause X21.3 requires the Project Manager to consult with the Client about the quotation. These consultations need to be concluded in sufficient time to allow the Project Manager to reply to the quotation within the period for reply (see clause 13.3 and Section 6.2). When making the reply, the Project Manager either accepts the quote or gives reasons for not accepting it. The reasons that can be given are unrestricted.

If the quotation is accepted, then clause X21.5 requires the Project Manager to change the Scope, the Prices, the Completion Date and any Key Dates as the quotation, and to accept the revised Programme. This change to the Scope does not constitute a compensation event as it is an agreed change to the relevant factors.

Alternatively, if the quotation is not accepted, then clause X21.4 expressly states that the Project Manager does not change the Scope as proposed by the Contractor. This clause is to clearly prevent the Client from taking advantage of the Contractor's ideas without giving the Contractor any reward for its initiative. Because of this express restriction, should the Scope be changed subsequently in the way proposed by the Contractor, the Client would be in breach of the Contract and a compensation event would arise under clause 60.1(18) (see Section 15.3.18), thereby facilitating compensation for the Contractor. There would also be a compensation event under clause 60.1(1) (see Section 15.3.1) for the actual change to the Scope, but other factors could be included in the 'breach' event.

20.10 X22: Early Contractor Involvement

20.10.1 Introduction

The concept of Early Contractor Involvement has been around for quite a time. It was referred to in the Banwell Report (*The Placing and Management of Contracts for Building and Civil Engineering Work*) published in 1964 and has featured in subsequent reports and learned discussions.

In the United Kingdom, the approach has been used widely by Highways England (formerly The Highways Agency) amongst others. Case studies on such projects tend to show that they are more likely to be finished on time, within the budget, with fewer defects and with higher client satisfaction.

In 2009 a book entitled *Early Contractor Involvement in Building Procurement: Contracts, Partnering and Project Management*, written by David Mosey, was published within the same overall series as this book.

The intent of the approach is to get the contractor's early input into construction methods, risks, timing and the 'buildability' of the scheme. By using the contractor's knowledge and experience in all these areas and then using that input to refine the design and programme, parties are able to deliver a more economical scheme than is often achieved by more traditional methods. The use of Option X22 is for those who are attempting to seek such advantages.

The use of Option X22 is limited to those contracts that are being operated under either Main Option C or Main Option E (see Section 3.5).

The basic premise with this Option is that the Contractor will get paid on a time charge basis for the work in Stage One and then will carry out Stage Two under either a target cost arrangement with an Activity Schedule (see Section 13.5.1), which is Main Option C or on a cost reimbursable basis, which is Main Option E.

20.10.2 Definitions

The first two definitions relate to monetary aspects; they also interact with each other, as will be explained later. The first of these is the term 'Budget', which clause X22.1(1) tells us is the items and associated amounts of money as stated in the Contract Data. The definition also identifies that the Budget can be changed later in the process. The second one is 'Project Cost' (X22.1(2)), which is the total paid by both the Client (see Section 5.2) and Others (see clause 11.2(12) and Section 5.11) for items within the Budget.

The third definition (X22.1(3)) does not provide a clear definition. Instead, it refers the user to the Scope where the meaning of 'Stage One' and 'Stage Two' will be set out. This approach, of using the Scope for the detail of the definition, allows the Client to set out what it wants to be included in each Stage rather than being tied to some predetermined allocation that may not suit the project requirements. It does, of course, mean that users of Option X22 will need to have more input in order to define what is going to be included in each Stage. There is some guidance as to what goes in the Scope at pages 57 and 58 of Volume 2 of the User Guide, but it must be said that it is not very helpful in respect of the Stages definitions.

The fourth and final definition is of the term 'Pricing Information'. This is stated to be the information that specifies how the Contractor prepares the assessment of the Prices for Stage Two. The location of this information is given in Contract Data Part Two. This information will often be part of the tender submitted by competing contractors in order for one to be selected as the preferred contractor, which is followed by the appointment under the ECC. This is a common factor in two-stage tendering processes.

20.10.3 Forecasts

The preparation, submission and acceptance of forecasts in relation to both Stages are an important function within the Option.

For Stage One, the Contractor is required to prepare and submit forecasts of the work to be done in that stage. The forecasts must be detailed. They are required to be submitted at the intervals stated in Contract Data Part One from the starting date (see Section 12.5.1) until the issue if a notice to proceed to Stage Two (see below). These requirements are all set out in clause X22.2(1).

In clause X22.2(2), the Project Manager is required to either accept the forecast submitted or notify the reasons for not accepting it. Two reasons for not accepting are included in the clause, these being that it does not comply with the Scope or includes work not required in Stage One. If the Project Manager does not accept a forecast, then clause X22.2(3) requires the Contractor to make a revised submission of the rejected forecast taking into account the Project Manager's reasons. If this is done correctly and all the reasons are addressed, then the revised forecast, by implication, should be accepted. However, if the revised forecast does not adequately cover the reasons given, or changes another aspect, then it could be rejected for a second time.

The provision and accuracy of these forecasts is directly relevant to the payment that will be made to the Contractor for the Stage One work. This is because clause X22.2(4) determines that the cost of any work that is not included in the accepted forecast is treated as Disallowed Cost.

In a similar fashion, clause X22.2(5) requires a separate forecast in respect of Project Cost (see definition at clause X22.1(2)). This forecast is also prepared by the Contractor, but this time in consultation with the Project Manager. As it is prepared together, there is no provision for acceptance or nonacceptance. Each forecast prepared is to be accompanied with an explanation of the changes from the previous forecast. The forecasts are also required to be submitted at the interval stated in Contract Data Part One, from the starting date (see Section 12.5.1) until Completions of the whole of the Works (see clause 30.2 and Section 12.5.2).

20.10.4 Proposals for and Moving to Stage Two

The aim of Stage One is to carry out design work and to develop the price for Stage Two, all so that the Client can decide whether to proceed to Stage Two, either with the Contractor as the builder or with another contractor or not at all. A major part of Option X22 is aimed at facilitating this exercise and transition.

Clause X22.3(1) starts the process by requiring the Contractor to submit its design proposals for Stage Two. These are submitted to the Project Manager for acceptance, all in accordance with the requirements for such a submission set out in the Scope. As clause X22.3(8) says that the Contractor completes any outstanding design during Stage Two, it is clear that it is not expected that the design will be completed during Stage One. This is consistent with the reference to 'design proposals', as this term is synonymous with several of the standard stages in the recognised and published work flow systems from professional bodies such as the Royal Institute of British Architects and other design discipline professional bodies. So Stage One could require the design to be advanced to a stage, as set out in the Scope, leaving the development and detailing of that design to something that can be built in Stage Two.

To use common parlance from the design and build market, the above equates to the Contractor developing the employer's requirements in Stage One and then the contractor's proposals and design are developed in Stage Two.

The submission of the Contractor's design proposals is accompanied by a forecast of any change to either the Project Cost or the Accepted Programme (clause X22.3(2)). Clause X22.3(3) gives the Project Manager three reasons for not accepting the design proposal submissions, these being:

- the submission does not comply with the Scope;
- the Client will be liable for unnecessary costs to Others; and
- that the Project Manager is not satisfied that the assessment of the Prices or any change thereto has been carried out correctly.

In the event that the submission is not accepted, then clause X22.3(4) requires that the Contractor resubmits the design proposal, taking account of the reason or reasons for it not being accepted previously.

Whilst carrying out the Stage One design tasks, the Contractor also seeks and obtains the approval or consent from any Others (see Section 5.11) as stated in the Scope (clause X22.3(6)). This could involve a wide range of issues such as planning permission, building control approval, party wall issues, access issues from local transport authorities and any other such matter that has been identified in the Scope to be obtained by the Contractor.

Clause X22.3(7) determines that any new Scope that is created by the Contractor during Stage One is classified as being Scope provided by the Contractor for its design. This provision acts to leave responsibility for all such Scope with the Contractor. The Contractor will, of course, require the protection of Option X15 (see Section 11.3) in order to limit its liability for design to reasonable skill and care and also to introduce the provision of professional indemnity insurance. No such protection or requirement is included within Option X22.

Turning to the financial aspects of the process clause X22.3(5) establishes that the Contractor uses the Pricing Information to assess the total of the Prices for Stage Two. If the Main Option to be used is C (but not E) then clause X22.3(9) the assessment of the Prices for Stage Two is presented in the form of revisions to the Activity Schedule (see Section 13.5.1). The Activity Schedule will always have been part of the ECC and will include the Stage One work as well as that to be carried out in Stage Two.

Once all of the Stage One processes have been carried out it becomes time to consider whether to proceed to Stage Two. This is dealt with in a clear manner in clause X22.5. Clause X22.5(1) sets out four requirements, all of which must be satisfied in order for the Project Manager to issue the notice to proceed. These four requirements are that:

- the Contractor has satisfied the requirement in clause X22.3(6) to obtain approvals and consents from Others;
- any changes to the Budget has been agreed by the Project Manager or if not agreed then assessed by the Project Manager;
- the total of the prices for Stage Two have been agreed by the Contractor and the Project Manager; and
- the Client has confirmed that it wishes to proceed with the works.

Should any of the above factors be absent, then the notice to proceed cannot be given.

In the event that the notice to proceed is not issued, for whichever reason, then clause X22.5(2) requires the Project Manager to issue an instruction which removes all of the work required in Stage Two from the Scope. The clause states that this instruction is not a compensation event, thereby removing any claim from the Contractor for any loss of profit or the like. In addition clause X22.5(3) allows the Client to appoint another contractor to carry out and complete Stage Two, but only if the reason that the notice to proceed was because of one of two reasons. The first of these is a failure by the Project Manager and the Contractor to agree the total of the Prices for Stage Two, which makes sense and requires no explanation. The second reason is that the Contractor failed to achieve the 'performance requirements' as stated in the Scope. I have emphasised the phrase 'performance requirements' because it is not a phrase that is used elsewhere in Option X22; so what does it mean? Volume 4 of the User Guide (NEC Panel 2017e) gives little help as it simply says the same thing but in different words. In the author's opinion, this must be referring to something along the lines of standards required by the Scope related to the various obligations of the Contractor set out, in the main, in clauses X22.2 and X22.3. For example, if the Scope requires that one of the approvals or consents is to obtain planning permission a standard which could be required is that there must be no more than, say, 10 conditions attached to the permission that is obtained. A failure to acquire planning position with 10 or less conditions would be failure of performance. There could, of course, be numerous different performance requirements for various aspects of the Contractor's obligations. Such performance requirements should always be objective, as these are then capable of measurement, rather than subjective.

20.10.5 Other matters

There are several other provisions in this Option that require comment and that have been included in this sub-section rather than creating a heading for each one.

The first one occurs at clause X22.4, which determines that the Contractor shall not remove and replace any key person (see Section 7.5.1) during Stage One unless either the Project Manager instructs the replacement (see clause 24.2 and Section 7.5.1) or the person concerned is no longer able to act in connection with the Contract. This is a sensible provision and can make sure that any key person who has influenced the Client's decision to enter into contract with the Contractor is retained for the key Stage of the process. The importance of such key people was recognised in the judgement in the UK courts in *Fitzroy Robinson Ltd v Mentmore Towers Ltd* [2009] EWHC 1552 (TCC). In this case, the design consultant had been appointed on the basis of the vision and involvement of one of the consultant's partners. The consultant had known that the involvement of the partner concerned was key to their appointment. The consultant also knew at that time that the partner concerned had resigned and would be leaving the practice. These circumstances were sufficient for the court to find that the consultant had been guilty of fraudulent misrepresentation. The second of the reasons that allows the replacement of a key person is necessary as there can be a number of genuine reasons why somebody might need to be replaced.

The second matter concerns changes to the Budget. Clause X22.6(1) sets out two distinct events which, if they occur, require the Project Manager and the Contractor to discuss different ways of dealing with changes to the Budget. Such changes are required to be practicable. The first of the two events arises if the Project Manager instructs a change in the Client's requirements stated in the Scope. This is referring specifically to a Client requirement stated in the Scope rather than any part of the Scope which may not be expressed as a Client requirement. Under the ECC, the Project Manager is always entitled to change the Scope (see clause 14.3 and Section 5.3), and this normally creates a compensation event under clause 60.1(1) (see Section 15.3.1). Whilst this clause does not say that such a instruction is not a compensation event, it is talking specifically about changing the Budget rather than the Prices, the Completion Date or any Key Date. It is suggested that these three factors may also need to be change but via the compensation event mechanism (see Chapters 15–17). The second event refers to the occurrence of additional events which are listed in Contract Data Part One. This class of occurrences will be very clear, as they will have to be included in that list. If not in that list, then they would more than likely fall within the first reason.

Clause X22.6(2) requires the Project Manager and the Contractor to assess and agree the change to the Budget within four weeks of the event which gives rise to such a change occurring. It also provides that if they cannot agree to the change, then the Project Manager makes the assessment and notifies the Contractor of it. Should this occur, the Contractor could take the assessment to dispute resolution under the relevant Option (see Chapter 19).

The final matter to consider is the provision at clause X22.7 for an incentive payment. Clause X22.7(1) determines that if the Project Cost (see Section 20.10.2) is less than the Budget (see Section 20.10.2), then the Contractor is paid the incentive payment. This payment is determined by applying the relevant percentage stated in Contract Data Part One to the difference between the two amounts. If Option C is the Main Option this calculation will be done after the addition or reduction, as the case may be, under the provision for the Contractor's Share at clauses 54.1 to 54.4 (see Section 13.7). Just as under clause 54 the Project Manager makes this calculation twice, once at Completion (as clause X22.7(2)) and once (as clause X22.7(3)) for inclusion in the final payment (see clause 53 and Section 13.6).

20.11 Y(UK)3: The Contracts (Rights of Third Parties) Act 1999

This is one of the three Secondary Options for use in the United Kingdom only (the other two being Y(UK)1 (see Section 13.10.4) and Y(UK)2 (see Section 13.9)) and is provided in order to allow the parties to operate a feature of The Contracts (Rights of Third Parties Act) 1999. The statute allows parties to opt out of its provisions, which if allowed to act would give the right to third parties to enforce terms of the Contract between the Client and the Contractor under certain conditions. By using this Secondary Option, clause Y3.2 blocks that right for any party not identified as a beneficiary in the Contract.

By making an entry in Contract Data Part One, the Client identifies those parties or class of persons or organisations who will qualify as beneficiaries and what terms of the Contract between the Parties each beneficiary will be able to enforce. This right to enforce an identified term is established by the application of clause Y3.1.

Wherever the entry in Contract Data Part One identifies a class of person or organisation rather than an individual person or organisation as the beneficiary, then clause Y3.3 requires that the Client notifies the Contractor of the identity of the actual beneficiary as soon as it is known. There could, of course, be several such beneficiaries within a class (e.g. a class identified as 'future tenants' in a multi-unit commercial development).

20.12 Z: Additional conditions of contract

Secondary Option Z is probably the most commonly used of the secondary options available. It could also be considered to be either the easiest or most difficult to write about. This is because it is simply a blank page upon which the user sets out any additional conditions that they wish to add to the ECC. In practice, this usually includes any changes to the existing core clauses as well as any completely new conditions to be incorporated in the contract. The key to the successful use of Option Z is to ensure that its contents are clear and preferably written in a style that is consistent with the clarity and simplicity (see Section 2.10) adopted by the drafting body. Other than that, it would appear that anything could be included.

However, users making amendments should always remember that the conditions are the vehicle for setting out the rights and obligations of the parties. They are not the vehicle for setting out detail of what the Contractor is required to do or for imposing constraints on how the work can be done. The vehicle for setting out such matters is the Scope (see definition at clause 11.2(16) and Section 22.2), and care should be taken to use this document and the conditions in the way intended. By keeping matters intended for the Scope in that document users will assist the application of the other conditions that make reference to the Scope, thereby preserving the integrity of the structure created within the core clauses.

Of historical note, in April 2007 the UK's Office of Government Commerce (OGC) issued a set of standard Z clauses for use by the public sector. Clause Z1 covered 'Official Secrets and Confidentiality' and clause Z2 'Security', both of which are accompanied by (somewhat brief) Guidance Notes. As well as for the ECC, versions of the Z clauses were also made available for the Engineering and Construction Short Contract, the Professional Services Contract and the Term Service Contract.

In June 2008 the NEC panel prepared and published a further Z clause for use with the OGC's model 'Fair Payment Charter'. This new clause was called 'Z3: Project Bank Account' and came with additional entries for Contract Data Parts One and Two, a Trust Deed, a Joining Deed, Guidance Notes and a Flow Chart.

These OGC Z clauses were available for download from the OGC website;[3] however, this has now been closed down and is stored in the National Archive, which whilst being accessible electronically over the internet does not easily give up its secrets.

The Project Bank Account Z Clause is, however, no longer necessary, as Option Y(UK)1 (see Section 13.10.4) now provides the necessary terms and conditions, together with Trust Deed and Joining Deed for the operation of such a facility. UK Government frameworks continue to include bespoke Z clauses for 'Official Secrets', 'Confidentiality' and 'Security', together with numerous other subjects.

20.13 Practical issues

20.13.1 Selection of Options

The Schedule of Options, which is set out at the first page of the ECC, details that certain Secondary Options should not be used either with certain main options or in combination with each other. There is nothing complicated about the combinations, and the reasons for most of the prohibitions are immediately obvious. In practice, users simply need to be aware of these restrictions and ensure that they do not create a conflict by ignoring them.

20.13.2 Electronic changes

From around 2007, initially in connection with the third edition and continuing with the fourth edition, there has been an increasing trend of employers, or their advisors acting

---

[3] www.ogc.gov.uk/About_OGC_news_7045.asp.

for them, making changes to the core clauses electronically within the text of the ECC. This appears to be facilitated by the increasing use of electronic versions of the document being acquired under licence. There is nothing wrong with this approach in principle and, indeed, it makes the practical use of the document far easier as all the changes that otherwise would be set out in Secondary Option Z (see Section 20.12) are plain to see within the clause that they apply to. This approach also means that the user will not forget about any such changes when applying the terms and conditions in practice.

However, what has on occasion been lacking in practice is any clear statement within the tender documentation that the copy of the contract provided, as part of the papers issued by the Client, has in fact been amended. This has resulted in several contractors tendering on what they believed to be an un-amended ECC when what they were actually submitting a price against was a heavily and often onerously amended contract, presenting a completely different level of risk than that which the contractor believed would apply. The discovery of such a position after a contract has been formed is not good ground upon which to build a spirit of mutual trust and cooperation. Another common factor in such tenders was that the changes that had been made to the standard ECC were not separately identified within the tender documents.

In the event that employers amend the ECC using electronic facilities, then, it is suggested that it would be good practice and show an intent from the outset to be open (thereby demonstrating a spirit of mutual trust and cooperation) to clearly state that the contract provided with the tender is based on the ECC and incorporates amendments. Provision of either a list setting out the amendments made or the provision of a version showing the changes (either simply using a standard software 'tracked changes' type function or the more sophisticated tools often referred to as giving a 'delta view') removes any doubt about what is being presented to the tendering contractors.

From the tendering contractor's point of view, they either need to ensure that every copy of an ECC presented to them as part of a tender is clearly annotated, as suggested above, or that they check by one means or another whether there are any amendments in the document. Only by taking these steps can they be fully sure about the basis on which they are being asked to provide a price.

Chapter 21
Completing the Contract Data

21.1 Introduction

Throughout the previous chapters, numerous references have been made to the Contract Data and entries that will either need to be made or referred to within it. The purpose of this chapter is to review those entries and, for each one in the blank version provided at the end of the black-covered edition of the ECC, to provide some practical guidance on what should be inserted and/or a cross-reference to the clause or clauses within the contract which rely on the data concerned. The green-covered versions of the ECC, which refer to a single Main Option, will not include all the entries referred to in this chapter, but all entries in those versions are covered by this chapter. Where the substance of the entry has been provided elsewhere in this guide, then a reference to the section where that comment has been made will be given instead of repeating what has been written previously.

Where an entry is reasonably self-explanatory, little guidance will be given. If there is complexity of some degree or associated potential pitfalls or complications, a greater degree of comment will be made unless covered elsewhere.

21.2 Purpose and form of the Contract Data

It is the contents of the Contract Data which provide all the detail that is required in order to allow the provisions contained in the core clauses and options to be operated. In other standard forms of construction contract, the part of the documentation that contains such information is often labelled as the 'Appendix', 'Contract Particulars' or something similar. The title is not actually significant as long as the clauses that rely on the information use the same title as is used in the documentation. As stated above, in the ECC this is achieved by references throughout the clauses to the Contract Data.

The purpose, therefore, is to clearly provide the detail referred to by the various clauses, without which the clause cannot be operated. The clauses make the reference by either:

- a direct reference to the Contract Data; or
- the use of *italic* lettering to indicate that there is detail in relation to the term set out in italics in the Contract Data (see clause 11.1 and Section 6.6.1).

A Practical Guide to the NEC4 Engineering and Construction Contract, First Edition. Michael Rowlinson.
© 2019 John Wiley & Sons Ltd. Published 2019 by John Wiley & Sons Ltd.

At the start of both parts of the Contract Data there is a clear statement reminding users that, in order for a complete contract to be created, all the entries must be fully completed for all of the options chosen. Without this exercise being fully completed, the contract will not be complete and the parties will almost certainly run into difficulties in administering the contract, which, in turn, will lead to conflict. Both the Client and the Contractor, therefore, need to carefully check that every required entry has been completed before any contract is formed. In practice, one of the more common issues that the author encounters is incomplete Contract Data.

The Contract Data is set out in two parts, the first of which is completed by the Client and provided to the Contractor with the tender documents and the second of which is completed by the Contractor and returned with its tender. This being the case, the first part acts to inform the Contractor of what the Client requires the Contractor to build and all the related conditions in respect of time, payment, compensation, insurances and other matters in the conditions. By completing the second part the Contractor is informing the Client of the Price which it requires to carry out the work, in relation to the appropriate main option, together with information for use in the pricing of compensation events and other details appertaining to people and design. The Contract Data therefore also acts as a communication medium between the parties before a contract is formed, and acts as a record of many important details that form part of the basis of the agreement between them should a contract come into existence.

Within NEC4 the entries to be made are each inserted into a box set out next to or below the relevant entry. If producing this information electronically, it will be possible to enlarge boxes as required. If producing the information by hand in a hard copy, then it may be necessary for some entries to refer to separate sheets that expand on the detail.

In the black-covered version, the blank Contract Data is set out at pages 78–95. In order to consider this in manageable sections, the required entries are broken down into groups under the following sections.

21.3 Contract Data Part One

21.3.1 Format

The format of the Contract Data in NEC4 is noticeably different to that in NEC3 albeit that the majority of the items are identical. The entries for each of the core clauses that require entries are set out below a heading for the core clause concerned. Each set of entries include the standard detail for the matters referred to in that core clause. In addition, where the use of a Main Option (see Section 3.5) requires further detail to be added to one of the core clauses then statements, all of which start with 'If', are included to provide the space for such information. Not all the main clauses have such entries.

Not all of the core clauses require entries in Contract Data Part One; those which have no entries are core clauses 7 and 9. The requirements for the other seven core clauses are discussed under the following sub-headings.

After the core clauses have been addressed separate sections are provided for each of the Secondary Options, under the appropriate heading.

The sub-headings below follow the order of all the sections with Contract Data Part One. It is only necessary to complete those entries which are relevant to the core clauses and the chosen Options.

21.3.2 Core Clause 1

The first set of entries requires the Client to identify which main option, which dispute resolution option and which secondary options apply. Separate boxes are provided for each of these three classes. It is reasonable to state that this is the most important of the entries as it is the one that determines the structure and content of the conditions and therefore the allocation of risk between the parties.

The next box requires nothing more than a simple description of the works that identifies what the Client wishes to have constructed. In practice, this entry does not need to be any more than one sentence long.

The next three entries identify the Client, the Project Manager and the Supervisor together with the address of each (see Sections 5.2–5.4). Each of these three entries is accompanied by three boxes, one each for name, address for communications and address for electronic communications. It is good practice in the case of the Project Manager and Supervisor to identify the individual who will fulfil this role as opposed to the name of the organisation which employs the individual concerned. The individuals named should be the people who will fulfil the role in a hands-on manner, notwithstanding that they may delegate some of their duties to others (see clause 14.2 and Section 5.3). By naming these individuals, the Client will from the outset be laying the foundations for the spirit required by clause 10.2. There is nothing that prevents an employee of the Client being named in either of these positions. In practice, there is nothing to stop the same person being named in both the roles of Project Manager and Supervisor.

The next entry is in practice probably the one to which most attention should be paid. It is here that the Scope is identified, which directly determines the extent of the Contractor's obligations (see clause 20.1 and Section 7.2). The many references throughout the ECC and this guide give testimony to the importance of this document and clear identification of its location. Should the parties make any amendments to the tender version during post-tender/pre-contract negotiations, these should be reflected in the final version which forms part of the Contract.

The following entry concerns the location of the Site Information (see clause 60.2 and Section 15.3.12), which serves to influence the Contractor's Price and the allocation of risk due to physical conditions.

The next three entries are all straightforward. The list below sets out the entries required and identifies the primary relevant clause:

- the boundaries of the site: 11.2(17) (Section 6.6.4);
- the language of the contract: 13.1 (Section 6.2); and
- the law of the contract: 12.2 (Section 6.6.8).

These are followed by an entry for the period of reply (see clause 13.3 and Section 6.2). This entry requires the standard period for reply to be included in the first box. This is then followed by two pairs of identical boxes. Each pair allows the Client to identify, firstly, a subject or class of communications followed by, secondly, a period of reply specifically for the subject or class. There is no restriction on what could be separated from the standard period and whether the period for reply stated is longer or shorter than the standard. That there are two such entries provided doesn't mean that any have to be used. Nor does it restrict the number of separate subjects or classes to a maximum of two. It is entirely up to the Client to insert whatever number of special classes or subjects that it wishes to create.

The final two boxes to be completed both concern the early warning mechanism at clause 15. The first entry requires the Client to list those matters that it has identified and that are to be included in the Early Warning Register (see clause 11.2(8) and Section 6.4). A practical alternative here is for the Client to draft the Early Warning Register, include any matters which it has identified and make reference to that draft in this entry. The Project Manager then has a starting point for use during administration of the project. The second box is used to identify the intervals at which early warning meetings will be held, as required by clause 15.2 (see Section 6.4).

21.3.3 Core Clause 2

Both classes of entry provided for in relation to core clause 2 are optional. The first class refers to Key Dates and Conditions (see clause 11.2(11) and Section 12.5.3). For any such combination of Key Date and related Condition, the key date and relevant condition to be met need to be adequately set out in this entry. This set of entries will only be required if the Client wishes to utilize that mechanism. If there are no Key Dates and related Conditions, then simply leave the entries blank.

The second class only applies if one of Option C, D, E or F is the chosen Main Option. Should one of these apply it is necessary to insert the maximum interval that the Contractor will be required to prepare and submit forecasts of the total Defined Cost (see clause 20.4 and Section 7.2).

21.3.4 Core Clause 3

The first two entries here concern dates connected to the starting of the works. The first is the starting date for the whole of the Contractor's responsibilities and the second is the date or dates upon which the Contractor will be allowed access to all or part of the Site (see clause 30.1 and Section 12.5.1).

The third entry requires the insertion of the period at which the Contractor is required to submit revised programmes as the third bullet point at clause 32.2 (see Section 12.3).

This is followed by an optional entry that applies when the Client wishes to state the Completion Date for the works. By including this as an optional statement, it allows the Client to either set the date it requires the works to be completed by (and in doing so, potentially impose inefficiencies that will be built into the Price) or to allow the tendering contractors to determine the most efficient combination of method and time (thereby enabling them to tender the lowest price). Which option the Client selects will depend on how important completion by a set date is to its business needs and goals.

The next item does not require an entry. It simply requires the Client to decide whether or not it is willing to take over the works (see clause 35 and Section 12.6.3) before Completion. Whichever way the Client has decided is indicated by deleting one of the two alternatives.

The final entry requires a period to be inserted for the submission of the first programme in the event that there is no programme identified in Contract Data Part Two. If the tendering contractors are required to submit a programme as part of their tender, then this statement will not be required. In the event that it is required, the most common period seen in practice is two weeks.

21.3.5 Core Clause 4

The first entry sets the period after the Contract Date (see clause 11.2(4) and Section 12.3) that the Contractor is required to submit a quality policy statement and a quality plan (see clause 40.2 and Section 9.2). This is likely to be in the region of two to four weeks.

The second entry determines what the period between Completion of the whole of the Works (see clause 30.2 and Section 12.5.2) and the defects date, being a period referred to in clauses 43.1, 43.2 and 44.3 (see Section 9.5). The extent of the period inserted here is important as it determines the duration of both the Contractor's obligations and rights under the ECC in respect of defects. The third entry establishes the defects correction period, which, as clause 44.2 (see Section 9.5), sets the time in which the Contractor must correct a defect subject to access being provided (see clause 44.4 and Section 9.5). The final entry, which does not necessarily need to be completed, provides the Client with a way of setting a different defects correction period for specified elements or sections of the works. For example, it may be that the Client decides to set a longer period for the correction of defects to the mechanical installation than for the rest of a building, or it decides to include a shorter period for any defect in the operating theatres of a hospital than for the rest of the unit.

21.3.6 Core Clause 5

For core clause 5 (Payment) a total of six entries are required; the first three are standard whilst the fourth is optional and the fifth and sixth relate to stated Main Options.

The first states the currency that payments will be made in; this only becomes complicated if Secondary Option X3 (see Section 13.3) is in use when further entries will be required (see Section 21.3.11). The second requires a period to be stated as the assessment interval (see clause 50.1 and Section 13.2.1). Users should note that if they insert four weeks, there will be 13 payments per year; if they insert five weeks there will only be 10 payments per year. For users who require 12 monthly payments to be made per year, then this entry needs to be completed to read 'monthly' or 'calendar month'. Alternatively, it could be amended to give the date each month that the assessment will be carried out. Such an amendment may require the wording of clause 50.1 to be considered via Option Z.

The third entry concerns the interest rate and requires the Client to state what percentage over the base rate of the bank (which must also be named) will be used to determine the rate of interest in the event that the provisions of clauses 51.2 and 51.3 (see Sections 13.2.2 and 13.2.3) should apply. The blank form states that this entry should not be less than 2%. Users in the United Kingdom should be aware of the judgement in *Yuanda (UK) Co Ltd v. WW Gear Construction Ltd* [2010] EWHC 720 (TCC) (13 April 2010) and are referred to the comments in Section 13.11.2 in connection with the percentage to be inserted here. Users in other jurisdictions are advised to consider any similar legislation or case law.

The optional statement at the fourth entry in this section relates to the period for payment and only applies if Secondary Option Y(UK)2 does not apply. (If Option Y(UK)2 applies see Section 13.9.) By electing to state a period other than three weeks in the box, the Client gains the ability to change the period stated in clause 51.2 as to when the payment is to be made.

The fifth set of entries only applies if main Option C or D is being used and requires the completion of the table that sets out the share ranges and the Contractor's share percentage for use with the Contractors share mechanism (set out at clauses 54.1–54.4 for Main Option C and clauses 54.5–54.8 for Main Option D). Full comment on this mechanism and the practical issues associated with it are given in Sections 13.7 and 13.8, respectively.

The final entry requires the inclusion of the exchange rate details for use in the operation of clause 50.7 (Main Option C and D) or clause 50.8 (Main Option E and F) (see Section 13.3). This entry requires both identification of the publication source of the exchange rates and the date which is to be used.

21.3.7 Core Clause 6

Core clause 6 (Compensation events) requires more thought and application than most of the preceding entries. The first sets of entries that are required all concern the compensation event at clause 60.1(13) (see 'Event 13' at Section 15.3.13). The need for and the content of these entries together with related issues are the subject of two of the Practical Issues identified at Sections 15.4.3 and 15.4.4, to which the reader is referred.

The remaining three entries all refer to optional positions. The first concerns Options A and B and requires the insertion of the value engineering percentage used in conjunction with clause 63.12 (see Section 17.2.6) following a Contractor's proposal under clause 16.1 (see Section 6.6.3).

The second of the three is to identify the method of measurement that has been used in the preparation of the bill of quantities (see clause 56.1 and Section 13.5.2). Unlike previous editions there is no prompt to make any reference to any amendments that have been made to that method of measurement. Users will need to consider how to include any such amendments when completing the Contract Data. It is suggested that a Z Clause entry would be required to introduce any such amendments. The amendments could be set out in the Scope with the Z Clause simply acting to incorporate those amendments and to make them effective.

The third and final optional entry allows the Client to list other events that will constitute compensation events as referred to in clause 60.1(21). See Section 15.3.21 for comment on such entries.

21.3.8 Core Clause 8

There are six sets of entries required in connection with core clause 8 (Liabilities and insurance), four sets of which are optional.

The first of these allows the Client the option of identifying additional matters that will be one of its liabilities under clause 80.1. The use of the Contract Data to identify additional liabilities in this way is referred to in the last bullet point at clause 80.1 and allows the Client to limit the Contractor's liabilities for any specified matters. Volume 2 of the User Guide gives an example of an additional liability, which might be covered by this method.

The second and third entries are mandatory and are in respect of the minimum limit of indemnity for two of the insurances listed in the Insurance Table at clause 83.3 (see Section 11.2.3), which needs to be stated in the Contract Data. Before entering into contract, the Contractor should confirm that the wording in these entries corresponds to the wording in its insurance policies. These entries require the minimum liability to be provided for any one event, whereas some insurance policies only provide cover in the aggregate over the insured period.

The fourth entry needs to be completed by the Client whenever it is free-issuing Plant and Materials to the Contractor for use in the works. This is necessary to provide the required level of cover referred to in the first line of the Insurance Table (see clause 83.3 and Section 11.2.3), which gives the clue that the entry is simply the replacement value of the free-issue goods.

The fifth and sixth sets of optional statements are those referred to by clause 83.2 (see Section 11.2.3) and are the mechanism by which the Client either reduces or increases the type of insurances to be provided by the Contractor. By using the first of these two entries, the Client removes the need for the Contractor to provide a type of insurance stated in the Insurance Table. By using the second entry, the Client can increase the types of insurance to be provided by the Contractor over those required by the Insurance Table. There is no reason why, in practice, both of these optional statements cannot be used at the same time as a way of varying the types of insurance cover that the Contractor is to provide.

21.3.9 Resolving and avoiding disputes (Options W1, W2 and W3)

The first statement to be completed is to identify the nature of the tribunal, a body that is referred to in all three of the dispute options. In all jurisdictions that the author is aware of, the choice will be between the Courts and Arbitration.

The next three entries will only become necessary when the tribunal has been stated to be Arbitration. In this case, further entries are required which state that arbitration rules or procedure will apply[1], where any arbitration is to be held (the Client should identify a general physical location such as 'London'), and who shall appoint the arbitrator, if the parties do not agree or the rules do not provide for such an event.

The next three sets of entries refer to Options W1 and W2 only. The first requires the name and contact addresses, both postal and electronic, for the Client's Senior Representatives (see clauses W1.1 and W2.1 and Sections 19.2 and 19.3). The Contractor's Senior Representatives are identified in Contract Data Part Two. The second gives the same information but for the Adjudicator if named in the Contract (see Sections 19.2 and 19.3). Whether or not the Adjudicator is named, the third of this set is used to identify the Adjudicator nominating body. This is far more important than naming an individual adjudicator in advance.

The final four entries all refer to Option W3 and relate to the Dispute Avoidance Board (DAB) (see Section 19.4). The first of the four requires the deletion of one of the two options in bold text so as to determine whether the DAB will consist of one or three members. There is no default provided so one must be deleted for the mechanism to be clear. The second of the four is used by the Client to identify its nomination to the DAB when it will consist of three members. The third sets the interval, in months, for visits to the Site by the DAB (see clause W3.1(5) and Section 19.4). The last entry is to identify the DAB nominating body, which is vital if the DAB is only to consist of one member and possibly necessary if the number of members is three.

21.3.10 Option X1

(See Section 13.10.1.) In order to make the Price adjustment for inflation mechanism work, there are three separate entries to complete. The third of these requires the preparer of the indices to be used to be identified. The identity of the preparer and the index to be used will then, in practice, partially facilitate the completion of the first entry by identifying the categories to which the proportions can be allocated within the index to be used. The second entry establishes the base date that will act as the baseline against which the inflation will be measured, and therefore the baseline for the Contractor to prepare its Price against.

21.3.11 Option X3

(See Section 13.3.) Where multiple currencies are to be used, then the Client needs to identify those items or activities that will be paid for in a currency other than the currency of the contract (see clause 51.1 and Section 13.3), which currency that will be and the maximum amount that will be paid in that currency. In practice, this last item could always be completed as 'unlimited' in order to ensure that all payments against the

---

[1] In practice, it is usual to refer to a standard published set of rules.

relevant item or activity would all be paid in the other currency. This may be of particular relevance when used in combination with Main Options C and D, where the Price for Work Done to Date (PWDD) is based on the expended Defined Cost (see clause 11.2(31) and Section 13.4.5). The other factor that is needed is the date and identity of publication of the exchange rates that will apply.

21.3.12 Options X5, X6 and X7

(See Section 12.7.) There are five different combinations that can apply under this heading depending on which of these three Secondary Options have been chosen. The key to the combinations is whether Option X5 (Sectional completion) is used or not. If X5 is used, there will always be a set of entries which contain an adequate description of each section of the works and a corresponding Completion Date for each section. If either or both of X6 (Bonus for early completion) and X7 (Delay damages) are used in conjunction with X5, then the entries for these two options will require the bonus or damages sum to be identified against each of the sections and also for the remainder of the works outside of the sections.

In the event that X5 is not used but either or both of X6 and X7 are, then the entry for either of the latter two is straightforward (simply being the appropriate sum in respect of the relevant option). Regardless of the combination of options used, any damages that are stated must be a genuine pre-estimate of the loss that the Client will incur as a result of the delay.

21.3.13 Option X8

(See Section 7.3.2.) Within this section of the Contract Data entries are required for the three different types of undertakings that may be required. For the undertakings to be given to Others (see Section 5.11) the beneficiaries, by name or class, must be identified. For undertakings to be given by Subcontractors it is necessary to identify those parts of the works which might be carried out by subcontractors and then which beneficiaries the undertakings will be provided to, by name or class. For undertakings to be given by Subcontractors to the Client, it is only necessary to identify those parts of the works that might be carried out by subcontractors.

In all three instances, if the entries are left blank, then this will indicate that no undertakings of that class are required.

21.3.14 Option X10

(See Section 7.4.) Of the three required entries, the first identifies the period after Contract Date within which the first Information Execution Plan (see Section 7.4.3) is required to be provided if one is not identified in Contract Data Part Two. It is suggested that between two and four weeks would be reasonable.

The second and third entries are related to professional indemnity insurance, identifying firstly the sum to be insured and secondly the period that such insurance should be maintained for following Completion of the whole of the works (see clause 30.2 and Section 12.5.2).

21.3.15 Option X12

(See Section 20.4.) The entries required in connection with the Partnering Option are far less than the study of this option would suggest. All that is required is the identity of the Promotor, the location of the Schedule of Partners, a statement as to what the Promotor's objective is and sufficient detail to identify what constitutes the Partnering Information. Everything else required for the operation of the Option is either included in the Partnering Information or is created by the Partners or the Core Group formed by the Partners under the Option.

21.3.16 Option X13

(See Section 20.5.) A single entry is required to establish the amount of the performance bond that the use of this option brings into the contract. The standard statement within the ECC does not suggest any unit for the amount of the performance bond; instead, it simply provides a blank space. This allows the Client the freedom to select how it would prefer the value of the bond to be determined. For example, the Client could from the outset state a specific sum of money or elect that the amount will be a stated percentage of the tendered total of the Prices. The only restriction is that however the amount is expressed, it must relate to something that will be known at the time the Contractor has to provide the bond, i.e. no later than four weeks after the Contract Date.

21.3.17 Option X14

(See Section 13.10.2.) In order to provide all the necessary information for the advance payment mechanism to work, three separate entries need to be considered plus one selection. The first entry requires the Client to state what amount will constitute the advance payment. This doesn't have to be stated as a sum of money; it could instead be stated as a percentage of something that is available at the time the payment is to be made, such as a percentage of (say) the tendered total of the Prices or the value of listed items of Plant, Materials or Equipment to be provided by the Contractor.

The second and third entries concern the latest time at which the repayment instalments shall start to be deducted from the regular assessments carried out pursuant to clause 50.1 (see Section 13.2) and the amount of each instalment, respectively. The amount of the instalments can be stated either as figures or as a percentage of the total advanced payment. This entry can be written so that where there are several repayment instalments to be made they are not all of the same value.

The final item requires a selection to determine whether a bond is required to give the Client some security over the advanced payment. The amount of the bond is not required to be stated, as clause X14.2 determines that it will be the amount of the advance payment.

21.3.18 Option X15

(See Section 11.3.) Of the three required entries the first identifies the period following Completion of the whole of the works (see clause 30.2 and Section 12.5.2) that the documents prepared by the Contractor for its design will have to be retained. This is known as the 'period for retention'. In this context, the word *retention* is not referring to the monetary sum retained under Option X16.

The second and third entries are related to professional indemnity insurance, identifying firstly the sum to be insured and secondly the period that such insurance should be maintained for following Completion of the whole of the works.

21.3.19 Option X16

(See Section 13.10.3.) The retention option requires two entries and one selection to be made. The first is the amount of the PWDD that retention will not be applied to, known as the retention-free amount. In practice, it is common to see this entry completed as 'nil'. Alternatively, either a fixed sum of money or a percentage of the tendered total of the Prices could be included should the Client decide that it does not want to apply the retention percentage to the whole of the PWDD. The second entry is the percentage of the PWDD over and above the retention-free amount that will be retained in accordance with this option.

The selection to be made determines whether the Contractor may or may not give the Client a retention bond instead of having a sum held (see clauses X41.1 and X14.2).

21.3.20 Option X17

(See Section 20.6.) Where this option is in use, as many or as few different performance levels can be stated each with the associated amount of damages for the failure to achieve each level of performance. The levels of performance must be identifiable and measurable and should be something over which the Contractor has significant influence as part of its obligations to Provide the Works (see clauses 11.2(15) and 20.1 and Section 7.2). This will normally mean that the Contractor will be responsible for designing the process to which the level of performance applies. As with delay damages (see Option X7 above), the amount of damages should be a genuine pre-estimate of the loss that will be incurred by the Client should the performance level not be achieved.

21.3.21 Option X18

(See Section 20.7.) Each of subclauses X18.2 to X18.6 requires an entry, the first four of which are the sums of money to which the Contractor's liability is limited for the four risks set out in clauses X18.2–X18.5. As stated in Section 20.7, it is outside the remit of this guide to discuss the law related to such limits of liability. Accordingly, users should take appropriate legal advice where such limitations are required. Such legal advice should include recommendations of the amount to be inserted against each of these four entries as appropriate for the particular project being considered. In practice, it is permitted and workable to state a proportion or multiplier of the tendered total or final total of the Prices instead of a fixed sum in some or all of these entries. This is a matter that should also be considered alongside the insurance requirements and cover that will be required by core clause 8 (see Chapter 11).

The final entry is the number of years after Completion that the Contractor's liability will cease to apply. In the United Kingdom, this is normally stated as 6 or 12 years to match the provisions of the Limitation Act 1980.

21.3.22 Option X20

(See Section 20.8.) In order to support the Key Performance Indicator (KPI) mechanism, two entries are required: the first identifies the location of the incentive schedule and the second sets the interval at which the Contractor is required to submit reports of its performance against that of the KPI. The actual entries themselves are not complicated. Where many users find difficulty in practice is identifying the KPIs to include in the incentive schedule, how to establish the baseline and how to measure the actual performance.

21.3.23 Option X22

(See Section 20.10.) A total of five entries are required to make this Option work. The first and last two all concern the Budget. The first sets out the Budget with a description and amount of money for each element, plus a total. Only four lines are provided, whereas in practice most projects would have far more elements within a Budget. Where this is the case it is suggested that the Budget is created in a separate document that is then referenced in place of the first set of boxes. The fourth entry allows additional events which would change the Budget to be identified. This is an important feature that gives flexibility to the Budget and could include things such as another aspect of Work, which the Client may decide to add or the risk in certain conditions (e.g. the finding of asbestos in an existing facility) occurs. These are not restrictive examples; many other possibilities will no doubt arise in practice. The fifth entry sets the budget incentive percentage, being the portion of the saving between the Budget and the Project Cost that the Contractor can earn.

The second and third entries are both periods for the preparation of forecasts. The second entry refers to that for the submission by the Contractor of forecasts of the total

Defined Cost. The third entry concerns the forecasts by the Contractor of the total Project Cost. There is no suggested or default period so it is open to the Client to insert its own requirements. In practice, these periods will be typically in the region of four weeks, five weeks or monthly. On occasions the total Project Cost forecast might be at longer intervals than the total Defined Cost forecast.

21.3.24 Option Y(UK)1

(See Section 13.10.4.) Option Y(UK)1's sole entry requires a selection that determines whether it the Contractor or the Client who will be required to pay any bank charges in connection with the running of the Project Bank Account and to receive the benefit of any interest paid on the deposits within the account.

21.3.25 Option Y(UK)2

(See Section 13.9.) The entry here is optional. Should the Client wish to change the period between the due date and the final date for payment from the 14 days stated in clause Y2.2 (2nd paragraph) to a different period (whether shorter or longer) that revised period is stated in the box provided.

21.3.26 Option Y(UK)3

(See Section 20.11.) The Contract Data requires two columns to be completed. The two columns are headed 'term' and 'beneficiary'. In the event that any third party is to have the right to enforce any term of the Contract against either or both of the Client or the Contractor then that third party is named in the 'beneficiary' column. The term or terms listed in the 'term' column will be the only ones that the named third party can enforce.

In contracts where both Option Y(UK)1 and Option Y(UK)3 are used, then an entry that requires no input is included. The 'beneficiary' named is 'Named Suppliers' thereby giving all such parties who come under that definition (see Section 13.10.4) the right to enforce the provisions of Option Y(UK)1. This right will be necessary to give the Named Suppliers the ability to enforce payment from the Project Bank Account when they are entitled to a payment but that payment has not been made.

21.3.27 Option Z

(See Section 20.12.) The entry for Option Z requires the additional conditions to be identified. What most users do is simply identify where the additional conditions are located, as they often occupy several or more pages. The potential for what could be included in Option Z is so large it is not a subject that can be covered in any book. Various references have been made in this Practical Guide to matters that could be incorporated in Option Z if the Client or the parties felt it was appropriate.

21.4 Contract Data Part Two

21.4.1 Format

The format of Contract Data Part Two is identical to that for Contract Data Part One albeit that there are far fewer entries to make and not all core clauses or Options will need to be considered.

When completing Part Two of the Contract Data as part of the tender process, the Contractor needs to ensure that all the relevant entries are completed. The Contractor should also ensure that the copy that has been provided for completion includes every relevant entry. The author has seen many incomplete versions issued with tenders, especially when the subcontract is being used (i.e. issued by contractors for subcontractors to complete).

21.4.2 Core Clause 1

A total of five subjects need to be considered and completed for core clause 1. The first of these is simply the Contractor's own name together with addresses for postal and electronic communications.

The second entry is for the fee percentage referred to in the definition of the Fee at clause 11.2(10). More detailed comment in respect of the Fee is given in Section 13.4.7.

The third entry gives the Contractor the opportunity at tender stage to define which other areas in addition to the Site should be designated as the Working Areas (see clause 11.2(20) and Section 6.6.4). The additional areas identified in this entry should be limited to areas that will only be used for this project and not for any other project being managed by the Contractor or any of its subcontractors or suppliers. The content of this entry will influence the costs that can be claimed as Defined Cost under all the main options and therefore also the matters that need to be included in the Fee.

The fourth requires the Contractor to identify the key people that it intends to use on the project, together with their relevant qualifications and experience. Which job roles the key people are to be identified for is often specified by the Client and serves to allow clause 24.1 (see Section 7.5.1) to be operated and, if necessary, administered.

Finally, in this group the Contractor is provided with the opportunity to list any items which it considers should be included in the Early Warning Register (see clause 11.2(8) and Section 6.4). Any such matters listed will be in addition to those identified in Contract Data Part One (see Section 21.3.2), and these two entries combined will be the Early Warning Register at the Contract Date. The purpose of this entry is to allow the Contractor to bring to the Client's attention any additional matters which constitute a risk to the project and which the Contractor has identified at that time. It is not the purpose of these entries to allow the Contractor to omit things, which under the ECC are at the Contractor's risk, from its tendered total of the Prices (see Section 21.4.5).

21.4.3 Core Clause 2

The only entry for this core clause requires the Contractor to identify the documents in which the Scope for its design can be found (see clause 21.1 and Sections 7.3 and 22.2.10). This Scope shows how the Contractor proposes to satisfy the Client's requirements set out in the Scope referred to in Contract Data Part One. This Scope will only need to specify and describe the works. There will be no need or right for the Contractor to identify any constraints; only the Client needs to consider what constraints are to be set. Where the Contractor has design responsibility (see clause 21.2 and Section 7.3), the information that will be referred to here would normally form a key part of the tender submission and be appraised carefully by the Client and its advisors before the offer was accepted.

21.4.4 Core Clause 3

The first of the two optional entries in core clause 3 refers to a key document that is requested by many employers as part of the tender submission. This is the tendering contractor's programme showing how it intends to carry out the work. If the Client requires a programme, or if the Contractor chooses to submit one that is subsequently accepted by referring to that programme in this entry within the Contract Data, it automatically becomes the first Accepted Programme (see clause 11.2(1) and Section 12.1). Accordingly, it should be compiled so that it includes all the information required by clause 31.2 (see Section 12.2). This is not always easy to achieve in practice, as some of the required information will often not be available (such as timings of the work by Others or even the start and access dates). Where this is the case, then a compromise needs to be made in order to facilitate the use of this method of establishing and accepting the first programme.

The second of the two entries may or may not be linked to the first. The Client has the option in Part One to state the Completion Date (see clause 11.2(3) and Section 12.5.2). If the Client does not exercise that option, then it can require the Contractor to state the Completion Date as part of its tender. It is often the case that, where the Client requires the Contractor to state the Completion Date in this way, it also requires the provision of a programme with the tender (hence the potential for a link with the previous entry). On the other hand, it is, of course, possible to ask the Contractor to state the Completion Date without also requiring the submission of a programme with its tender albeit that there seems little practical sense behind such a position.

21.4.5 Core Clause 5

The four entries for core clause 5 all concern optional statements in respect of one or more of the main options and no more than two of these can apply at any one time.

Unlike the other optional statements, these statements will have to be used when the main option referred to is in use; if this rule is not followed, then the information supporting the conditions will be incomplete and the parties will encounter difficulties in administering the contract. The first two of these statements act to identify either the activity schedule where Main Option A or C applies or the bill of quantities where Main Option B or D applies. This requirement to identify the bill of quantities in the data supplied by the Contractor is somewhat odd as, under Main Options B and D, it is the Client who prepares the bill of quantities and takes the risk on the accuracy and completeness of the information within that document. It would appear that the reason that this entry is required can only be to identify by some reference the copy of the bill of quantities that contains the Contractor's rates and prices, and that the Contractor should provide such a reference and include it here.

The third of these four statements requires the tendered total of the Prices to be inserted, i.e. the Contractor's tender sum, and is only used when Main Options A, B, C or D apply. This will not generally be the total that the Contractor will be paid for Providing the Works, as the total of the Prices stated here will be subject to adjustment as a result of the compensation event mechanism. In any event, the amount paid to the Contractor is the amount due as calculated under clause 50.3 (see Section 13.4.2). No entry is required here for Main Options E and F as the basis of a tender under these two options will not result in such a figure.

The fourth statement is only used when Main Option F applies and consists of a list of work activities that the Contractor will carry out (as opposed to those activities being carried out by the subcontractors) and the price, whether it is a lump sum or unit rate for each of those work activities. This entry is required to give meaning to the requirements of clause 20.2 (see Section 7.2) and, in particular, the exception at the end of that clause.

21.4.6 Resolving and avoiding disputes (Options W1, W2 and W3)

Two entries are required for this section. The first, which refers to Options W1 and W2, requires the name and contact addresses, both postal and electronic, for the Contractor's Senior Representatives (see clauses W1.1 and W2.1 and Sections 19.2 and 19.3). The Client's Senior Representatives are identified in Contract Data Part One.

The second entry refers to Option W3 and relates to the DAB (see Section 19.4). It is used by the Contractor to identify its nomination to the DAB when it will consist of three members. Whether the DAB will consist of one or three members is identified in the same section in Contract Data Part One.

21.4.7 Option X10

(See Section 7.4.) The single entry required for this Option is to identify the Information Execution Plan (see Section 7.4.2) that has been prepared by the Contractor and which will be the first accepted Information Execution Plan.

21.4.8 Option X22

(See Section 20.10.) Where this Option applies, two types of entry are required from the Contractor. The first requires the Contractor to identify the key people that it intends to use for Stage One together with their relevant qualifications and experience. These are, possibly, different people to those who will have been identified in the entries required for Core Clause 1 (see Section 21.4.2). These Stage One key people are referred to in clause X22.4.

21.4.9 Option Y(UK)1

(See Section 13.10.4.) Where this Option applies, the Contractor is required to give two pieces of information. The first is to identify which bank the Contractor proposes to use as the project bank; this simply requires the name of the chosen institution. The second one requires the Contractor to list all those subcontractors and suppliers who it proposes should be classified as Named Suppliers (see Section 5.9) and therefore receive payments direct from the project bank account. Given that the Scope (see Section 22.2.44) may contain restrictions or other stipulations concerning who may or must be included in this entry, the Contractor must make sure that it takes note of any such issues.

21.4.10 Data for the Schedule of Cost Components

Data entries for the Schedule of Cost Components are only required when one of Main Options C, D or E has been selected. There are five sets of entries to complete, which are all required to facilitate the proper operation of the Schedule of Cost Components (SCC) (see Section 14.2). In practice, this is often one of the areas of the Contract Data that is not properly or fully completed. Contractors, in particular, are advised to ensure that this set of entries is fully completed and that they fully understand the significance of each one.

Of the five entries, the first two relate to Equipment under section 2. The first is information referred to by item 23 of the SCC and requires the Contractor to identify items of Equipment that it intends to purchase for use on the project, together with the associated time-related charge and unit of time to which that time-related charge is applicable (i.e. £X per hour/day/week/month as appropriate). This entry does not require either the purchase price of the items listed or the anticipated depreciation in value to be identified, both being factors in the application of item 23 of the SCC. The intent of this item within the SCC was to cover large specialist equipment; users should note that the wording would also act to pick up small items that the Contractor may purchase. In the event that an item of equipment to which the rules in this item apply is missed from the list when the contract is formed, users should also note that the item in the SCC contains wording that allows, with the Project Manager's agreement, any additional items to be added and treated as if they had always been included.

The second item relating to Equipment provides the information needed by item 24 of the SCC. Here the Contractor needs to list any special Equipment with appropriate details and a rate. This is the entry in which to consider any equipment that cannot be hired or purchased on the open market. This typically includes anything which the Contractor or its potential subcontractors have constructed specifically for the type of work that they will be undertaking. In practice, most projects do not include any entries for such special equipment. As for the previous entry, items not included can be subsequently added with the Project Manager's agreement.

The third item to be considered for the SCC concerns section 6, manufacture and fabrication. Again, practical experience shows that this is not used very often; the applicability of this section of the SCC is limited due to the specific requirements contained within it and since most contractors subcontract this type of work. When it is required, the entry requires the categories of people to be identified together with the appropriate hourly rate, these being the rates referred to by item 61 of the SCC.

The last two entries relate to the cost of design carried out outside the Working Areas as section 7 of the SCC. The first requires the class and related hourly rates for each type of design employee, as referred to in item 71 of the SCC to be listed. Users should appreciate that this is for design carried out outside the Working Areas. Where design is carried out in the Working Areas, the designers carrying out such work are classified as people under section 1 of the SCC and would be costed under those rules. The second entry requires the Contractor to list those categories of design employees for which it will require reimbursement of any travelling expenses between their normal place of work and the Working Areas, in the event that it is necessary for those individuals to travel to the Working Areas in order to carry out their work. This cost is recoverable under item 72 of the SCC.

21.4.11 Data for the Short Schedule of Cost Components

This data is only required when one of either Main Option A or B has been selected. There are seven entries to complete that are all required to facilitate the proper operation of the Short Schedule of Cost Components (SSCC) (see Section 14.3). In practice, this is often one of the areas of the Contract Data that is not properly or fully completed. Contractors in particular are advised to ensure that this set of entries is fully completed and that they fully understand the significance of each one.

The first entry refers to the People Rates (see clause 11.2(28) and Section 14.3) which is referred to in item 11 of the SSCC. It requires the category of person (e.g. site manager, working supervisor, bricklayer, roofer etc.), the unit of cost (e.g. hour, day, shift etc.) and the rate, in the currency of the contract, per unit. The black book only provides four sets of entries, which will be far too few for the vast majority of projects. It is suggested that this information is set out in a separate document, which is referenced here. It should also be remembered that the list of categories and associated information can be expanded as required during the project (see clause 63.16 and Section 14.3).

The second and third entries should be considered together and relate to item 21 of the SSCC. Here the tendering contractor should insert the proper title of a published list of rates for Equipment (see definition at clause 11.2(9)) together with a percentage adjustment, as discussed at Section 14.3. Tendering contractors should remember that this information will form part of their tender and, accordingly, will be assessed as part of the tender adjudication process by the Client and its advisors. Apart from the commercial aspects of these entries, users should not encounter any practical difficulties with them.

The fourth entry also relates to Equipment and is provided in order to allow any items that the Contractor anticipates it will use in Providing the Works but that are not included in the published list to be identified with the rate which will apply (see item 22 of the SSCC). This entry requires the Contractor to consider its intended method of working (which would be required anyway if it is preparing a programme[2] for inclusion in the tender) and to compare the Equipment it intends to use against the published list. Any item not in the published list is then included here. The percentage adjustment that applies to the published list does not apply to the rates quoted here.

The fifth item to be considered for the SSCC concerns section 6, manufacture and fabrication. Again, practical experience shows that this is not used very often; the applicability of this section of the SSCC is limited due to the specific requirements contained within it and since most contractors subcontract this type of work. When it is required, the entry requires the categories of people to be identified together with the appropriate hourly rate, these being the rates referred to by item 61 of the SSCC. This entry is exactly the same as the third entry for the SCC.

The last two entries relate to the cost of design carried out outside the Working Areas as section 7 of the SSCC. The first requires the class and related hourly rates for each type of design employee, as referred to in item 71 of the SSCC to be listed. Users should appreciate that this is for design carried out outside the Working Areas. Where design is carried out in the Working Areas, the designers carrying out such work are classified as people under section 1 of the SSCC and would be costed under those rules. The second entry requires the Contractor to list those categories of design employees for which it will require reimbursement of any travelling expenses between their normal place of work and the Working Areas, in the event that it is necessary for those individuals to travel to the Working Areas in order to carry out their work. This cost is recoverable under item 72 of the SSCC. These two entries are exactly the same as the last two entries for the SSCC.

21.5 Practical issues

21.5.1 Failing to fully complete the Contract Data

It is common in practice to find that the parties have failed to fully complete the Contract Data. The consequence of this failure is clearly stated at the start of both Parts: unless the Contract Data is fully completed as required to provide all the information necessary to facilitate the administration and application of the chosen options, then the contract formed between the parties will be incomplete. Not only does this create difficulties for

---

[2] See the penultimate bullet point at clause 31.2 and Section 12.2.4.

both parties but can easily result in disputes which generally only result in an outcome that is unsatisfactory for all those involved.

One of the common causes of this position that is often found in practice, particularly in relation to Contract Data Part Two, is that whoever prepares the draft document does not include all the entries that are required. While employers and/or their representatives tend to be reasonably proficient at completing Part One of the Contract Data, there are far too many occurrences of the template that the tendering contractors are asked to fill in being incomplete. Tendering contractors should be aware of this and check that everything that is required for the options selected is completed before submitting the tender, including adding those items that may have been missed.

Before entering into any contract, it is good practice for the parties to sit down together and review the whole of the Contract Data to ensure that all the necessary information has been included and that the contract they are about to enter into will be complete.

21.5.2 Using a previous project as the template

One of the causes that often contributes to the practical issue above is that users, and in particular employers and their representatives, start the exercise of preparing the documents for a new project by using a previous project as a template. This is very common and has primarily been encouraged in practice by the availability of computers and the perceived simplicity and speed gained by simply renaming a previous file and then editing it for the next job. The result in the majority of cases is that one or more entries that require editing are not actually considered, and the Contract Data is issued with incorrect details. Such examples have been seen with incorrect names and addresses contained in them, an incorrect description of the work or a reference to the Site Information for the previous scheme that was in a different area of the country.

There is only one foolproof way of avoiding such mistakes: use a blank template every time. By all means, refer back to previous schemes to remind yourself of the nature of the entry required, but overwriting a previous template is a recipe for error making and is in the interest of no one.

Chapter 22
The Supporting Documents: Need and Content

22.1 Introduction

This chapter considers the supporting documents (some of which may be electronic documents) that are created in order to make the procedures work, which have not been considered elsewhere in this guide. The full list of potential supporting documents is set out below and, where a document has been considered elsewhere, that location is identified:

- Scope;
- Site Information;
- Accepted Programme (see Chapter 12);
- Quality Policy Statement (see Section 9.2);
- Quality Plan (see Section 9.2);
- Activity Schedule (see Section 13.5.1);
- Bill of Quantities (see Section 13.5.2);
- Parent Company Guarantee (see Section 20.3);
- Undertakings (see Section 7.3.2);
- Information Execution Plan (see Section 7.4);
- Project Model (see Section 7.4);
- Information Model (see Section 7.4);
- Information Model Requirements (see Section 7.4);
- Schedule of Partners (see Section 20.4);
- Partnering Information (see Section 20.4);
- Partnering timetable (see Section 20.4);
- Performance bond (see Section 20.5);
- Advanced payment bond (see Section 13.10.2);
- Incentive schedule (see Section 20.8);
- Budget (see Section 20.10);
- Pricing Information (see Section 20.10); and
- Additional Conditions of Contract (see Section 20.12).

A Practical Guide to the NEC4 Engineering and Construction Contract, First Edition. Michael Rowlinson.
© 2019 John Wiley & Sons Ltd. Published 2019 by John Wiley & Sons Ltd.

Those documents in the list above that are not identified as having been considered elsewhere, i.e. the Scope and the Site Information, are the subject of the following sections.

22.2 Scope

22.2.1 Introduction

The following sections consider the Scope (as defined at clause 11.2(16)) firstly at a high level and then at a more practical level. Each of the following sections refers to a specific part of the contents by reference to the clause in the ECC, which needs that information to be effective. Practical considerations will be included as appropriate.

22.2.2 Importance and purpose

In practical terms, the Scope is the most important supporting document under the ECC, no matter which Main Option has been selected. Scope, which is carefully and comprehensively drafted prior to the tender being issued, will provide the Contractor with more information upon which to base its Price. The more information that is available to the Contractor, especially when it is in a well-presented form, the more accurately the Contractor can determine the price (thereby negating uncertainty and the consequent risk allowances that have to be included to ensure that the Price is sufficient). The result is that the tenders received by the Client are more competitive and more representative of the final outturn cost, providing of course that the Client does not instigate wholesale change to the project once work commences.

Further, the more accurate the Scope at the Contract Date, the fewer the number of instructions that will be required to correct omissions, errors or conflicts within the document, hence resulting in fewer compensation events.

The purpose of the Scope is established by a reference to the straightforward but broadly written definition of the term at clause 11.2(16). This definition clearly sets out that the Scope consists firstly of information which specifies and describes the works and secondly of information which sets out any constraints that apply to how the Contractor is allowed to carry out the works. This is done by reference to the term Provide the Works (see clauses 11.2(15) and 20.1 and Section 7.2), which is the crux of the Contractor's primary obligation under the ECC. In simple terms, the Contractor is obliged to build everything described and specified in the Scope and to carry out that work within the constraints set out in that document. If something is not specified and described, then the Contractor does not provide it. If a constraint is not stated, then the Contractor is not bound by it.

22.2.3 Location

As well as outlining the content, the Scope also identifies that it is located in two places. The first and most important of these is in the documents identified in Contract Data

Part One. This entry is but a blank space in which the Client must ensure that all the various documents and requirements that make up the Scope are identified. In practice, the most effective way to do this is to identify all the information that needs to be included and, in a structured way with a clear contents list, collect together as much of the information as possible which makes up the Scope to the exclusion of any surplus information. As this document will inevitably include drawings, the bound Scope can sensibly be completed with a list of all drawings within it which serves to incorporate the drawings by reference. That said, it is of course necessary to make copies of all the drawings available to the Contractor.

It should also be noted that this first reference is also referring to the space in Contract Data Part Two where, in '1 General', the Contractor is required to identify the Scope that it has produced in connection with any design as required by clause 21.1 (see Section 7.3). This makes the documents somewhat circular because the Contractor's design obligations are set out in the Client's version of the Scope (see Section 22.2.10). Where such a requirement exists, the Contractor then produces 'Scope' to specify and describe how it will satisfy that requirement. The general requirements for the Contractor's Scope are the same as the Client's Scope, except that the Contractor cannot use its Scope to impose obligations on the Client. The Contractor can only specify and describe how it is going to satisfy the requirements set out in the Client's Scope.

The second location of Scope is in an instruction given in accordance with the Contract. Such an instruction can only be given by the Project Manager under the provisions of clause 14.3 (see Section 5.3) and will more often than not result in a compensation event arising.

22.2.4 General contents: Specification

Specifications should be included for all elements of the construction in sufficient detail to inform the Contractor of the full extent of what is required. These should include all the information which the industry sector concerned needs to enable a successful project to be constructed, and they should cover both materials and workmanship. Any requirements for information from vendors of materials that are to be provided to the Client should also be stated, as well as a list of any spares to be handed over by the Contractor for use in regular maintenance.

In practice, this aspect of the Scope does not appear to present too many problems other than conflicts between different design disciplines both specifying the same part of the finished construction and the possibility that something has been missed altogether. The remedy to this type of problem is better design coordination, usually achieved by appointing a lead designer to manage and coordinate all aspects of the design. It is the lead designer who is best placed to prepare this part of the Scope, as it is essentially a function of the design process.

22.2.5 General contents: Drawings

Following the same guidelines as for the specifications considered above, all necessary and relevant drawings including (but not limited to) general arrangements, location, working, production and detailed drawings need to be identified and included in the Scope. The amount, need and detail will vary depending on the nature of the works to be constructed. The drawings should be supported as necessary with other information such as reinforcement bending schedules, which amplify the information.

In practice, drawings often make up the bulk of the physical mass of the Scope. As a result of their physical size, it is often difficult to bind them into a document which brings all the information together; they are therefore often provided loose. This can lead to problems in keeping track of which revision of a drawing formed part of the Scope at the Contract Date (see clause 11.2(4) and Section 12.3) and which is the current revision that the Contractor has been instructed to work to by the Project Manager. The inclusion of drawing registers within the Scope is a practical solution to this issue. If the Project Manager re-issues the drawing register every time an instruction is issued to the Contractor to work to a revised drawing, it can act as both the transmission notice and become the current drawing register to which everyone should work.

22.2.6 General contents: Constraints

The word *constraints* as used in the definition of Scope creates a position for the necessary inclusion within the document of all details that will affect how the Contractor delivers the project. From this, it can be seen that the concept of what constitutes a constraint is extremely wide to the extent that it is impractical to produce a definitive list of everything that needs to be considered for inclusion. Nonetheless, a list of possible examples of matters to be considered in this class is set out below in order to present users with ideas around which to formulate this section of the Scope for a project. Matters to consider include, but are definitely not limited to:

- restrictions on access;
- working hours;
- noise limits;
- environmental issues;
- the sequencing of the works;
- delivery and storage requirements of materials before fixing;
- provision of facilities for the Client, Project Manager and Supervisor; and
- timing and notice provisions for tests and inspections.

When considering the constraints to be included in the Scope, users should remember that if something that will act as a constraint on the Contractor is not included in the document, then it can be added later by means of an instruction from the Project Manager. However, such an instruction will constitute a compensation event. The more comprehensive the constraints at the Contract Date, the more certain are the parties of their respective obligations and expectations. Further, the total of the Prices (see clauses 11.2(32), 11.2(33), 11.2(34) and 11.2(35)) and planned Completion (see clause 31.2 and Section 12.2) will be more accurate as they will be less subject to change resulting from the need to introduce constraints that had not been included in the Scope.

22.2.7 The Construction (Design and Management) Regulations 2015

In the United Kingdom, in order to comply with The Construction (Design and Management) Regulations 2015 the Client needs to appoint a Principal Designer (see Section 5.14) and a Principal Contractor (see Section 5.15). The identity of both of these people or organizations needs to be included in the Scope, as there is no alternative place provided in the Contract Data for these entries.

In addition, and under the same heading, the Client could include in the Scope reference to the Pre-Construction Phase Health and Safety File. This entry could then lead into any constraints that the Client may wish to impose in respect of the Construction Phase Plan, such as content, style, delivery method and time. Similar matters could also be included in respect of the final Health and Safety file.

22.2.8 The work to be done by Completion: Clause 11.2(2)

In order to remove any subjectivity about whether the Contractor has achieved the position known as Completion, a full and concise statement of the works that must be completed by this date should be included. This statement can cover not only the physical work required, but also any incidental work including (but not limited to) operation and maintenance manuals, as built drawings, health and safety files, testing and commissioning certificates, etc.

22.2.9 Use of a communication system: Clause 13.2

Whether this entry is used is optional, as clause 13.2 provides for the making of communications without such a system. Should the Client decide to use one of the systems that are available in the market then the system to be used must be identified in the Scope. It would be good practice to include details of how the system to be used will be managed, and by whom. Any training requirements and how they can be obtained, together with any user protocol that is to be followed, should also be included.

It is also worth considering and specifying what will happen to the data in the communication system at the end of the project–who will hold the data and how users will access that data after Completion (see Section 12.5.2), the issue of the Defects Certificate (see Section 9.6) and the making of the final payment (see clause 53 and Section 13.6).

22.2.10 Contractor's design responsibilities: Clause 21.1

The extent of the Contractor's design responsibilities, if any, needs to be stated. It is suggested that in practice the Scope should contain a clear heading under which is set out details of those elements of the works that the Contractor shall design. In the event that the Contractor does not have any design responsibility, then this same heading should be employed with a statement to the effect that the Contractor has no design responsibility. This approach provides clarity between the parties, which is an advantage to them both.

On the other hand, it is common to find that there are one or more statements scattered throughout the specification which, by words that are not clearly highlighted, pass design responsibility to the Contractor. Such entries are often difficult for a contractor to find during a limited tender period, and result in the Contractor taking on obligations that have not been properly understood prior to the Contract Date. Such occurrences are unhelpful and only serve to create an atmosphere of mistrust, which is contrary to the requirements of clause 10.2.

22.2.11 Procedures for submission of design: Clauses 21.2 and 23.1

Should there be any requirements for the submission of design information, such as format, recipients, etc. that may be required and that are different to the basic provisions for communications (see clause 13 and Section 6.2), then these should be included. Care should be taken to ensure that any submission and reply procedure included does not create a conflict with clause 21.2 or any other relevant core clause.

The same procedure, or a different one if required, could be used in relation to the submission of temporary works design under clause 23.1.

22.2.12 Purposes for which the Client may require to use and copy the Contractor's design: Clause 22.1

Should the Client wish to expand the purposes for which it can use and copy the Contractor's design set out in clause 22.1, then any additional such purposes must be included.

22.2.13 Details of Others and their works: Clause 25.1

Where the Client will be having other work done either on the Site or within the Working Areas by Others (see clause 11.2(12) and Section 5.11), details need to be included setting out the extent of the cooperation required, any interfaces between the operations of the respective parties and details of the extent of sharing that is required. If the work of the Others requires coordination with that of the Contractor, then sufficient information needs to be provided to allow the Contractor to prepare its programme (see contents list at clause 31.2 and Section 12.2). In the event that such information is not fully comprehensive, then it is possible that a compensation event will arise as identified at clause 60.1(5) (see Section 15.3.5).

22.2.14 Services and other things to be provided by the Client and Contractor to each other: Clause 25.2

The wording at clause 25.2 is one of several examples throughout the ECC where the drafting presents users with great flexibility. By doing so, however, it gives no guidance as to what can be covered. Here, the answer to that conundrum is that the drafter of the Scope simply needs to cover every matter which the Client is going to provide to the Contractor and every matter which the Client requires the Contractor to provide for use by the Client. The only exclusion to these requirements would be the requirement for the Contractor to provide the permanent works specified and described elsewhere in the Scope. That exclusion narrows the contents required by this clause to anything that is of a temporary nature.

Provisions by the Client might include matters such as materials issued free of charge by him to the Contractor, the provision and free use of water or electricity, the free use of existing onsite office facilities for the Contractor's site set-up or any other such facility. Requirements from the Contractor might include the provision of adequate personal protective equipment, office space, computer equipment, stationary, etc. for the Project Manager and Supervisor. Beyond these common examples, the possibilities are limitless. The only controlling factor is that determined by the entries included in the Scope.

22.2.15 Matters relating to Subcontractors: Clause 26

The Scope can be used by the Client to place constraints on the Contractor as to how it can subcontract parts of the works. This could include:

- lists of acceptable subcontractors for particular elements of the works;
- a statement of any work which the Contractor will not be allowed to subcontract;
- a statement of any work which the Contractor will be required to subcontract; and
- any other relevant restrictions or requirements linked to the use of subcontractors.

The extent and content of such statements and requirements are not limited except that they should not conflict with any of the provisions of clause 26. Should this occur, then either the statements need to be reconsidered, which is the preferable option, or an amendment to clause 26 should be considered via Secondary Option Z.

22.2.16 Health and safety requirements: Clause 27.4

The health and safety provisions with which the Contractor is obliged to comply are included. It is not necessary to include health and safety requirements that are imposed on the Contractor by the law. For example, in the United Kingdom it is not necessary in practice to state in the Scope that the Contractor needs to ensure that all employees wear a hard hat. This requirement is set down in statute which cannot be opted out of and is therefore implied into the contract by that statute.

The requirements that need to be included will be those over and above the law of the country in which the Site is located. Most industry sectors and many large employers have this information as well as recognised safety levels and procedures to hand. It is these to which the ECC is referring as being necessary to include.

22.2.17 Any additional information to be included in the programme: Clause 31.2

The final bullet point at clause 31.2 requires the Contractor to show any other information required by the Scope on every programme submitted for acceptance. Given the comprehensive coverage of the rest of clause 31.2, it is often difficult to think of other things that might be required. Users may consider including a requirement to clearly show the critical path or paths through the project. Another matter that users could consider for inclusion includes the type of planning software to be used and the format of programmes to be submitted.

A more controversial suggestion would be to include the effect of notified compensation events. By including this requirement, the effect of matters that are accepted as being a compensation event are shown on the programme earlier, which many consider to be something that provides a more accurate programme at each revision date. This is controversial, as the acceptance of a programme by the Project Manager showing the effect of a notified compensation event for which the quotation had not been accepted would be considered by many as indicating that the Project Manager accepted the delay shown to arise as a result of the compensation event in question. That could well be the case unless the Parties created a position whereby any such acceptance did not act as any form of agreement to the effect of any such compensation event. No doubt, any party that considered adopting this suggestion would need to consider this position carefully before deciding to go ahead and implement it.

22.2.18 Identify reasons why use of the works before Completion will not constitute take over: Clause 35.2

To create the position where by use of the works before Completion (see clause 30.2 and Section 12.5.2) does not constitute take over under clause 35 (see Section 12.6.3) the Client can identify such reasons as will exclude elements of the works from the deemed take over in such circumstances. To do this the Client needs to identify the parts of the works that it may need to use before Completion and set out the reasons for its use.

22.2.19 Requirements for the Contractor's quality management system: Clause 40.1

Should the Client want to ensure that the Contractor's quality management system includes any particular requirements, then they should be set out in the Scope. If nothing is included, then the Contractor is free to operate the system as it sees fit. This requirement could be filled by simply stating the relevant British Standard or International Standard.

22.2.20 Descriptions of tests and inspections to be carried out: Clause 41.1

As the ECC only requires the tests and inspections required by the Scope to be carried out, it is necessary for all such requirements to be included. Those required by the applicable law are implied and therefore do not need to be identified. Rather than spreading such requirements throughout the specification section of the document, it is better in practice to include a full and detailed test and inspection plan. This plan not only identifies what is required but also sets out who is to carry out each test. The identity of the tester does not need to be confined to just the Supervisor and the Contractor, but could also include the Client, Project Manager or someone covered by the term 'Others'.

22.2.21 Materials, facilities and samples to be provided by the Contractor and the Client for tests: Clause 41.2

The separate requirement to identify which party is to supply something in connection with any test can also be included in a full test and inspection plan, thereby combining the requirements of clauses 41.1 and 41.2 into one comprehensive section.

22.2.22 Plant and materials to be inspected or tested before delivery to the Working Areas: Clause 42.1

This requirement is a further matter that could be incorporated into a detailed test and inspection plan. Details contained in this respect could be linked to requirements that must be fulfilled before such Plant and Materials would be marked by the Supervisor in order to comply with the requirements of clauses 70.1 and 71.1 (see Section 10.2).

22.2.23 Details required in each application for payment submitted by the Contractor: Clause 50.2

Use of this entry allows the Client to specify the type and level of detail that is to be included by the Contractor with every application for payment submitted by the Contractor. There is no prescribed minimum, but users should remember that this is only about interim payments. It does not impact the assessment of compensation events (see Chapter 17) and does not form part of the requirements in relation to the final payment (see clause 53 and Section 13.6). If no requirements are stated, then the Contractor would still have to provide sufficient information to satisfy what the industry would recognise as a reasonable standard. Such a requirement should not cause experienced construction professionals any problems.

22.2.24 Preparation of Equipment, Plant and Materials for marking by the Supervisor: Clause 71.1

See Section 22.2.22 regarding clause 42.1.

22.2.25 Statement of any materials from excavation and demolition to which the Client will have title: Clause 73.2

Unless attention is given to the requirements of this inclusion within the Scope, the Contractor will retain title of all materials that arise from either excavations or demolitions carried out by the Contractor. That being the case, should the Client wish to retain the title of any excavated materials or anything from any demolition, then the Scope will need to detail what items are to remain in the Client's ownership. As the Contractor will own any materials not so identified, the Contractor would be able to remove them from the site for disposal and retain any monies received by selling them on.

22.2.26 Any acceptance or procurement procedures to be followed by the Contractor: Clause 11.2(26) (Options C, D and E) and Clause 11.2(27) (Option F)

The requirement to include these matters in the Scope is established by a reference to such matters in the definition of the term 'Disallowed Cost' found in these four Main Options. Without such requirements being identified, the Project Manager would find it impossible to apply this factor in calculating the Price for Work Done to Date under any of these options. The definition is worded such that the detail of what can be set out is not limited by the clause; the Client is therefore free to create whatever constraints it requires in this respect.

22.2.27 Details of records to be kept by the Contractor: Clause 52.2 (Options C, D and E) and Clause 52.3 (Option F)

The requirement at clauses 52.2 and 52.3 (see Section 13.5.3) for the Contractor to keep records is one that, on the face of it, refers simply to records associated with Defined Cost. Users may choose to limit the ambit of anything they add to the Scope to match this intent, but they should also recognise that the wording is drafted such that it could refer to any records whatsoever that the Client or Project Manager might wish to have access to. Such matters could cover any aspect of the works or the way in which the Contractor has carried out the works. The requirements are limitless, but whatever is required will have to be clearly set out in the Scope.

22.2.28 Ultimate holding company guarantee: Clause X4.1

In the event that this Secondary Option has been included, in order to be operative the form of the ultimate holding company guarantee must be set out in the Scope.

This requirement brings out two practical issues. First, the Client will need to ensure that the wording of the guarantee is drafted and included in the Scope when it is sent out to the tendering contractors. This is relatively straightforward.

The second problem is one for the Contractor and can, depending on the attitude of the Client, be more problematic. In the event that the wording of the guarantee is not (for whatever reason) acceptable to the Contractor, then it is vital that the Contractor brings this to the attention of the Client before any contract is formed between the two parties. If the Contractor does not then, once a contract is formed (part of the basis of which would be the Scope), the Contractor would become obliged to provide the guarantee in the form set out therein. In order to prevent this from occurring, the Contractor needs to raise any queries in respect of the wording either prior to or as part of its tender. If the Client is an organisation that, for any reason (such as compliance with local authority or EU regulations), will not accept any tender which is not fully compliant with the tender documents, then raising such a point is where the problem arises. On the other hand, if the Client is open to discussion about such matters, then this is not problematic but just something that needs to be raised.

22.2.29 Undertakings to the Client or Others: Clause X8.4

The comments and issues surrounding these undertakings are the same as for the ultimate holding company guarantee above, to which the reader should refer. The added complexity is that there are up to three types of undertakings to be considered and that these documents are more complex.

22.2.30 Transfer of rights: Clause X9.1

Should the Client wish to relax any aspect of its ownership of the material prepared for the design, then any such exemptions must be identified. If none are stated, then the client will have complete ownership should the Option be in use.

22.2.31 Information Modelling: Clause X10.1

Sub-clause (4) refers to the Information Model Requirements being set out in the Scope. It is therefore necessary to set out the requirements for the Information Model (see clause X10.1(3)) or any change thereto. This should consider matters such as, for example, format, minimum requirements in relation to the data and methods of transmission.

22.2.32 Performance Bond: Clause X13.1

The comments and issues surrounding this Bond are generally the same as for the parent company guarantee above, to which the reader should refer. In addition, the Project Manager is given the role of checking the commercial strength of the bank or insurer proposed as the provider of the bond, as this is a reason for not accepting the proposed bond provider. It is likely that the Project Manager will need specialist advice in order to facilitate the discharge of this responsibility with accuracy and fairness.

22.2.33 Advanced Payment Bond: Clause X14.2

The comments and issues surrounding this bond are the same as for the performance bond above, to which the reader should refer.

22.2.34 The Contractor's design: Clause X15.3

If Option X9 does not apply but the Client wishes to restrict the use of the Contractor's design material, then the restrictions to be complied with need to be detailed.

22.2.35 The Contractor's design: Clause X15.4

It is necessary to identify the form of the drawings, specifications, reports and other documents related to the design that are required to be kept for the period of retention. This could be electronically (identifying the file format or formats), in hard copy or both (or a mixture thereof). It is also a good idea to specify any requirements for backups or second copies and for these to be kept at a separate location.

22.2.36 Retention Bond: Clause X16.3

The comments and issues surrounding this bond are the same as for the performance bond above, to which the reader should refer.

22.2.37 Details of performance tests: Clause X17.1

While there is no actual reference to the Scope in this clause, it would be good practice to include an entry in connection with it. In order to establish whether the performance level that the Contractor is required to exceed, in order to avoid any such damages being charged by the Client, tests to establish such factors need to be included in the Scope. The actual requirement to carry out these tests is established by reference to clause 41.1, and it is that clause which would require the tests and inspections linked to this Secondary Option to be performed.

22.2.38 Early Contractor Involvement: Clause X22.1(3)

To give effect to the definition of the terms 'Stage One' and 'Stage Two', it is necessary to detail what these will consist of. As a prime element of Stage One is design, the extent of design required could be linked to a recognised Plan of Work such as that issued by RIBA (in the UK). By stating the stages to be completed, it would be clear how far the design process was expected to be completed in each stage. Other requirements relating to health and safety, environmental and other issues should also be included.

22.2.39 Early Contractor Involvement: Clause X22.2(2)

This sub-clause refers to the Scope in one of the two reasons given for not accepting a forecast of Defined Cost that has been submitted by the Contractor. In order to allow this reason to function, it is necessary to set out any requirements related to the submission of a forecast of Defined Cost. The requirements stated could include any relevant aspect of the forecast. The more obvious requirements would be the content and format. The format could be prescribed by the inclusion of a template in the Scope. As to content, this could set the nature and level of information to be included in the forecast. Any other requirement outside of these two obvious areas could also be stated; there is no limit.

22.2.40 Early Contractor Involvement: Clause X22.3(1)

This sub-clause refers to the submission of the Contractor's design for Stage Two (delivered during Stage One) and requires that submission to be made in accordance with the procedure set out in the Scope. See Section 22.2.11 above for further comment.

22.2.41 Early Contractor Involvement: Clause X22.3(3)

In the same way that clause X22.2(2) (see Section 22.2.39) refers to a reason not to accept a submission due to a failure to comply to the requirements set out in the Scope, this clause includes the same reason for rejecting a design submission. Similar content would be relevant but related to a design submission rather than to a forecast.

22.2.42 Early Contractor Involvement: Clause X22.3(6)

This clause requires the Contractor to obtain approvals and consents from Others (see clause 11.2(12) and Section 5.11). What has to be obtained for both of these factors must be stated clearly in the Scope. Matters that could be included could relate to subjects such as planning permission, building control consent, party wall agreements and over-sailing rights for tower cranes. These are but a few of the matters that might be on a project. This is one of the entries that the compiler of the Scope will need to use their experience and knowledge and also seek input from other members of the team in order to cover everything that might be needed.

22.2.43 Early Contractor Involvement: Clause X22.5(3)

This is another clause within Option X22 that makes reference to a requirement in the Scope as part of a reason for either not accepting a submission or for a following action not been taken. In this case, the clause refers to a reason for a notice to proceed to Stage 2 not having been issued because the Contractor did not achieve the performance requirements in the Scope. There is no clause anywhere in Option X22 as to what these performance requirements might be. Nonetheless, if the Client states any such performance requirements, in order for them to be effective they will have to be objective (i.e. capable of being measured). As there are no clues as to what they could be, it follows that they could be anything, whether related to time, cost or quality or to something else that is important to the Client.

22.2.44 Requirements for the Project Bank Account: Clause Y1.6

This clause makes reference to the Scope in relation to a reason for the Project Manager not accepting a proposed from the Contractor to add a Supplier to the Named Suppliers for the purposes of Option Y(UK)1.

Having considered Option Y(UK)1 and its intent, it would appear that the use of this restriction would be to limit the Suppliers who could become Named Suppliers to persons or organisations of stated classes, e.g. persons or organisations who would be carrying out certain trades or parts of the Works identified within the Scope. That said, given the open nature of the wording of clause Y1.6, it would be open for the Client and its advisors to make whatever stipulations that they considered necessary for the project in question.

22.3 Site Information

The Site Information (as defined at clause 11.2(18)) only has a role to play within the ECC in judging what physical conditions the Contractor should have determined to have more than a small chance of encountering and that it would be reasonable for the Contractor to have allowed for such circumstances. This judgement is only relevant when considering whether a compensation event as described at clause 60.1(12) (Section 15.3.12) has arisen or not. The reference to the Site Information in connection with this judgement is made at clause 60.2 with a secondary consideration at clause 60.3.

It can be seen that the purpose of the Site Information is to inform the Contractor of what conditions it should expect to encounter, and therefore allow for within its Price. These conditions act as a benchmark against which to determine whether a compensation event has arisen, should the actual conditions encountered be materially different. As the term 'physical conditions' is wider than the term 'ground conditions' (see Event 12 at Section 15.3.12), the information contained within the Site Information must include details of not just the ground but of anything else of a physical nature that would influence the way in which a contractor could carry out the works. For the majority of construction contracts, the type of matters to be considered and included as appropriate would include:

- geotechnical information relating to the ground;
- surveys of the existing topography including size and levels of existing features;
- maps and details of existing underground services;
- maps and details of existing overhead services;
- as built drawings of any existing structures;
- surveys of critical aspects of existing structures (such as the asbestos surveys that are a statutory obligation in the UK);
- information on groundwater and other hydrographical information;
- ecological surveys identifying the possible presence of any protected species; and
- references to any relevant publicly available information.[1]

In most instances, the above list will cover everything that can be provided. Any additional information about physical conditions that would affect the way the work was done should also be included.

22.4 Practical issues

22.4.1 'The Scope is in' and/or 'The Site Information is in'

In Contract Data Part One the entries for both the Scope and the Site Information start with the words 'The [Scope][Site Information] is in'. The intent is that the entry that follows identifies all the documents that contain the relevant information. The entries are phrased in this way so that it is possible to refer to documents as containing that type of information under both entries. For example, a general arrangement drawing of the works might also show the existing services. The general arrangement is Scope, while the location of the existing services is Site Information. There is nothing wrong with this in principle as, in this example, all the information on the drawing is information of one type or the other.

In practice, the problem with this potentially loose approach to identifying what constitutes either Scope or Site Information is when a document that is referred to as including such information also includes something that the Client does not intend to form part of either type of information. As such superfluous information is in a document that is identified as containing one or both types of information, then the Contractor can only safely assume that the superfluous information applies and take account of such information when pricing the project. In such circumstances, the Client may find itself paying more for the project than if the contents of the Scope and Site Information had been more precisely identified. The problem here can be avoided by careful preparation and compartmentation of the information between the two types to the exclusion of anything that is superfluous.

---

[1] See the second bullet point at clause 60.2 and Section 15.3.12.

22.4.2 Using 'constraints'

The inclusion within the Scope of 'constraints' provides the Client with a powerful tool that can be employed to control and exert influence not only over the actual appearance and construction of the project but also over how it is constructed and who constructs part of it. The use of this term allows the Client to impose what in essence are restrictions on what the Contractor can do but, when used in a positive way, these restrictions can act to the benefit of both the Client and the Contractor. This benefit is achieved by using the constraints to inform the Contractor of how the Client would like the scheme to be constructed in any way that fits within the use of such a mechanism.

For example, if the Client required the scheme to be constructed in a certain order, it could use a constraint to direct the Contractor in this way. If the Client required a certain element to be constructed by a particular subcontractor, it could impose a constraint to ensure that only that subcontractor could be used for that element. The way the term is included in the definition at clause 11.2(16) opens the door for the Client and its advisors to be creative about how they use such constraints to create the operating environment for the project. This should be seen as a positive mechanism in that it enables such matters to be detailed, thereby acting as a clear method of communication that also (for the protection of both parties) imposes contractual rights and obligations.

22.4.3 Omission of the physical conditions compensation event

The physical conditions compensation event at clause 60.1(12) is one of the more common omissions from the list of compensation events. This changes the allocation of risk between the Client and the Contractor in the contract. This omission also results in some employers deciding that the provision of Site Information to the tendering contractors in such circumstances is not necessary. It is suggested that, from a practical point of view, the non-provision of Site Information where the compensation event has been omitted can only serve to push up the Prices tendered by contractors.

For projects where the physical conditions compensation event has been omitted, knowledge of the physical conditions that may be encountered becomes even more important to a tendering contractor. As the amended contract requires the contractor to take the risk of the physical conditions, the more information it has about what can be expected, the more accurately the Contractor can price the works. Where the Site Information is not provided, then the tendering contractors are left with little or no basis upon which to judge the risk. The tendency is then to increase the price they tender by the inclusion of risk allowances against the unknown.

The outcome of this analysis is that when an employer decides to change the risk allocation in this way, it will still be in the Client's financial interest to provide the tendering contractors with a reasonable level of Site Information.

It should also be remembered that in the United Kingdom, the need to provide some information about existing structures, such as an asbestos survey, is a statutory requirement. As such information constitutes Site Information, the Client would be obliged to provide such information even if the compensation event had been deleted.

22.4.4 Guidance on writing the Scope

In January 2012, approximately 10 months after the first edition of this book's predecessor (*A Practical Guide to the NEC3 Engineering and Construction Contract*) was published, the NEC Panel published a document entitled 'Works Information Guidance' (NEC Panel 2012) (Works Information being the term for what is now called Scope). This was the first time that the NEC Panel had issued formal guidance on how to prepare and what to put in the Works Information. At that time, the document was a free download accessible through the NEC website. This has subsequently been removed and is no longer available.

When the NEC3 suite of contracts was reprinted in April 2013, one of the new documents produced at that time is entitled 'How to … write the ECC Works Information Scope' (NEC Panel 2013b). Although the author has not checked these two documents against each other, the chapter and sub-section titles are the same across the pair, giving rise to the thought that the 2013 document is the formal published version of the 2012 download.

Within the NEC4 suite, Volume 2 of the User Guide (NEC Panel 2017c) includes a chapter on the Scope dealing with several aspects of this important document, including advice on how to draft it.

Whilst the reader of this book has the guidance at Section 22.2 above, it would be amiss to ignore all guidance that was available to assist the user in preparing what, as I have said previously, is the most important document in the contract. For that reason, I must recommend the contents of the User Guide to readers.

Bibliography

Construction Industry Council (CIC) (2013). Building Information Model (BIM) Protocol. In: 1e. London: CIC.

Latham, M. (1994). *Constructing the Team*. Final Report of the Government/Industry Review of Procurement and Contractual Arrangements in the UK Construction Industry. London: HMSO.

NEC Panel (2012). *Engineering and Construction Contract; Works Information Guidance*. London: Thomas Telford.

NEC Panel (2013a). *NEC3 Engineering and Construction Contract*. London: Thomas Telford.

NEC Panel (2013b). *NEC3 How to Write the ECC Works Information*. London: Thomas Telford.

NEC Panel (2013c). *NEC3 How to Write the PSC Scope*. London: Thomas Telford.

NEC Panel (2013d). *NEC3 How to Write the TSC Service Information*. London: Thomas Telford.

NEC Panel (2013e). *NEC3 How to Use the ECC Communication Forms*. London: Thomas Telford.

NEC Panel (2013f). *NEC3 How to Use the PSC Communication Forms*. London: Thomas Telford.

NEC Panel (2013g). *NEC3 How to Use the TSC Communication Forms*. London: Thomas Telford.

NEC Panel (2013h). *NEC3 How to Use BIM with NEC3 Contracts*. London: Thomas Telford.

NEC Panel (2013i). *NEC3 Professional Services Contract*. London: Thomas Telford.

NEC Panel (2013j). *NEC3 Engineering and Construction Subcontract*. London: Thomas Telford.

NEC Panel (2013k). *NEC3 Engineering and Construction Contract, Guidance Notes*. London: Thomas Telford.

NEC Panel (2017a). *NEC4 Engineering and Construction Contract*. London: Thomas Telford.

NEC Panel (2017b). *NEC4 Establishing a Procurement and Contract Strategy, Volume 1*. London: Thomas Telford.

NEC Panel (2017c). *NEC4 Preparing an Engineering and Construction Contract, Volume 2*. London: Thomas Telford.

NEC Panel (2017d). *NEC4 Selecting a Supplier, Volume 3*. London: Thomas Telford.

NEC Panel (2017e). *NEC4 Managing an Engineering and Construction Contract, Volume 4*. London: Thomas Telford.

NEC Panel (2017f). *NEC4 Supply Contract*. London: Thomas Telford.

NEC Panel (2017g). *NEC4 Supply Short Contract*. London: Thomas Telford.

NEC Panel (2017h). *NEC4 Dispute Resolution Service Contract*. London: Thomas Telford.

NEC Panel (2017i). *NEC4 Engineering and Construction Subcontract*. London: Thomas Telford.

NEC Panel (2017j). *NEC4 Engineering and Construction Short Subcontract*. London: Thomas Telford.

NEC Panel (2017k). *NEC4 Professional Services Contract*. London: Thomas Telford.

NEC Panel (2017l). *NEC4 Professional Services Short Contract*. London: Thomas Telford.

NEC Panel (2017m). *NEC4 Engineering and Construction Short Contract*. London: Thomas Telford.

Rowlinson, M. (2017a). NEC4: What can we expect? Civil Engineering Surveyor, April 2017, 23 (and http://www.michael-rowlinson.co.uk/wp-content/uploads/2018/03/NEC4-What-can-we-expect.pdf).

Rowlinson, M. (2017b). NEC4: Changes to the family, Part 1 Civil Engineering Surveyor, September 2017, 29 (and http://www.michael-rowlinson.co.uk/wp-content/uploads/2018/03/NEC4-Changes-to-the-family-Part-1.pdf).

Rowlinson, M. (2017c). NEC4: Changes to the family, Part 2 Civil Engineering Surveyor, October 2017, 23 (and http://www.michael-rowlinson.co.uk/wp-content/uploads/2018/03/NEC4-Changes-to-the-family-Part-2.pdf).

Rowlinson, M. (2017d). NEC4: Changes to the family, Part 3 Civil Engineering Surveyor, November 2017, 44 (and http://www.michael-rowlinson.co.uk/wp-content/uploads/2018/03/NEC4-Changes-to-the-family-Part-3.pdf).

Rowlinson, M. (2018a). NEC4: Changes to the family, Part 4. Civil Engineering Surveyor, December 2017/January 2018, 52 (and http://www.michael-rowlinson.co.uk/wp-content/uploads/2018/03/NEC4-Changes-to-the-family-Part-4.pdf).

Rowlinson, M. (2018b). NEC4: Changes to the family, Part 5. Civil Engineering Surveyor, February 2018, 43 (and http://www.michael-rowlinson.co.uk/wp-content/uploads/2018/03/NEC4-Changes-to-the-family-Part-5.pdf).

The Construction Task Force (1998). *Rethinking Construction*. HMSO.

Further reading

Eggleston, B. (2006). *The NEC3 Engineering and Construction Contract: A Commentary*, 2e. Blackwell Science.

Gracia, P. and Gracia, D. (2010). Dispute resolution under NEC3 contracts: the Peter Pan conditions destined never to grow up. *Arbitration* **76** (1): 79–85.

Pickavance, K. (2000). *Delay and Disruption in Construction Contracts*, 2e. LLP.

Rowlinson, M. (2006). Subcontracting under the NEC3. Civil Engineering Surveyor, July/August 2006, 28 (and http://www.michael-rowlinson.co.uk/wp-content/uploads/2013/01/SUBCONTRACTING-UNDER-THE-NEC3-ENGINEERING-AND-CONSTRUCTION-CONTRACT1.pdf).

Rowlinson, M. (2009a). Approving and removing people under NEC3. Civil Engineering Surveyor, March 2009, 40 (and http://www.michael-rowlinson.co.uk/wp-content/uploads/2013/01/Approving-and-Removing-People-under-NEC3-Contracts1.pdf).

Rowlinson, M. (2009b). The role of the Supervisor under the ECC. Civil Engineering Surveyor, June 2009, 33 (and http://www.michael-rowlinson.co.uk/wp-content/uploads/2013/01/The-Role-of-the-Supervisor-under-the-ECC.pdf).

Appendix 1
Tables of Clause Numbers, Case Law and Statutes

Table A1.1 Table of clause numbers. Note that clause number in bold indicates the main section where the clause is considered.

Clause Number	Main Option (if applicable)	Section (in this guide)
10.1		Chapter 4; also applies to all clauses and therefore all sections in this guide whether expressly mentioned or not
10.2		
11.1		1.6, **6.6.1**, 10.2, 12.7.1, 13.4.6, 13.7, 14.3, 15.2.2
11.2(1)		6.4, **12.1**, 12.3
11.2(2)		6.4, 9.5, **12.5.2**, 12.7.2, 12.7.3, 12.8.1, 19.5.2, 22.2.8
11.2(3)		9.5, 12.2.1, 12.4.4, 12.4.5, **12.5.2**, 12.6.3, 12.6.4, 12.7.2, 12.7.3, 12.8.1, 12.8.3, 15.3.13
11.2(4)		9.2, **12.3**, 15.3.22, 15.3.25, 19.5.5, 20.2, 20.3, 20.5, 22.2.5
11.2(5)		**6.6.6**, 18.2.3
11.2(6)		**9.4**, 15.3.1
11.2(7)		5.4, **9.6**, 11.2.1, 11.2.3, 20.6
11.2(8)		**6.4**, 6.5.6
11.2(9)		6.6.6, 7.3, 11.2.1, 12.2.4, **14.2.2**
11.2(10)		**13.4.7**, 13.11.6, 14.2, 17.2.1
11.2(11)		5.3, 6.4, 12.2.1, 12.4.4, **12.5.3**, 12.6.4, 15.3.4, 17.5.3
11.2(12)		**5.11**, 7.3.2, 12.2.1, 15.3.5, 18.2.3, 20.10.2, 22.2.13
11.2(13)		**5.2**, **5.5**, 11.2.3
11.2(14)		6.6.6, 8.2, 11.2.1, 11.4.1, **14.2.3**, 18.5.1
11.2(15)		5.5, **7.2**, 8.2, 9.2, 10.2, 11.2.1, 12.2.1, 12.2.2, 16.5.4, 22.2.2
11.2(16)		5.3, 6.3.3, 6.4, 7.2, 7.3.3, 13.4.5, 15.3.1, 20.3, 20.12, **22.2**, 22.4.2
11.2(17)		**6.6.4**, 15.3.12
11.2(18)		6.6.4, 6.7.2, 15.3.12, 22.3
11.2(19)		5.10, **8.2**, 14.2.4, 14.2.7
11.2(20)		**6.6.4**, 6.7.2, 18.5.1
11.2(21)	A, C	**13.5.1**, 17.2.5
11.2(22)	B, D	**13.5.2**, 17.2.5

(*Continued*)

Table A1.1 (Continued)

Clause Number	Main Option (if applicable)	Section (in this guide)
11.2(23)	A, B	3.5, 13.10.1, 14.1, **17.2.4**, 17.6.1, 18.6.1
11.2(24)	C, D, E	3.5, 7.2, **13.4.5**, 13.10.1, 14.1, 14.2.3, 16.5.4, 17.2.4, 17.6.1, 18.6.1
11.2(25)	F	3.5, 7.2, **13.4.6**, 17.6.1, 18.6.1
11.2(26)	C, D, E	3.5, 6.4, **13.4.5**, 13.4.6, 13.11.5, 14.2.3, 14.2.4, 14.4, 22.2.26
11.2(27)	F	3.5, 6.4, **13.4.6**, 22.2.26
11.2(28)	A, B	14.3, 14.5.2, 18.6.2
11.2(29)	A	3.5, 12.4.2, **13.4.3**, 13.5.1
11.2(30)	B	3.5, **13.4.4**, 13.5.2, 13.7, 15.3.22
11.2(31)	C, D, E, F	3.5, **13.4.5**
11.2(32)	A, C	3.5, 6.4, **13.4.3**, 13.5.1, 13.7, 22.2.6
11.2(33)	B, D	3.5, 6.4, **13.4.4**, 22.2.6
11.2(34)	E, F	3.5, 6.4, **13.4.5**, **13.4.6**, 22.2.6
11.2(35)	D	**13.7**, 15.3.22, 22.2.6
12.1		**6.6.8**
12.2		**6.6.8**, 18.2.1, 18.2.4
12.3		**6.6.8**
12.4		**6.6.8**
13.1		**6.2**, 6.3.1, 6.3.4, 19.1, 22.2.11
13.2		**6.2**, 6.3.4, 16.6.2, 19.1, 22.2.9, 22.2.11
13.3		5.13, **6.2**, 6.3.4, 12.3, 15.3.6, 16.5.2, 16.10.7, 19.1, 20.9, 22.2.11
13.4		**6.2**, 6.3.4, 12.3, 13.10.4, 15.3.9, 16.3.2, 19.1, 22.2.11
13.5		**6.2**, 6.3.4, 16.6.5, 19.1, 22.2.11
13.6		**6.2**, 6.3.4, 12.5.2, 19.1, 22.2.11
13.7		**6.2**, 6.3.2, 6.3.4, 9.5, 9.8.1, 16.2, 16.6.6, 16.10.1, 19.1, 22.2.11
13.8		**6.2**, 6.3.4, 12.8.1, 15.3.9, 19.1, 22.2.11
14.1		**6.6.2**, 7.2, 7.3
14.2		**5.3**, 6.6
14.3		**5.3**, 6.6, 9.3, 12.6.4, 15.3.4, 20.10.5, 22.2.3
14.4		**5.2**, 6.6
15.1		**6.4**, 6.5.2, 9.3, 12.4.1, 13.4.5, 16.3.1, 16.4.1, 17.5.1, 20.4.5
15.2		5.11, **6.4**, 6.5.2, 12.2.1, 16.4.1, 17.5.1, 20.4.5
15.3		**6.4**, 6.5.2, 6.5.7, 16.4.1, 17.5.1, 20.4.5
15.4		**6.4**, 6.5.2, 16.4.1, 17.5.1, 20.4.5
16.1		**6.6.3**, 20.9
16.2		6.2, **6.6.3**
16.3		6.2, **6.6.4**, 15.3.9
17.1		**6.6.5**, 9.3, 9.4, 15.3.1, 16.3.1, 17.5.2
17.2		**6.6.5**, 15.3.1, 16.3.1
18.1		**6.6.6**, 18.2.3
18.2		**6.6.6**, 18.2.3

Table A1.1 (Continued)

Clause Number	Main Option (if applicable)	Section (in this guide)
18.3		**6.6.6**, 18.2.3
19.1		**6.6.7**, 6.7.3, 15.3.19, 18.2.4, 18.5.1
20.1		5.5, **7.2**, 18.2.3, 18.4, 22.2.2
20.2	F	6.6.4, **7.2**
20.3	C, D, E, F	**7.2**
20.4	C, D, E, F	**7.2**, 7.5.2, 12.2.1, 16.9.2, 16.10.6
21.1		6.6.2, **7.3**, 7.6.3, 7.6.4, 11.3, 12.2.2, 22.2.3
21.2		6.2, **7.3**, 7.6.4, 9.4, 12.2.2, 15.3.9, 22.2.11
21.3		**7.3**, 7.6.4, 12.2.2
22.1		**7.3**, 7.6.4, 22.2.10, 22.2.12
23.1		6.2, **7.3**, 15.3.9, 22.2.11
24.1		6.2, 7.5, **7.5.1**, 15.3.9
24.2		7.5, **7.5.1**, 13.4.5, 20.10.5
25.1		7.5, **7.5.2**, 12.2.1, 18.2.3, 20.4.5, 22.2.13
25.2		7.4, **7.5.2**, 7.6.5, 11.4.1, 12.2.3, 13.4.2, 15.3.3, 18.2.2, 22.2.14
25.3		7.5, 7.5.2, **12.5.3**, 13.4.2
26.1		7.5, **8.3**, 8.5.2, 14.4, 20.4.5, 22.2.15
26.2		6.2, 7.5, **8.3**, 8.5.1, 8.5.2, 15.3.9, 18.2.3, 20.4.5, 22.2.15
26.3		4.2.3, 6.2, 7.5, **8.3**, 8.5.1, 8.5.2, 8.6, 15.3.9, 18.7.1, 20.4.5, 22.2.15
26.4	C, D, E, F	7.5, **8.4**, 8.5.1, 8.5.2, 13.4.5, 20.4.5, 22.2.15
27.1		7.5, **7.5.2**, 18.2.3, 20.4.5
27.2		7.5, **7.5.2**
27.3		5.3, 7.5, **7.5.2**, 16.2, 16.8, 16.9.1, 16.10.3, 17.2.1
27.4		7.5, **7.5.2**, 12.2.2, 18.2.3, 22.2.16
28.1		7.5, **7.5.3**
29.1		7.5, **7.5.4**
29.2		7.5, **7.5.4**
30.1		6.6.4, 7.2, **12.5.1**, **12.5.2**, 12.7.3
30.2		**12.5.2**, 12.6.3, 13.10.1, 16.10.4, 20.10.3, 22.2.18
30.3		**12.5.3**
31.1		**12.3**
31.2		6.6.4, 7.5.2, 7.6.2, 8.5.1, **12.2**, 12.2.6, 12.3, 12.4.2, 12.4.3, 12.4.5, 12.4.7, 12.5.1, 12.6.1, 12.8.1, 15.3.5, 16.7.1, 17.3, 17.6.1, 22.2.6, 22.2.13, 22.2.17
31.3		6.2, **12.3**, 12.4.3, 12.8.1, 15.3.9, 16.7.1, 19.2.2
31.4	A	**12.2.5**
32.1		**12.3**, 12.8.1, 16.7.1, 16.10.6, 17.3, 17.6.1
32.2		12.2.1, **12.3**, 16.7.1, 17.6.2
33.1		5.2, 6.6.4, 12.2.1, **12.6.1**, 12.8.1, 15.3.2, 18.2.2
34.1		**12.6.2**, 12.8.2, 15.3.4, 18.2.2, 18.2.4
35.1		9.5, 11.2.1, **12.6.3**, 12.7.2, 15.3.15, 22.2.18

(Continued)

Table A1.1 (Continued)

Clause Number	Main Option (if applicable)	Section (in this guide)
35.2		9.5, 11.2.1, **12.6.3**, 12.7.2, 15.3.15, 22.2.18
35.3		9.5, 11.2.1, **12.6.3**, 12.7.2, 15.3.15, 22.2.18
36.1		**12.6.4**, 16.6.1, 20.4.6
36.2		**12.6.4**, 20.4.6
36.3		**12.6.4**, 20.4.6
40.1		**9.2**, 9.8.6, 22.2.19
40.2		6.2, **9.2**, 9.8.6, 15.3.9
40.3		**9.2**, 9.8.6
41.1		5.4, **9.3**, 22.2.20, 22.2.21
41.2		5.4, **9.3**, 15.3.16, 22.2.21
41.3		5.4, **9.3**, 13.5.3
41.4		5.4, **9.3**
41.5		5.4, **9.3**, 15.3.11
41.6		5.4, **9.3**, 13.4.2
41.7	C, D, E	5.4, **9.3**
42.1		5.4, **9.3**, 22.2.22
43.1		**9.5**, 9.8.5, 15.3.10, 16.2
43.2		**9.5**
44.1		**9.5**, 20.6
44.2		**9.5**, 20.6
44.3		5.4, **9.5**, 9.6, 9.8.4, 13.6, 16.4.2, 20.6, 20.8
44.4		**9.5**
45.1		**9.7**, 15.3.1
45.2		**9.7**, 15.3.1, 20.6
46.1		9.5, 9.6, **9.7**, 13.4.2, 15.3.1, 20.6
46.2		9.5, 9.6, **9.7**, 13.4.2, 15.3.1, 20.6
50.1		10.3.4, 12.2.1, **13.2.1**, 13.2.2, 13.10.2, 13.10.4, 16.3.1
50.2		10.3.4, **13.2.1**, 13.4.1, 13.4.2, 13.10.4, 22.2.23
50.3		7.6.5, 9.7, 10.3.4, 12.5.3, **13.4.2**, 13.4.5, 13.4.8, 13.5.1, 13.6, 13.10.4, 13.11.1, 13.11.2, 18.6.1, 20.4.7
50.4		10.3.4, **13.4.8**, 13.9, 13.11.1
50.5		10.3.4, 13.2.1, **13.2.2**
50.6		10.3.4, **13.2.2**
50.7	C, D	10.3.4, **13.3**
50.8	E, F	10.3.4, **13.3**
50.9	C, D, E, F	**10.3.4**
51.1		5.2, 8.6, 10.3.4, **13.2.2**, 13.3, 13.6, 13.8.4, 13.9, 13.10.4, 18.2.2, 18.6.1
51.2		5.2, 8.6, 10.3.4, **13.2.2**, **13.2.3**, 13.4.2, 13.8.4, 13.9, 13.10.4, 13.11.7, 15.3.18, 18.2.2
51.3		5.2, 8.6, 10.3.4, **13.2.3**, 13.4.2, 13.10.4, 15.3.18, 16.7.1

Table A1.1 (Continued)

Clause Number	Main Option (if applicable)	Section (in this guide)
51.4		5.2, 8.6, 10.3.4, **13.2.3**, 13.10.4, 15.3.18
51.5		5.2, 8.6, 10.3.4, **13.2.4**, 13.10.2, 13.10.4, 13.11.3
52.1		**13.4.7**, 14.2.7, 14.5.1, 14.5.2
52.2	C, D, E	**13.5.3**, 14.4, 22.2.27
52.3	F	**13.5.3**, 22.2.27
52.4	C, D, E, F	**13.5.3**
53.1		13.4.2, **13.6**, 13.7, 13.9, 13.11.8, 18.6.1, 20.10.5, 22.2.9, 22.2.23
53.2		13.4.2, **13.6**, 13.9, 13.11.8, 20.10.5, 22.2.9, 22.2.23
53.3		13.4.2, **13.6**, 13.9, 13.11.8, 20.10.5, 22.2.9, 22.2.23
53.4		13.4.2, **13.6**, 13.9, 13.11.8, 20.10.5, 22.2.9, 22.2.23
54.1	C	3.5, 6.6.3, 10.3.4, 13.3, 13.5.1, **13.7**, 16.10.4, 18.6.2, 20.10.5
54.2	C	3.5, 6.6.3, 10.3.4, 13.3, 13.4.2, 13.5.1, **13.7**, 16.10.4, 18.6.2, 20.10.5
54.3	C	3.5, 6.6.3, 10.3.4, 13.3, 13.5.1, **13.7**, 16.10.4, 18.6.2, 20.10.5
54.4	C	3.5, 6.6.3, 10.3.4, 13.3, 13.5.1, **13.7**, 16.10.4, 18.6.2, 20.10.5
54.5	D	3.5, 6.6.3, 10.3.4, 13.3, 13.5.1, **13.7**, 16.10.4, 18.6.2
54.6	D	3.5, 6.6.3, 10.3.4, 13.3, 13.4.2, 13.5.1, **13.7**, 16.10.4, 18.6.2
54.7	D	3.5, 6.6.3, 10.3.4, 13.3, 13.5.1, **13.7**, 16.10.4, 18.6.2
54.8	D	3.5, 6.6.3, 10.3.4, 13.3, 13.5.1, **13.7**, 16.10.4, 18.6.2
55.1	A	**13.5.1**
55.2	C	**13.5.1**
55.3	A	**13.5.1**
55.4	A	6.2, **13.5.1**, 15.3.9
56.1	B, D	**13.5.2**
60.1(1)		7.2, 9.3, 15.3, **15.3.1**, 15.3.3, 15.3.4, 15.3.5, 15.3.22, 16.2, 16.10.2, 16.10.4, 17.2.6, 17.5.2, 20.9, 20.10.5
60.1(2)		6.6.4, 12.6.1, 15.3, **15.3.2**, 16.3.1, 17.6.1
60.1(3)		5.2, 7.5.2, 7.6.5, 12.6.1, 15.3, **15.3.3**, 15.4.1, 16.3.1, 17.6.1
60.1(4)		12.6.2, 12.6.4, 15.3, **15.3.4**, 16.2
60.1(5)		5.2, 6.6.4, 15.3, **15.3.5**, 16.3.1, 17.6.1, 22.2.13
60.1(6)		15.3, **15.3.6**, 16.3.1
60.1(7)		6.6.4, 10.2, 15.3, **15.3.7**, 16.2
60.1(8)		15.3, **15.3.8**, 16.2
60.1(9)		8.3, 12.8.1, 15.3, **15.3.9**, 16.3.1
60.1(10)		9.5, 9.8.5, 15.3, **15.3.10**, 16.2
60.1(11)		15.3, **15.3.11**, 16.3.1
60.1(12)		6.6.4, 15.3, **15.3.12**, 16.3.1, 16.10.4, 22.3, 22.4.3
60.1(13)		12.2.2, 15.3, **15.3.13**, 15.4.3, 16.3.1
60.1(14)		15.3, **15.3.14**, 16.3.1
60.1(15)		15.3, **15.3.15**, 16.2

(*Continued*)

Table A1.1 (Continued)

Clause Number	Main Option (if applicable)	Section (in this guide)
60.1(16)		15.3, **15.3.16**, 16.3.1
60.1(17)		15.3, **15.3.17**, 16.2, 17.2.6, 17.3, 17.4
60.1(18)		6.4, 12.6.1, 15.2.3, 15.3, **15.3.18**, 15.4.1, 16.3.1, 20.9
60.1(19)		6.6.7, 6.7.3, 15.3, **15.3.19**, 16.3.1, 18.2.4
60.1(20)		15.3, **15.3.20**, 16.2
60.1(21)		15.3, **15.3.21**, 16.2, 16.3.1
60.2		6.6.4, **15.3.12**, 15.4.2, 22.3
60.3		**15.3.12**, 22.3
60.4	B, D	3.5, 13.4.4, 13.5.2, 15.3, **15.3.22**, 15.3.23, 16.3.1, 17.2.6
60.5	B, D	3.5, 13.4.4, 13.5.2, 15.3, **15.3.23**, 16.3.1
60.6	B, D	3.5, 13.4.4, 13.5.2, 15.3, **15.3.24**, 16.2, 17.2.6
60.7	B, D	13.4.4, **13.5.2**
61.1		6.3.2, 12.4.6, **16.2**, 16.3.1, 16.6.1, 16.8, 16.10.1, 16.10.3, 16.10.7, 17.4
61.2		6.3.2, **16.2**, 16.6.2, 17.5.1
61.3		12.4.6, 16.1.1, **16.3.1**, 16.3.2, 16.5.3, 16.6.1, 16.10.3, 16.10.7, 17.4, 17.6.4, 20.2
61.4		12.4.6, 12.6.2, 15.2.2, 15.3.4, **16.3.2**, 16.4.1, 16.5.3, 16.6.1, 16.6.2, 16.6.4, 16.6.6, 16.7.4, 16.10.2, 16.10.7, 17.5.1, 19.2.2, 19.2.7, 19.3.7
61.5		6.4, 13.4.5, **16.4.1**, 17.5.1
61.6		15.3.17, 16.4.3, 16.5.1, 16.6.3, 16.10.4, 17.2.2, 17.3, 17.3.1, **17.4**
61.7		16.3.2, **16.4.2**, 16.6.6, 16.10.7
62.1		16.5.3, 16.5.4, **16.6.1**
62.2		15.2.2, **16.5.1**, 16.6.2, 16.10.7
62.3		8.6, 16.5.2, 16.5.3, 16.6.1, 16.6.2, **16.6.3**, 16.6.4, 16.6.6, 16.7.1, 16.7.3, 16.8, 16.10.4, 16.10.5, 16.10.7, 17.5.1, 17.6.4
62.4		16.6.2, **16.6.3**, 16.8
62.5		**16.6.5**, 16.7.1, 16.7.3, 16.10.4, 16.10.5
62.6		**16.6.6**, 16.7.4, 16.9.1, 16.10.7, 19.2.2, 19.2.7, 19.3.7
63.1		14.1, **17.2.1**, 17.6.1, 17.6.4
63.2		**17.2.5**, 17.4
63.3		**17.2.6**, 20.2
63.4		16.1.1, **17.2.6**
63.5		12.2.1, 12.3, 12.4.5, 13.4.5, **17.3**, 17.5.3, 17.6.1
63.6		15.2.3, **16.3.1**
63.7		6.4, 16.4.1, **17.5.1**
63.8		12.2.2, 12.3, 12.4.6, **17.2.2**, **17.3.1**, 17.4
63.9		12.3, **17.2.3**, **17.3.2**
63.10		9.3, 9.4, **17.5.2**

Table A1.1 (Continued)

Clause Number	Main Option (if applicable)	Section (in this guide)
63.11		15.3.4, **17.5.3**
63.12	A, B	6.6.3, **17.2.6**
63.13	C, D	6.6.3, **17.2.6**
63.14	A, C	**16.9.2**
63.15	B, D	**16.9.2**
63.16	A, B	**14.3**
63.17	F	**16.9.2**
64.1		16.5.2, **16.7.1**, 16.7.2, 16.9.1, 16.10.7, 17.2.2, 17.3.1, 17.4, 17.6.4
64.2		**16.7.2**
64.3		**16.7.3**, 16.7.4, 16.10.7, 17.6.4
64.4		**16.7.4**, 16.9.1, 16.10.7, 19.2.2, 19.2.7, 19.3.7
65.1		15.3.20, **16.8**
65.2		15.3.20, **16.8**
65.3		15.3.20, **16.8**
66.1		16.6.2, **16.9.1**, 16.9.2
66.2		**16.9.2**, 16.10.6
66.3		15.3.17, **17.4**
70.1		**10.2**, 18.5.1, 18.6.1, 22.2.22
70.2		**10.2**, 18.5.1, 18.6.1, 18.7.1
71.1		**10.2**, 22.2.22, 22.2.24
72.1		6.6.4, **10.2**
73.1		6.6.4, **10.2**, 10.3.3, 15.3.7
73.2		**10.2**, 10.3.4, 14.2.5, 22.2.25
74.1		**10.2**, 10.3.5
80.1		6.6.4, 7.6.5, **11.2.1**, 11.4.1, 12.2, 12.6.3, 15.3.14
81.1		7.6.5, **11.2.1**, 11.4.1
82.1		**11.2.2**, 13.4.2
82.2		**11.2.2**, 13.4.2, 15.3.14
82.3		**11.2.2**
83.1		**11.2.3**
83.2		**11.2.3**, 12.2.1
83.3		**11.2.3**, 11.3
84.1		**11.2.3**, 12.2.1
84.2		**11.2.3**
84.3		**11.2.3**
85.1		**11.2.3**, 13.4.2
86.1		**11.2.3**, 11.4.2, 12.2.1
86.2		**11.2.3**, 11.4.2
86.3		**11.2.3**, 11.4.2, 13.4.2
90.1		11.2.3, 13.6, 13.9, 18.2, **18.4**, 19.2.7
90.2		**18.2**, 18.5.1, 18.6.1

(Continued)

Table A1.1 (Continued)

Clause Number	Main Option (if applicable)	Section (in this guide)
90.3		18.2, **18.5.1**, 18.6.1
90.4		**18.4**
91.1		**18.2.1**, 18.6.1, 18.7.1
91.2		8.3, **18.2.3**, 18.6.1, 20.3
91.3		12.8.2, **18.2.3**, 18.6.1
91.4		13.8.4, **18.2.2**
91.5		**18.2.4**
91.6		12.6.2, **18.2.2**, **18.2.3**, **18.2.4**, 18.6.1
91.7		**18.2.4**, 18.5.1, 18.6.1
91.8		6.6.6, **18.2.3**
92.1		**18.5.1**
92.2		6.6.4, **18.5.1**
93.1		15.3.18, **18.6.1**
93.2		15.3.18, **18.6.1**
93.3	A	**18.6.2**
93.4	C	**18.6.2**
93.5	D	**18.6.2**
93.6	C, D	13.4.2, **18.6.2**
W1		3.1, 5.6, 5.7, 5.8, 12.2.1, 13.6, 13.11.8, 14.3, 16.10.7, 19.1, **19.2**, 19.3, 19.4, 19.5.1, 19.5.2, 19.5.3, 19.5.4
W2		2.4, 3.1, 5.6, 5.7, 5.8, 12.2.1, 13.6, 13.11.8, 14.3, 16.10.7, 19.1, **19.3**, 19.4, 19.5.1, 19.5.2, 19.5.3, 19.5.4
W3		3.1, 5.8, 5.9, 6.6.4, 12.2.1, 13.6, 13.11.8, 14.3, 16.10.7, 19.1, **19.4**, 19.5.1, 19.5.3, 19.5.4
X1		3.1, 13.4.2, **13.10.1**
X2		3.1, 6.6.4, 15.3.21, 15.3.25, 16.3.1, 17.2.6, **20.2**
X3		3.1, **13.3**
X4		3.1, 3.2, 6.2, 15.3.9, 18.2.3, **20.3**, 20.5, 22.2.28
X5		3.1, 9.5, 12.4.4, **12.7.1**, 12.7.2, 12.7.3
X6		3.1, **12.7.2**, 12.8.4, 13.4.2
X7		3.1, **12.7.3**, 12.8.3, 12.8.4, 13.4.2, 17.6.2
X8		3.1, 3.2, **7.3.2**, 22.2.29
X9		3.1, 7.3.1, **7.3.3**, 22.2.30
X10		3.1, 3.2, 6.2, **7.4**, 11.3, 12.2.1, 15.3.9, 22.2.31
X11		3.1, **18.3**, 18.5.2, 18.6.3
X12		3.1, 3.2, 13.4.2, 15.3.26, 16.3.1, 17.2.6, **20.4**
X13		3.1, 3.2, 6.2, 13.10.3, 15.3.9, 18.2.3, **20.5**, 22.2.32
X14		3.1, 3.2, 6.2, 13.4.2, **13.10.2**, 15.3.9, 15.3.27, 16.3.1, 18.6.1, 22.2.33
X15		3.1, 7.3.1, 8.6, **11.3**, 12.2.1, 15.3.28, 16.3.1, 20.10.4, 22.2.34, 22.2.35

Table A1.1 (Continued)

Clause Number	Main Option (if applicable)	Section (in this guide)
X16		3.1, 3.2, 6.3, 13.4.2, **13.10.3**, 14.2.4, 15.3.9, 18.6.1, 22.2.36
X17		3.1, 13.4.2, **20.6**, 22.2.37
X18		3.1, **20.7**
X20		3.1, 3.2, 12.2.1, 13.4.2, 20.4.7, **20.8**
X21		3.1, 6.2, **20.9**
X22		3.1, 3.2, 6.2, 12.2.1, 13.4.2, 15.3.9, **20.10**, 22.2.38, 22.2.39, 22.2.40, 22.2.41, 22.2.42, 22.2.43
Y(UK)1		2.4, 3.1, 3.2, 5.2, 5.12, 6.2, **13.10.4**, 13.11.7, 15.3.9, 20.11, 22.2.44
Y(UK)2		3.1, 8.6, 12.7.2, 12.7.3, 13.2.2, 13.8.4, **13.9**, 13.10.4, 13.11.7, 15.3.29, 16.3.1, 18.2.2, 18.5.2, 18.6.1, 20.11
Y(UK)3		3.1, 7.3.2, 20.4.1, **20.11**
Z		3.1, 8.5.2, 13.5.3, 14.5.2, 19.1, 19.5.4, **20.12**, 20.13.2

Note that clause number in bold indicates the main section where the clause is considered.

Table A1.2 Table of cases.

Case Law Citation	Section
Anglian Water Services Ltd v Laing O'Rourke Utilities Ltd [2010] EWHC 1529 (TCC)	2.10
Balfour Beatty Building Ltd v Chestermount Properties Ltd (1993) 62 BLR 1	17.3
Birse Construction Ltd v St David Ltd [1999] BLR 194	4.3
Cavendish Square Holding BV v El Makdessi and ParkingEye Ltd v Beavis [2015] UKSC 67	12.7.3
Compass Group UK and Ireland Ltd (trading as Medirest) v Mid Essex Hospital Services NHS Trust [2012] EWHC 781 (QB)	4.4.5
(1) Costain Ltd, (2) O'Rourke Civil Engineering Ltd, (3) Bachy Soletanche Ltd, (4) Emcor Drake & Scull Group plc v (1) Bechtel Ltd, (2) Mr Fady Bassily [2005] EWHC 1018 (TCC)	5.3
Fitzroy Robinson Ltd v Mentmore Towers Ltd [2009] EWHC 1552 (TCC)	20.10.5
Gilbert-Ash (Northern) Ltd v Modern Engineering (Bristol) Ltd [1974] AC 689	13.11.1
Hancock v Brazier [1966] 2 All ER 901	15.2.3
Hart Investments v Fidler & Ors [2007] BLR 30	19.3.6
Henry Boot Construction (UK) Ltd v Malmaison Hotel 70 ConLR 32	12.4.5
Humber Oil Trustees Ltd v Harbour and General Public Works (Stevin) Ltd (1991) 59 BLR 1	15.3.12
London Borough of Merton v Stanley Hugh Leach Limited (1985) 32 BLR 51	5.2
Northern Ireland Housing Executive v Healthy Buildings (Ireland) Limited [2017] NIQB 43	4.3, 17.6.4
Yuanda (UK) Co Ltd v WW Gear Construction Ltd [2010] EWHC 720 (TCC) (13 April 2010)	13.11.2, 21.3.6

Table A1.3 Table of statutes.

Statute	Section
The Bribery Act 2010	6.6.6
The Construction (Design and Management) Regulations 2015	5.14
The Contracts (Rights of Third Parties) Act 1999	7.3.2
The Data Protection Act 1998	14.5.2
The Housing Grants, Construction and Regeneration Act 1996	3.1, 5.7, 13.9, 13.10.4, 13.11.7, 15.3.29, 18.2.2, 18.6.1, 19.1, 19.2, 19.3, 19.3.2, 19.3.4, 19.3.5, 19.3.6, 19.5.1
The Sale of Goods Act 1979	7.3.1
The Scheme for Construction Contracts (England and Wales) Regulations 1998 SI 649	13.9
The Scheme for Construction Contracts (England and Wales) Regulations 1998 (Amendment) (England) Regulations 2011 SI 2333	13.9
The Late Payment of Commercial Interest (Debts) Act 1998	13.11.2
The Limitation Act 1980	11.3, 16.3.2, 16.6.6, 19.2.8, 19.3.8, 19.4.3
The Local Democracy, Economic Development and Construction Act 2009	5.7, 13.9, 13.10.4, 13.11.7, 15.3.12, 18.6.1, 19.1, 19.3

Appendix 2
Tables of Client's, Project Manager's, Supervisor's, Contractor's, Senior Representatives, Adjudicator's, Dispute Avoidance Board and Tribunals Actions

Table A2.1 Client's actions (note that the order of the clauses follows the printed order of the black-covered book).

Clause	Role	Mandatory or Discretionary
10.1	shall act as stated in this contract	Mandatory
10.2	act in a spirit of mutual trust and cooperation	Mandatory
13.1	communicate in a form which can be read, copied and recorded	Mandatory
14.4	replace the Project Manager after notifying the Contractor of the name of the replacement	Discretionary
14.4	replace the Supervisor after notifying the Contractor of the name of the replacement	Discretionary
16.1	consult with the Contractor and the Project Manager about a proposal to change the Scope made by the Contractor	Mandatory
16.2	consider a Contractor's proposal	Discretionary
22.1	use and copy the Contractor's design for any purpose connected with construction, use, alteration, extension or demolition of the works	Discretionary
25.2	provide services and other things to the Contractor as stated in the Scope	Mandatory
28.1	notifies the Contractor before transferring benefit of the contract or any rights under it	Mandatory

(Continued)

Table A2.1 (Continued)

Clause	Role	Mandatory or Discretionary
28.1	does not transfer any benefit to a party that does not intend to work in a spirit of mutual trust and cooperation	Mandatory
29.1	keep information confidential	Mandatory
29.2	consent to the Contractor publicizing the works	Discretionary
33.1	allows access to and use of each part of the Site to the Contractor which is necessary for the work included in the contract on or before the later of its access date and the date of access shown on the Accepted Programme	Mandatory
35.1	take over the works no later than two weeks after Completion	Mandatory
35.2	use any part of the works before Completion has been certified	Discretionary
41.2	provide materials, facilities and samples for tests and inspections as stated in the Scope	Mandatory
44.4	allows access to and use of a part of the works which he has taken over if they are needed for correcting a Defect	Discretionary
51.1	makes interim payments to the Contractor	Mandatory
53.1	makes the final payment to the Contractor	Mandatory
53.2	makes the final payment to the Contractor if he agrees with the Contractor's assessment and the Project Manager has not made the required assessment	Mandatory
53.3 (W1)	refers a dispute about the assessment of the final amount due to the Senior Representatives within four weeks of the assessment being issued	Discretionary
53.3 (W1)	refers any issues not agreed by the Senior Representatives to the Adjudicator within three weeks of the list of issues not agreed being produced or when it should have been produced	Discretionary
53.3 (W1)	refers to the tribunal its dissatisfaction with a decision of the Adjudicator as to the final assessment of the amount due within four weeks of the decision being made	Discretionary
53.3 (W2)	refers a dispute about the assessment of the final amount due to the Senior Representatives within four weeks of the assessment being issued	Discretionary
53.3 (W2)	refers any issues not agreed by the Senior Representatives to the Adjudicator within three weeks of the list of issues not agreed being produced or when it should have been produced	Discretionary
53.3 (W2)	refers to the tribunal its dissatisfaction with a decision of the Adjudicator as to the final assessment of the amount due within four weeks of the decision being made	Discretionary

Table A2.1 (Continued)

Clause	Role	Mandatory or Discretionary
53.3 (W3)	refers a dispute about the assessment of the final amount due to the Dispute Avoidance Board	Discretionary
53.3 (W3)	refers to the *tribunal* its dissatisfaction with the recommendation of the Dispute Avoidance Board within four weeks of the recommendation being made	Discretionary
80.1	carry the liabilities listed	Mandatory
82.2	pay any costs which have been paid by the Contractor and which are the Client's liability	Mandatory
83.1	provides insurances as stated in the Contract Data	Mandatory
84.2	ensure that insurance policies include a waiver by the insurers of their subrogation rights against directors and other employees of every insured except where there is fraud	Mandatory
84.3	comply with the terms and conditions of the insurance policies	Mandatory
85.1	insure a risk which the Contract requires the Contractor to insure if the Contractor does not submit a required certificate	Discretionary
86.3	pay the Contractor the cost of any insurance taken out by the Contractor where the Contract requires the Employer to take out such insurance but for which the Employer does not submit a required policy or certificate	Mandatory
90.1	notify the Project Manager if it wishes to terminate the Contractor's obligation to Provide the Works	Discretionary
90.2	terminate for a reason identified in the Termination Table	Discretionary
91.1	terminate for one of the insolvency based reasons (R1–R10)	Discretionary
91.2	terminate if the Project Manager has notified that the Contractor has defaulted in one of the specified ways and not put the default right within four weeks (R11–R13)	Discretionary
91.3	terminate if the Project Manager has notified that the Contractor has defaulted in one of the specified ways and not stopped defaulting within four weeks (R14–R15)	Discretionary
91.5	terminate if the Parties have been released under the law from further performance of the whole of the contract (R17)	Discretionary
91.6	terminate if an instruction to stop or not to start any substantial work or all work has not been countermanded within thirteen weeks under the stated circumstances (R18 and R20)	Discretionary
91.7	terminate if a prevention event (clause 19) occurs (R21)	Discretionary

(*Continued*)

Table A2.1 (Continued)

Clause	Role	Mandatory or Discretionary
91.8	terminate if the Contractor has carried out a Corrupt Act (R22)	Discretionary
92.1	complete the works and use any Plant and Materials to which it has title	Discretionary
92.2 (P2)	instruct the Contractor to leave the Site, remove any Equipment, Plant and Materials from the Site and assign the benefit of any subcontract and other contract related to the performance of the Contract	Discretionary
92.2 (P3)	use any Equipment to which the Contractor has title to complete the works	Discretionary
W1.1(1)	refer a dispute to the Senior Representatives	Discretionary
W1.1(2)	notify the Senior Representatives, the Contractor and the Project Manager of the nature of the dispute to be resolved	Mandatory
W1.1(2)	submits it statement of case within one week of the notification	Mandatory
W1.1(3)	put into effect the issues agreed by the Senior Representatives	Mandatory
W1.1(4)	do not disclose any evidence, statement or discussion in any subsequent proceedings	Mandatory
W1.2(1)	appoint the Adjudicator under the NEC Dispute Resolution Service Contract	Mandatory (joint)
W1.2(3)	if the Adjudicator resigns or is unable to act, jointly choose a new adjudicator	Mandatory (joint)
W1.2(3)	if a replacement adjudicator is not chosen jointly ask the Adjudicator nominating body to choose one	Discretionary
W1.3(1)	notify and refer disputes not agreed by the Senior Representatives to the Contractor and the Project Manager	Discretionary
W1.3(3)	include with his referral information to be considered by the Adjudicator	Mandatory
W1.3(3)	provide any further information to the Adjudicator within four weeks of the referral	Discretionary
W1.3(3)	extend periods referred to by agreement	Discretionary
W1.3(5)	provide any further information to the Adjudicator as instructed	Mandatory
W1.3(5)	comply with any instruction of the Adjudicator	Mandatory
W1.3(6)	copy any communication to the Adjudicator to the Contractor at the same time	Mandatory
W1.3(9)	proceed as if the matter disputed was not disputed	Mandatory

Table A2.1 (Continued)

Clause	Role	Mandatory or Discretionary
W1.4(1)	do not refer any dispute to the tribunal unless it has first been referred to the Adjudicator	Mandatory
W1.4(2)	notify the Contractor that it intends to refer a dispute to the tribunal within 4 weeks of the Adjudicator's decision	Discretionary
W1.4(3)	notify the Contractor that it intends to refer a dispute to the tribunal where the Adjudicator has not notified the decision within the time allowed	Discretionary
W1.4(3)	do not refer any dispute to the tribunal unless it is less than four weeks since the Adjudicator should have notified the decision	Discretionary
W1.4(6)	do not call the Adjudicator as a witness in tribunal proceedings	Mandatory
W2.1(1)	refer a dispute to the Senior Representatives	Discretionary
W2.1(1)	if a dispute is not resolved by the Senior Representatives refer that dispute to the Adjudicator	Discretionary
W2.1(2)	notify the Senior Representatives, the Contractor and the Project Manager of the nature of the dispute to be resolved	Mandatory
W2.1(2)	submits it statement of case within one week of the notification	Mandatory
W2.1(3)	put into effect the issues agreed by the Senior Representatives	Mandatory
W2.1(4)	do not disclose any evidence, statement or discussion in any subsequent proceedings	Mandatory
W2.2(1)	refer a dispute to the Adjudicator	Discretionary
W2.2(3)	appoint the Adjudicator under the NEC Adjudicator's Contract	Mandatory (joint)
W2.2(5)	if the Adjudicator is not identified in the Contract Data jointly choose a new adjudicator or ask the Adjudicator nominating body to choose an adjudicator	Mandatory
W2.2(5)	if the Adjudicator resigns or is unable to act, jointly choose a new adjudicator or ask the Adjudicator nominating body to choose an adjudicator	Mandatory
W2.2(7)	do not refer a dispute that is substantially the same as one already decided by the Adjudicator	Mandatory
W2.3(1)	give notice to the Contractor before referring a dispute to the Adjudicator	Mandatory
W2.3(1)	if the Adjudicator is named in the Contract Data send a copy of the notice of adjudication to the Adjudicator when it is issued	Mandatory

(Continued)

Table A2.1 (Continued)

Clause	Role	Mandatory or Discretionary
W2.3(1)	if a named adjudicator does not notify whether he is able to act or not within three days of the notice act as if the Adjudicator had resigned	Discretionary
W2.3(2)	within seven days of giving a notice of adjudication refer the dispute to the Adjudicator, providing the information relied on including any supporting documents and provide a copy of all such information and documents to the other party	Mandatory
W2.3(2)	provide the Adjudicator with any further information within fourteen days of the referral	Mandatory
W2.3(2)	extend the fourteen day period if agreed with the Adjudicator and other party	Discretionary
W2.3(4)	provide any further information to the Adjudicator as instructed	Mandatory
W2.3(4)	comply with any instruction of the Adjudicator	Mandatory
W2.3(6)	copy any communication to the Adjudicator to the Contractor at the same time	Mandatory
W2.3(8)	if the Client is the referring party, agree to the extension of the period for the Adjudicator to reach the decision by up to fourteen days	Discretionary
W2.3(8)	if Client is the other party, agree to any longer extension of the period for the Adjudicator to reach the decision	Discretionary
W2.3(9)	proceed as if the matter disputed was not disputed	Mandatory
W2.3(10)	if the Adjudicator does not make the decision and notify to the parties in the time provided agree an extension of the period for the Adjudicator to reach the decision	Discretionary
W2.3(10)	if the Adjudicator does not make the decision and notify to the parties in the time provided and an extension to the period is not agreed then act as if the Adjudicator had resigned	Discretionary
W2.4(1)	do not refer any dispute to the tribunal unless it has first been decided by the Adjudicator	Mandatory
W2.4(2)	notify the Contractor that it intends to refer a dispute to the tribunal within four weeks of the Adjudicator's decision	Discretionary
W2.4(5)	do not call the Adjudicator as a witness in tribunal proceedings	Mandatory
W3.1(1)	jointly select the third member of the Dispute Avoidance Board	Mandatory

Table A2.1 (Continued)

Clause	Role	Mandatory or Discretionary
W3.1(2)	appoint the Dispute Avoidance Board under the NEC Adjudicator's Contract	Mandatory (joint)
W3.1(4)	chose a new member of the Dispute Avoidance Board	Discretionary
W3.1(4)	ask the Dispute Avoidance Board nominating body to nominate a replacement member	Discretionary
W3.1(5)	agree that a site visit by the Dispute Avoidance Board is not required	Discretionary
W3.1(5)	request that the Dispute Avoidance Board makes an extra visit	Discretionary
W3.1(6)	propose the agenda for a site visit by the Dispute Avoidance Board	Mandatory
W3.2(2)	refer a potential dispute to the Dispute Avoidance Board	Mandatory
W3.2(4)	make materials available to the Dispute Avoidance Board	Mandatory
W3.3(1)	do not refer any dispute to the tribunal unless it has first been referred to the Dispute Avoidance Board	Mandatory
W3.3(2)	notify the Contractor that it intends to refer a dispute to the tribunal within four weeks of the Dispute Avoidance Board's recommendation	Discretionary
W3.3(5)	do not call a member of the Dispute Avoidance Board as a witness in tribunal proceedings	Mandatory
X7.2	repay any overpaid damages with interest if the Completion Date is changed	Mandatory
X8.5	prepares the undertakings and send them to the Contractor	Mandatory
X10.1(3)	provide the Project Information and similar information for inclusion in the Information Model	Mandatory
X10.7(1)	be responsible for the Client's liabilities	Mandatory
X11.1	terminate the Contractor's obligations for a reason not identified in the Termination Table	Discretionary
X14.1	make an advanced payment to the Contractor of the amount stated in the Contract Data	Mandatory
X16.3	agree to a retention bond in place of retention	Discretionary
X20.5	add a Key Performance Indicator and associated payment to the Incentive Schedule	Discretionary
X22.5(1)	confirm that Stage Two is to proceed	Discretionary
X22.5(3)	appoint another contractor to complete the Stage Two works	Discretionary
Y1.6	signs the Joining Deed for any new Named Supplier accepted by the Project Manager	Mandatory
Y1.8	make payment to the project bank account of the amount which is due to be paid under the contract	Mandatory

(*Continued*)

Table A2.1 (Continued)

Clause	Role	Mandatory or Discretionary
Y1.9	signs the Authorisation and submit it to the project bank no later than one day before the final date for payment	Mandatory
Y1.13	sign the Trust Deed before the first assessment date	Mandatory
Y2.2	make payment to the Contractor by the final date for payment	Mandatory
Y2.3	notify the Contractor of his intention to pay less by notifying the Contractor how much he considers is due and the basis on which that sum is calculated no later than seven days before the final date for payment	Discretionary
Y2.4	make payment to the Contractor after terminating for one of the listed Reasons unless one of the reasons given not to make such a payment applies	Mandatory

Table A2.2 Project Manager's actions (note that the order of the clauses follows the printed order of the black-covered book).

Clause	Role	Mandatory or Discretionary
10.1	shall act as stated in this contract	Mandatory
10.2	act in a spirit of mutual trust and cooperation	Mandatory
13.1	communicate in a form which can be read, copied and recorded	Mandatory
13.3	reply to a communication within the period for reply or other period stated in the contract	Mandatory
13.4	reply to a communication submitted or resubmitted to him by the Contractor for acceptance	Mandatory
13.4	if a reply is not acceptance then state his reasons	Mandatory
13.5	extend the period for reply to a communication by agreement and notify the Contractor of the extension which has been agreed	Discretionary
13.6	issue his certificates to the Employer and the Contractor	Mandatory
13.7	communicate notifications separately from other communications	Mandatory
13.8	withhold acceptance of a communication	Discretionary
14.2	delegate any of his actions	Discretionary
14.2	cancel any delegation	Discretionary
14.2	take an action which he has delegated	Discretionary
14.3	give an instruction to the Contractor which changes the Scope or a Key Date	Discretionary

Table A2.2 (*Continued*)

Clause	Role	Mandatory or Discretionary
15.1	give an early warning as soon as he becomes aware of any matter which could • increase the total of the Prices; • delay Completion; • delay meeting a Key Date; or • impair the performance of the works in use	Mandatory
15.1	enter early warning matters in the Early Warning Register	Mandatory
15.2	prepares and issues the first Early Warning Register	Mandatory
15.2	instruct the Contractor to attend a first early warning meeting	Mandatory
15.2	instruct other people to attend an early warning meeting if the Contractor agrees	Discretionary
15.3	at an early warning meeting cooperate as required by the contract	Mandatory
15.4	revise the Early Warning Register to record the decisions made at each early warning meeting	Mandatory
15.4	issues the revised Early Warning Register to the Contractor	Mandatory
15.4	issue an instruction if a decision at an early warning meeting needs a change to the Scope	Mandatory
16.1	consult with the Client and the Contractor about a proposal to change the Scope made by the Contractor	Mandatory
16.2	accept or not accept a proposal from the Contractor to change the Scope	Mandatory
16.3	accept or not accept a proposal from the Contractor to add an area to the Working Areas	Mandatory
17.1	notify the Contractor as soon as he becomes aware of an ambiguity or inconsistency in or between the documents which are part of the contract	Mandatory
17.1	state how an ambiguity or discrepancy should be resolved	Mandatory
17.2	notify the Contractor as soon as he becomes aware that the Scope includes an illegal or impossible requirement	
17.2	if the Scope does include an illegal or impossible requirement, give an instruction to change the Scope accordingly	Mandatory
19.1	if an event occurs which constitutes Prevention give an instruction to the Contractor stating how to deal with the event	Mandatory
21.2	accept or not accept the Contractor's design	Mandatory

(*Continued*)

Table A2.2 (Continued)

Clause	Role	Mandatory or Discretionary
23.1	instruct the Contractor to submit particulars of his design for Equipment for acceptance	Discretionary
23.1	accept or not accept the Contractor's design of an item of Equipment	Mandatory
24.1	accept or not accept a replacement key person	Mandatory
24.2	instruct the Contractor to remove an employee	Discretionary
25.2	assess any cost incurred by the Client as a result of the Contractor not providing the services and other things which he is to provide	Mandatory
25.3	decide whether the work meets the Condition for a Key Date	Mandatory
25.3	assess the additional cost which the Client has paid or will incur within four weeks of the meeting of a Condition for a Key Date, where the Condition was not met by the Key Date stated	Mandatory
26.2	accept or not accept a proposed Subcontractor	Mandatory
26.3	accept or not accept the proposed conditions of contract for a Subcontractor	Mandatory
27.2	notify the Contractor of Others to whom access to work being done and to Plant and Materials being stored for the contract is to be provided	Mandatory
30.2	decide the date of Completion	Mandatory
30.2	certify Completion within one week of Completion	Mandatory
31.3	accept or not accept a programme submitted to him	Mandatory
31.3	reply to a notification given by the Contractor regarding a failure to reply to a programme	Discretionary
32.2	instruct the Contractor to submit a revised programme	Discretionary
34.1	instruct the Contractor to stop or not to start any work	Discretionary
34.1	instruct the Contractor to start or re-start any work that has been instructed to be stopped or not started or remove that work from the Scope	Mandatory
35.3	certify the date upon which the Client takes over any part of the works and its extent within one week of the date	Mandatory
36.1	propose an acceleration to achieve Completion before the Completion Date to the Contractor	Discretionary
36.1	if the Contractor is prepared to consider a proposed acceleration instruct the Contractor to submit a quotation for an acceleration to achieve Completion before the Completion Date including any changes to the Key Dates to be included in the quotation	Discretionary
36.1	accept or not accept a quotation for acceleration	Mandatory

Table A2.2 (Continued)

Clause	Role	Mandatory or Discretionary
36.3	change the Prices, the Completion Date and any Key Dates when a quotation for an acceleration is accepted	Mandatory
40.2	accept or not accept a quality policy statement or quality plan	Mandatory
40.3	instruct the Contractor to correct a failure to comply with the quality plan	
41.6	assess the cost incurred by the Client in repeating a test or inspection after a Defect is found	Mandatory
44.4	arrange for the Client to allow the Contractor access to and use of a part of the works which he has taken over if they are needed for correcting a Defect	Mandatory
45.1	propose to the Contractor that the Scope should be changed so that a Defect does not have to be corrected	Discretionary
45.2	accept or not accept a quotation in respect of accepting a Defect	Mandatory
46.1	assess the cost to the Client of having an uncorrected Defect corrected by other people (access provided)	Mandatory
46.2	assess the cost to the Contractor of correcting an uncorrected Defect (access not provided)	Mandatory
50.1	assess the amount due at each assessment date	Mandatory
50.1	decide the first assessment date to suit the procedures of the Parties	Mandatory
50.5	retain one quarter of the Price for Work Done to Date until the Contractor has submitted a first programme for acceptance	Mandatory
50.6	correct any wrongly assessed amount due in a later payment certificate	Mandatory
51.1	certify payment within one week of each assessment date	Mandatory
51.1	give the Contractor details of how the amount due has been assessed	Mandatory
53.1	makes an assessment of the final amount due	Mandatory
53.1	certifies a final payment	Mandatory
53.1	gives the Contractor details of how the amount due has been assessed	Mandatory
61.1	notify the Contractor of a compensation event which arises from the Project Manager or the Supervisor giving an instruction or notification, issuing a certificate or changing an earlier decision	Mandatory
61.2	instruct the Contractor to submit quotations	Mandatory
61.4	reply to a notification of a compensation event by the Contractor	Mandatory

(Continued)

Table A2.2 *(Continued)*

Clause	Role	Mandatory or Discretionary
61.4	decide whether an event notified by the Contractor is a compensation event	Mandatory
61.4	instruct the Contractor to submit quotations (if event notified by contractor is a compensation event)	Mandatory
61.4	reply to a notification given by the Contractor regarding a failure to reply to the notification of a compensation event	Discretionary
61.5	decide whether the Contractor gave an early warning which an experienced Contractor could have given and notify if necessary	Mandatory
61.6	state assumptions about a compensation event if he decides that the effects are too uncertain to be forecast reasonably	Discretionary
61.6	notify a correction if any assumption is later found to be wrong	Mandatory
61.7	do not notify a compensation event after the defects date	Mandatory
62.1	discuss with the Contractor different ways of dealing with a compensation event	Mandatory
62.1	instruct the Contractor to submit alternative quotations	Discretionary
62.3	reply to the submission of a quotation within two weeks	Mandatory
62.4	explain reasons for instructing a revised quotation	Mandatory
62.5	extend the time allowed in relation to the quotation procedure	Discretionary
62.5	inform the Contractor of the agreed extension	Mandatory
62.6	reply to a notification given by the Contractor regarding a failure to reply to a quotation	Discretionary
63.2	agree rates or lump sum prices to assess the change to the Prices	Discretionary
63.11	if a change to the Scope makes the description of the Condition for a Key Date incorrect, correct the description	Mandatory
64.1	assess a compensation event if one of the four trigger events occurs	Mandatory
64.2	use his own assessment of the programme if stated conditions occur	Mandatory
64.3	notify the Contractor of his assessment of a compensation event and give the Contractor details of it within the period allowed for the Contractor's submission	Mandatory
64.4	reply to a notification given by the Contractor regarding a failure to assess a compensation event	Discretionary
65.1	instruct the Contractor to submit quotations for a proposed instruction or a proposed changed decision	Discretionary

Table A2.2 (*Continued*)

Clause	Role	Mandatory or Discretionary
65.1	state the date by which the proposed instruction may be given	Mandatory
65.2	reply to a quotation for a proposed change	Mandatory
65.3	if the quotation for a proposed instruction is not accepted, issue the instruction, notify it as a compensation event and instruct the submission of a quotation	Discretionary
72.1	allow Equipment to be left in the works	Discretionary
73.1	instruct the Contractor how to deal with an object of value or of historical or other interest found within the Site	Mandatory
84.1	accept or not accept the Contractor's insurance certificates	Mandatory
87.1	submit policies and certificates for insurances provided by the Employer to the Contractor for acceptance before the starting date and afterwards as the Contractor instructs	Mandatory
90.1	issue a termination certificate if the reason given complies with the contract	Mandatory
90.3	implement the procedures for termination immediately after issuing a termination certificate	Mandatory
91.2	notify that the Contractor has defaulted in one of the specified ways	Mandatory
91.2	notify the Employer that the Contractor has not put a specified default previously notified right within four weeks	Mandatory
91.3	notify the Employer that the Contractor has defaulted in one of the specified ways	Mandatory
91.3	notify the Employer that the Contractor has not stopped defaulting in a previously specified way within four weeks	Mandatory
55.3 (Opt A)	accept or not accept a revision of the Activity Schedule	Mandatory
63.16 (Opt A)	agree to add a new People Rate	Discretionary
63.16 (Opt A)	assess a required new People Rate if the Project Manager and the Contractor do not agree	Mandatory
60.6 (Opt B)	give an instruction to correct mistakes in the Bill of Quantities which are departures from the rules for item descriptions	Mandatory
60.6 (Opt B)	give an instruction to correct mistakes in the Bill of Quantities which are departures from the rules for the division of the work into items in the method of measurement	Mandatory

(*Continued*)

Table A2.2 (Continued)

Clause	Role	Mandatory or Discretionary
60.6 (Opt B)	give an instruction to correct mistakes in the Bill of Quantities which are due to ambiguities or inconsistencies	Mandatory
63.16 (Opt B)	agree to add a new People Rate	Discretionary
63.16 (Opt B)	assess a required new People Rate if the Project Manager and the Contractor do not agree	Mandatory
20.4 (Opt C)	consult with the Contractor as required in connection with the Contractor's forecasts of the total Defined Cost for the whole of the works	Mandatory
26.4 (Opt C)	agree that the pricing information for a proposed subcontract does not need to be submitted	Discretionary
41.7 (Opt C)	when assessing the cost incurred by the Employer in repeating a test or inspection after a Defect is found do not include the Contractor's cost of carrying out the repeat test or inspection	Mandatory
50.9 (Opt C)	review the Defined Cost records made available for inspection by the Contractor and notify the Contractor of his findings	Mandatory
50.9 (Opt C)	review any further Defined Cost records requested from the Contractor and notify the Contractor of his findings	Mandatory
52.4 (Opt C)	inspect the Contractor's records	Discretionary
54.1 (Opt C)	assess the Contractor's share	Mandatory
54.3 (Opt C)	make a preliminary assessment of the Contractor's share at Completion	Mandatory
54.4 (Opt C)	make a final assessment of the Contractor's share at the final payment	Mandatory
93.4 (Opt C)	assess the Contractor's share after termination	Mandatory
93.6 (Opt C)	add or deduct the Contractor's share from the amount due after termination	Mandatory
20.4 (Opt D)	consult with the Contractor as required in connection with the Contractor's forecasts of the total Defined Cost for the whole of the works	Mandatory
26.4 (Opt D)	agree that the pricing information for a proposed subcontract does not need to be submitted	Discretionary
41.7 (Opt D)	when assessing the cost incurred by the Employer in repeating a test or inspection after a Defect is found do not include the Contractor's cost of carrying out the repeat test or inspection	Mandatory
50.9 (Opt D)	review the Defined Cost records made available for inspection by the Contractor and notify the Contractor of his findings	Mandatory

Table A2.2 (Continued)

Clause	Role	Mandatory or Discretionary
50.9 (Opt D)	review any further Defined Cost records requested from the Contractor and notify the Contractor of his findings	Mandatory
52.3 (Opt D)	inspect the Contractor's records	Discretionary
54.5 (Opt D)	assess the Contractor's share	Mandatory
54.7 (Opt D)	make a preliminary assessment of the Contractor's share at Completion	Mandatory
54.8 (Opt D)	make a final assessment of the Contractor's share at the final payment	Mandatory
60.6 (Opt D)	give an instruction to correct mistakes in the Bill of Quantities which are departures from the rules for item descriptions	Mandatory
60.6 (Opt D)	give an instruction to correct mistakes in the Bill of Quantities which are departures from the rules for the division of the work into items in the method of measurement	Mandatory
60.6 (Opt D)	give an instruction to correct mistakes in the Bill of Quantities which are due to ambiguities or inconsistencies	Mandatory
93.5 (Opt D)	assess the Contractor's share after termination	Mandatory
93.6 (Opt D)	add or deduct the Contractor's share from the amount due after termination	Mandatory
20.4 (Opt E)	consult with the Contractor as required in connection with the Contractor's forecasts of the total Defined Cost for the whole of the works	Mandatory
26.4 (Opt E)	agree that the pricing information for a proposed subcontract does not need to be submitted	Discretionary
41.7 (Opt E)	when assessing the cost incurred by the Employer in repeating a test or inspection after a Defect is found do not include the Contractor's cost of carrying out the repeat test or inspection	Mandatory
50.9 (Opt E)	review the Defined Cost records made available for inspection by the Contractor and notify the Contractor of his findings	Mandatory
50.9 (Opt E)	review any further Defined Cost records requested from the Contractor and notify the Contractor of his findings	Mandatory
52.4 (Opt E)	inspect the Contractor's records	Discretionary
20.4 (Opt F)	consult with the Contractor as required in connection with the Contractor's forecasts of the total Defined Cost for the whole of the works	Mandatory

(*Continued*)

Table A2.2 (Continued)

Clause	Role	Mandatory or Discretionary
26.4 (Opt F)	agree that the pricing information for a proposed subcontract does not need to be submitted	Discretionary
50.9 (Opt F)	review the Defined Cost records made available for inspection by the Contractor and notify the Contractor of his findings	Mandatory
50.9 (Opt F)	review any further Defined Cost records requested from the Contractor and notify the Contractor of his findings	Mandatory
52.4 (Opt F)	inspect the Contractor's records	Discretionary
63.17 (Opt F)	agree the change to the Prices and any change to the Completion date or a Key Date for any compensation event which affects work to be done by the Contractor	Mandatory
63.17 (Opt F)	if the Project Manager and Contractor do not agree the assessment of a compensation event which affects work to be done by the Contractor then carry out the assessment	Mandatory
W1.1(3)	put into effect any issue agreed by the Senior Representatives	Mandatory
W1.3(2)	agree to an extension of the time for notifying or referring a dispute before the notice or referral is due	Discretionary
W1.3(2)	notify the Contractor of the extension that has been agreed	Mandatory
W1.3(9)	proceed as if the matter disputed was not disputed	Mandatory
W2.1(3)	put into effect any issue agreed by the Senior Representatives	Mandatory
W2.3(9)	proceed as if the matter disputed was not disputed	Mandatory
X4.2	accept or reject an alternative guarantor proposed by the Contractor	Mandatory
X7.3	if the Employer has taken over part of the works, assess the benefit to the Employer of taking over the part of the works as a proportion of the benefit to the Employer of taking over the whole of the works	Mandatory
X10.3	give an early warning of any matter which could adversely affect the creation or use of the Information Model	Mandatory
X10.4(2)	accept or reject an Information Execution Plan within two weeks	Mandatory
X10.4(2)	reply to a notification given by the Contractor regarding a failure to reply to the submission of an Information Execution Plan	Discretionary
X10.4(3)	instruct the submission of a revised Information Execution Plan	Discretionary

Table A2.2 (Continued)

Clause	Role	Mandatory or Discretionary
X13.1	accept or not accept the bank or insurer which will provide the performance bond	Mandatory
X14.2	accept or not accept the bank or insurer which will issue the advanced payment bond	Mandatory
X16.3	accept or not accept the bank or insurer which will issue the retention bond	Mandatory
X21.2	accept or reject a Contractor's proposal to make a change	Discretionary
X21.3	consult with the Contractor about a consultation for a proposed change	Mandatory
X21.3	accept or reject a Contractor's quotation for a change and give reasons if rejected	Mandatory
X21.4	do not change the Scope unless the quote is accepted	Mandatory
X21.5	change the Scope, the Prices and the Completion Date and the Key Dates if the quote is accepted	Mandatory
X22.2(2)	accept or reject a forecast	Mandatory
X22.2(5)	consult with the Contractor about forecasts being prepared by the Contractor	Mandatory
X22.3(1)	accept or reject the Contractor's design proposal for Stage Two	Mandatory
X22.3(3)	if a Contractor's design proposal for Stage Two is rejected give reasons	Mandatory
X22.4	instruct the change of a key person during Stage One	Discretionary
X22.5(1)	issue a notice to proceed to Stage Two	Discretionary
X22.5(2)	issue an instruction to remove Stage Two from the Scope	Discretionary
X22.6(1)	discuss with the Contractor different ways of dealing with changes to the Budget	Mandatory
X22.6(2)	agree changes to the Budget with the Contractor	Mandatory
X22.6(2)	assess changes to the Budget if agreement is not reached with the Contractor	Mandatory
X22.7(2)	make a preliminary assessment of the budget incentive at Completion of the whole of the works	Mandatory
X22.7(3)	make a final assessment of the budget incentive and include it in the final amount due	Mandatory
Y1.4	accept or not accept details of the banking arrangements for the project Bank Account	Discretionary
Y1.6	accept or not accept the addition of a Supplier to the Named Suppliers	Discretionary
Y1.9	pass the Authorisation, after receipt, to the Employer (implied action)	Mandatory

Table A2.3 Supervisor's actions (note that the order of the clauses follows the printed order of the black-covered book).

Clause	Role	Mandatory or Discretionary
10.1	shall act as stated in this contract	Mandatory
10.2	act in a spirit of mutual trust and cooperation	Mandatory
13.1	communicate in a form which can be read, copied and recorded	Mandatory
13.3	reply to a communication within the period for reply or other period stated in the contract	Mandatory
13.6	issue his certificates to the Project Manager, the Client and the Contractor	Mandatory
14.2	delegate any of his actions	Discretionary
14.2	cancel any delegation	Discretionary
41.3	notify the Contractor of each of his tests and inspections	Mandatory
41.3	notify the Contractor of the results of each of his tests and inspections	Mandatory
41.3	watch any test done by the Contractor	Discretionary
41.5	do his tests and inspections without causing unnecessary delay to the work or a payment	Mandatory
43.1	instruct the Contractor to search for a Defect	Discretionary
43.2	notify the Contractor of each Defect as soon as he finds it	Mandatory
44.3	issues the Defects Certificate at the later of the defects date and the end of the last defects correction period	Mandatory
71.1	mark Plant and Materials which is outside the Working Areas as being for this contract	Discretionary

Table A2.4 Contractor's actions (note that the order of the clauses follows the printed order of the black-covered book).

Clause	Role	Mandatory or Discretionary
10.1	shall act as stated in this contract	Mandatory
10.2	act in a spirit of mutual trust and cooperation	Mandatory
13.1	communicate in a form which can be read, copied and recorded	Mandatory
13.3	reply to a communication within the period for reply or other period stated in the contract	Mandatory
13.4	where the Project Manager has not accepted a communication resubmit that communication within the period for reply taking account of the reasons given by the Project Manager	Mandatory

Table A2.4 (Continued)

Clause	Role	Mandatory or Discretionary
13.5	agree to the extension of the period for reply	Discretionary
15.1	give an early warning as soon as he becomes aware of any matter which could • increase the total of the Prices; • delay Completion; • delay meeting a Key Date; or • impair the performance of the works in use	Mandatory
15.1	give an early warning of a matter which could increase the total of his cost	Discretionary
15.2	attend early warning meetings	Mandatory
15.2	instruct other people to attend a risk reduction meeting if the Project Manager agrees	Discretionary
15.3	at a risk reduction meeting cooperate as required by the contract	Mandatory
16.1	propose a change to the Scope in order to reduce the amount that the Client pays	Discretionary
16.1	consult with the Client and the Project Manager about a proposal to change the Scope made by the Contractor	Mandatory
16.3	submit a proposal for adding an area to the Working Areas to the Project Manager	Discretionary
17.1	notify the Project Manager as soon as he becomes aware of an ambiguity or inconsistency in or between the documents which are part of the contract	Mandatory
17.2	notify the Project Manager as soon as he considers that the Scope requires him to do anything which is illegal or impossible	Mandatory
18.1	do not do a Corrupt Act	Mandatory
18.2	take action to stop a Corrupt Act of a Subcontractor or a supplier	Mandatory
18.3	include equivalent provisions in subcontracts and contracts for the supply of Plant and Materials and Equipment	Mandatory
20.1	Provide the Works in accordance with the Scope	Mandatory
21.1	design the parts of the works which the Scope states the Contractor is to design	Mandatory
21.2	submits particulars of its design as the Scope requires to the Project Manager for acceptance	Mandatory
21.2	does not proceed with the relevant work until the Project Manager has accepted its design for that work	Mandatory
21.3	submit its design for acceptance in parts if the design of each part can be assessed fully	Discretionary

(Continued)

Table A2.4 (Continued)

Clause	Role	Mandatory or Discretionary
22.1	obtain equivalent rights for the Client to use any design prepared by subcontractors	Mandatory
23.1	submits particulars of the design of an item of Equipment to the Project Manager for acceptance if the Project Manager instructs him to	Mandatory
24.1	either employs each key person named to do the job for him stated in the Contract Data or employs a replacement person who has been accepted by the Project Manager	Mandatory
24.1	submits the name, relevant qualifications and experience of a proposed replacement person to the Project Manager for acceptance	Mandatory
24.2	where the Project Manager has instructed the removal of an employee, arrange that, after one day, the employee has no further connection with the work included in this contract	Mandatory
25.1	cooperates with Others in obtaining and providing information which they need in connection with the works	Mandatory
25.1	cooperates with Others and shares the Working Areas with them as stated in the Scope	Mandatory
25.2	provide services and other things as stated in the Scope	Mandatory
25.2	pay any cost incurred by the Employer as a result of not providing the services and other things which he is to provide as assessed by the Project Manager	Mandatory
25.3	pay the additional cost which the Employer has paid or will incur as a result of the failure to satisfy the Condition for a Key Date by the Key Date as assessed by the Project Manager	Mandatory
26.1	if it subcontracts work, be responsible for Providing the Works as if it had not subcontracted	Mandatory
26.2	submit the name of each proposed Subcontractor to the Project Manager for acceptance	Mandatory
26.2	do not appoint a proposed Subcontractor until the Project Manager has accepted it	Mandatory
26.3	submit the proposed conditions of contract for each subcontract to the Project Manager for acceptance subject to exceptions	Mandatory
27.1	obtain approval of its design from Others where necessary	Mandatory
27.2	provide access to work being done and to Plant and Materials being stored for this contract for the Project Manager, the Supervisor and Others notified to it by the Project Manager	Mandatory

Table A2.4 (*Continued*)

Clause	Role	Mandatory or Discretionary
27.3	obey an instruction which is in accordance with this contract and is given to it by the Project Manager or the Supervisor	Mandatory
27.4	act in accordance with the health and safety requirements stated in the Scope	Mandatory
28.1	notify the Client if it intends to transfer the benefit of this contract or any rights under it	Mandatory
29.1	keep information confidential	Mandatory
29.2	only publicise the works with the Client's agreement	Mandatory
30.1	do not start work on the Site until the first access date	Mandatory
30.1	do the work so that Completion is on or before the Completion Date	Mandatory
30.3	do the work so that the Condition stated for each Key Date is met by the Key Date	Mandatory
31.1	if a programme is not identified in the Contract Data, submit a first programme to the Project Manager for acceptance within the period stated in the Contract Data	Mandatory
31.2	show on each programme submitted for acceptance all of the things stated in the Contract	Mandatory
31.3	notify the Project Manager that the Project Manager has failed to reply to the submission of a programme within the necessary period	Discretionary
32.1	show on each revised programme the three things stated in the Contract	Mandatory
32.2	submit a revised programme to the Project Manager for acceptance as required by the Contract	Mandatory
36.1	propose an acceleration to achieve Completion before the Completion Date to the Project Manager	Discretionary
36.1	provide a quotation for acceleration	Mandatory
36.2	submit details of his assessment with each quotation for acceleration	Mandatory
40.1	operate a quality management system which complies with the requirements of the Scope	Mandatory
40.2	submit a quality policy statement and quality plan to the Project Manager for acceptance	Mandatory
40.2	submit any change to the quality plan to the Project Manager for acceptance	Mandatory
40.3	correct a failure to comply with the quality plan	Mandatory
41.2	provide materials, facilities and samples for tests and inspections as stated in the Scope	Mandatory

(*Continued*)

Table A2.4 (Continued)

Clause	Role	Mandatory or Discretionary
41.3	notify the Supervisor of each of its tests and inspections before it starts and afterwards notify the Supervisor of the results	Mandatory
41.3	notify the Supervisor in time for a test or inspection to be arranged and done before doing work which would obstruct the test or inspection	Mandatory
41.3	allow the Supervisor to watch any test done by the Contractor	Mandatory
41.4	if a test or inspection shows that any work has a Defect, correct the Defect and repeat any test or inspection	Mandatory
41.6	pay the amount incurred by the Employer in repeating a test or inspection after a Defect is found as assessed by the Project Manager	Mandatory
42.1	do not bring to the Working Areas those Plant and Materials which the Scope states are to be tested or inspected before delivery until the Supervisor has notified that they have passed the test or inspection	Mandatory
43.2	until the defects date, notify the Supervisor of each Defect as soon as it finds it	Mandatory
44.1	corrects a Defect whether or not the Supervisor notifies him of it	Mandatory
44.2	corrects a notified Defect before the end of the defect correction period	Mandatory
45.1	propose to the Project Manager that the Scope should be changed so that a Defect does not have to be corrected	Discretionary
45.2	submit a quotation for reduced Prices or an earlier Completion Date or both to the Project Manager for acceptance where a proposal to accept a Defect has been accepted	Mandatory
46.1	pay the amount of the cost to the Employer of having the Defect corrected by other people assessed by the Project Manager (where access was given but Defect was not corrected)	Mandatory
46.2	pay the amount of the cost to the Contractor of having the Defect corrected assessed by the Project Manager (where access was not given)	Mandatory
50.2	submit an application for payment to the Project Manager before each assessment date	Mandatory
51.1	pay to the Employer any amount shown on a payment certificate which is a reduction from the previously certified amount	Mandatory

Table A2.4 (Continued)

Clause	Role	Mandatory or Discretionary
52.1	include all costs which cannot be recovered as Defined Cost in the Fee	Mandatory
53.2	issue an assessment of the final amount due if the Project Manager has not issued it	Discretionary
53.3 (W1)	refers a dispute about the assessment of the final amount due to the Senior Representatives within four weeks of the assessment being issued	Discretionary
53.3 (W1)	refers any issues not agreed by the Senior Representatives to the Adjudicator within three weeks of the list of issues not agreed being produced or when it should have been produced	Discretionary
53.3 (W1)	refers to the tribunal its dissatisfaction with a decision of the Adjudicator as to the final assessment of the amount due within four weeks of the decision being made	Discretionary
53.3 (W2)	refers a dispute about the assessment of the final amount due to the Senior Representatives within four weeks of the assessment being issued	Discretionary
53.3 (W2)	refers any issues not agreed by the Senior Representatives to the Adjudicator within three weeks of the list of issues not agreed being produced or when it should have been produced	Discretionary
53.3 (W2)	refers to the tribunal its dissatisfaction with a decision of the Adjudicator as to the final assessment of the amount due within four weeks of the decision being made	Discretionary
53.3 (W3)	refers a dispute about the assessment of the final amount due to the Dispute Avoidance Board	Discretionary
53.3 (W3)	refers to the *tribunal* its dissatisfaction with the recommendation of the Dispute Avoidance Board within four weeks of the recommendation being made	Discretionary
61.3	notify the Project Manager of an event which has happened or which he expects to happen if he believes that it is a compensation event and it has not been notified by the Project Manager	Mandatory
61.4	agree a longer period for the Project Manager's notification in respect of a Contractor's compensation event notification if requested to do so by the Project Manager	Discretionary
61.4	notify the Project Manager that the Project Manager has failed to reply to the notification of a compensation event within the necessary period	Discretionary
61.6	take note of any assumption stated by the Project Manager in connection with a compensation event	Mandatory

(Continued)

Table A2.4 (Continued)

Clause	Role	Mandatory or Discretionary
61.7	do not notify a compensation event after the defects date	Mandatory
62.1	discuss with the Project Manager different ways of dealing with a compensation event	Mandatory
62.1	submit required quotations to the Project Manager	Mandatory
62.1	submit alternative quotations for other methods of dealing with the compensation event	Discretionary
62.2	submit details of his assessment with each quotation	Mandatory
62.2	if the programme for the remaining work is altered by the compensation event, include alterations to the Accepted Programme with his quotation	Mandatory
62.3	submit quotations within three weeks of being instructed to do so by the Project Manager	Mandatory
62.4	submit revised quotations within three weeks of being instructed to do so by the Project Manager	Mandatory
62.5	agree to extend the time allowed in relation to the quotation procedure	Discretionary
62.6	issue a notification to the Project Manager reminding the Project Manager that a quotation has not been replied to	Discretionary
63.2	agree rates and lump sums to assess the change to the Prices	Discretionary
63.7	assess the effect of a compensation event as if an early warning had been given if the Project Manager has issued a notice under clause 61.5	Mandatory
64.4	issue a notification to the Project Manager reminding the Project Manager that a Project Manager's assessment has not been issued	Discretionary
65.1	do not put a proposed instruction or a proposed changed decision into effect	Mandatory
65.2	submit a quotation for a proposed instruction within three weeks of being instructed to do so	Mandatory
70.1	pass whatever title it has to Plant and Materials which are outside the Working Areas to the Employer once the Supervisor has marked them as for the contract	Mandatory
70.2	pass whatever title it has to Plant and Materials to the Employer once they are brought within the Working Areas	Mandatory
72.1	remove Equipment from the Site when it is no longer needed (unless the Project Manager allows it to be left in the works)	Mandatory
73.1	notify the Project Manager when an object of value or of historical or other interest is found within the Site	Mandatory

Table A2.4 (Continued)

Clause	Role	Mandatory or Discretionary
73.1	do not move an object of value or of historical or other interest is found within the Site without an instruction from the Project Manager	Mandatory
74.1	use material provided by the Client to provide the Works	Mandatory
74.1	make material provided available to a subcontractor	Discretionary
81.1	carry the liabilities listed	Mandatory
82.1	pay any costs which have been paid by the Client and which are the Contractor's liability	Mandatory
83.2	provide insurances as stated in the Insurance Table (except where stated in the Contract Data to be provided by the Employer)	Mandatory
83.2	provide additional insurances as stated in the Contract Data	Mandatory
84.1	submit certificates which state that the insurance required by this contract is in force at the required intervals	Mandatory
84.2	ensure that insurance policies include a waiver by the insurers of their subrogation rights against the Parties and the directors and other employees of every insured except where there is fraud	Mandatory
84.3	comply with the terms and conditions of the insurance policies	Mandatory
85.1	pay the Client the cost of any insurance taken out by the Client where the Contract requires the Contractor to take out such insurance but for which the Contractor does not submit a required policy or certificate	Mandatory
86.1	accept the insurance policies and certificates submitted on behalf of the Client if they comply with the contract	Mandatory
87.3	insure a risk which the Contract requires the Client to insure if the Client does not submit a required certificate	Discretionary
90.1	notify the Project Manager if it wishes to terminate its obligation to Provide the Works	Discretionary
90.2	terminate only for a reason identified in the Termination Table	Discretionary
90.3	implement the procedures for termination immediately after the Project Manager has issued a termination certificate	Mandatory
90.4	do no further work necessary to Provide the Works after a termination certificate has been issued	Mandatory
91.1	terminate for one of the insolvency based reasons (R1–R10)	Discretionary
91.4	terminate if the Client has not paid an amount due within thirteen weeks of the date it should have been paid (R16)	Discretionary

(Continued)

Table A2.4 (Continued)

Clause	Role	Mandatory or Discretionary
91.5	terminate if the Parties have been released under the law from further performance of the whole of the contract (R17)	Discretionary
91.6	terminate if an instruction to stop or not to start any substantial work or all work has not been countermanded within thirteen weeks under the stated circumstances (R19 and R20)	Discretionary
92.2 (P2)	if instructed to by the Client leave the Site, remove any Equipment, Plant and Materials from the Site and assign the benefit of any subcontract and other contract related to the performance of the Contract	Mandatory
92.2 (P3)	promptly remove any Equipment from the Site when instructed to do so by the Client	Mandatory
92.2 (P4)	leave the Working Areas and remove the Equipment	Mandatory
31.4 (Opt A)	provide information which shows how each activity on the Activity Schedule relates to the operations on each programme which he submits for acceptance	Mandatory
55.3 (Opt A)	submit a revision of the Activity Schedule if he changes the planned method of working	Mandatory
63.16 (Opt A)	agree to add a new People Rate	Discretionary
60.7 (Opt B)	assume that the Bill of Quantities is correct	Mandatory
63.13 (Opt B)	agree to not compile a new bill item in accordance with the method of measurement	Discretionary
63.16 (Opt B)	agree to add a new People Rate	Discretionary
20.3 (Opt C)	advise the Project Manager on the practical implications of the design of the works and on subcontracting arrangements	Mandatory
20.4 (Opt C)	prepare forecasts of the total Defined Cost for the whole of the works in consultation with the Project Manager and provide an explanation of any changes	Mandatory
26.4 (Opt C)	submit the pricing information for each proposed subcontract unless the Project Manager has agreed that no submission is required	Mandatory
50.9 (Opt C)	notify the Project Manager when a part of the Defined Cost has been finalized and make the necessary records available for inspection	Mandatory
50.9 (Opt C)	provide any further records requested by the Project Manager	Mandatory
52.2 (Opt C)	keep the stated records	Mandatory
52.4 (Opt C)	allow the Project Manager to inspect at any time within working hours the accounts and records which it is required to keep	Mandatory
54.2 (Opt C)	pay its share of any excess of the Price for Work Done to date over the total of the Prices	Mandatory

Table A2.4 (Continued)

Clause	Role	Mandatory or Discretionary
20.3 (Opt D)	advise the Project Manager on the practical implications of the design of the works and on subcontracting arrangements	Mandatory
20.4 (Opt D)	prepare forecasts of the total Defined Cost for the whole of the works in consultation with the Project Manager and provide an explanation of any changes	Mandatory
26.4 (Opt D)	submit the pricing information for each proposed subcontract unless the Project Manager has agreed that no submission is required	Mandatory
50.9 (Opt D)	notify the Project Manager when a part of the Defined Cost has been finalized and make the necessary records available for inspection	Mandatory
50.9 (Opt D)	provide any further records requested by the Project Manager	Mandatory
52.2 (Opt D)	keep the stated records	Mandatory
52.3 (Opt D)	allow the Project Manager to inspect at any time within working hours the accounts and records which it is required to keep	Mandatory
54.6 (Opt D)	pay its share of any excess of the Price for Work Done to date over the total of the Prices	Mandatory
60.7 (Opt D)	assume that the Bill of Quantities is correct	Mandatory
63.13 (Opt D)	agree to not compile a new bill item in accordance with the method of measurement	Discretionary
20.3 (Opt E)	advise the Project Manager on the practical implications of the design of the works and on subcontracting arrangements	Mandatory
20.4 (Opt E)	prepare forecasts of the total Defined Cost for the whole of the works in consultation with the Project Manager and provide an explanation of any changes	Mandatory
26.4 (Opt E)	submit the pricing information for each proposed subcontract unless the Project Manager has agreed that no submission is required	Mandatory
50.9 (Opt E)	notify the Project Manager when a part of the Defined Cost has been finalized and make the necessary records available for inspection	Mandatory
50.9 (Opt E)	provide any further records requested by the Project Manager	Mandatory
52.2 (Opt E)	keep the stated records	Mandatory
52.4 (Opt E)	allow the Project Manager to inspect at any time within working hours the accounts and records which it is required to keep	Mandatory
20.2 (Opt F)	manage its design, the provision of Site services and the construction and installation of the works	Mandatory

(Continued)

Table A2.4 (Continued)

Clause	Role	Mandatory or Discretionary
20.2 (Opt F)	subcontract its design, the provision of Site services and the construction and installation of the works except work which the Contract Data states that it will do	Mandatory
20.3 (Opt F)	advise the Project Manager on the practical implications of the design of the works and on subcontracting arrangements	Mandatory
20.4 (Opt F)	prepare forecasts of the total Defined Cost for the whole of the works in consultation with the Project Manager and provide an explanation of any changes	Mandatory
26.4 (Opt F)	submit the pricing information for each proposed subcontract unless the Project Manager has agreed that no submission is required	Mandatory
50.9 (Opt F)	notify the Project Manager when a part of the Defined Cost has been finalized and make the necessary records available for inspection	Mandatory
50.9 (Opt F)	provide any further records requested by the Project Manager	Mandatory
52.2 (Opt F)	keep the stated records	Mandatory
52.3 (Opt F)	allow the Project Manager to inspect at any time within working hours the accounts and records which it is required to keep	Mandatory
63.17 (Opt F)	if work which it is to do itself is affected by a compensation event agree the change to the price for the work and any change to the Completion Date and Key Dates	Discretionary
W1.1(1)	refer a dispute to the Senior Representatives	Discretionary
W1.1(2)	notify the Senior Representatives, the Client and the Project Manager of the nature of the dispute to be resolved	Mandatory
W1.1(2)	submits it statement of case within one week of the notification	Mandatory
W1.1(3)	put into effect the issues agreed by the Senior Representatives	Mandatory
W1.1(4)	do not disclose any evidence, statement or discussion in any subsequent proceedings	Mandatory
W1.2(1)	appoint the Adjudicator under the NEC Dispute Resolution Service Contract	Mandatory (joint)
W1.2(3)	if the Adjudicator resigns or is unable to act, jointly choose a new adjudicator	Mandatory (joint)
W1.2(4)	refer a subcontract dispute to the Adjudicator if the matter disputed is the same	Discretionary
W1.3(1)	notify and refer disputes not agreed by the Senior Representatives to the Contractor and the Project Manager	Discretionary

Table A2.4 (*Continued*)

Clause	Role	Mandatory or Discretionary
W1.3(2)	agree to an extension of the time for notifying or referring a dispute before the notice or referral is due	Discretionary
W1.3(3)	include with his referral information to be considered by the Adjudicator	Mandatory
W1.3(3)	provide any further information to the Adjudicator within four weeks of the referral	Discretionary
W1.3(3)	extend periods referred to by agreement	Discretionary
W1.3(5)	provide any further information to the Adjudicator as instructed	Mandatory
W1.3(5)	comply with any instruction of the Adjudicator	Mandatory
W1.3(6)	copy any communication to the Adjudicator to the Contractor at the same time	Mandatory
W1.3(9)	proceed as if the matter disputed was not disputed	Mandatory
W1.4(1)	do not refer any dispute to the tribunal unless it has first been referred to the Adjudicator	Mandatory
W1.4(2)	notify the Client that it intends to refer a dispute to the tribunal within 4 weeks of the Adjudicator's decision	Discretionary
W1.4(3)	notify the Client that it intends to refer a dispute to the tribunal where the Adjudicator has not notified the decision within the time allowed	Discretionary
W1.4(3)	do not refer any dispute to the tribunal unless it is less than four weeks since the Adjudicator should have notified the decision	Discretionary
W1.4(6)	do not call the Adjudicator as a witness in tribunal proceedings	Mandatory
W2.1(1)	refer a dispute to the Senior Representatives	Discretionary
W2.1(1)	if a dispute is not resolved by the Senior Representatives refer that dispute to the Adjudicator	Discretionary
W2.1(2)	notify the Senior Representatives, the Contractor and the Project Manager of the nature of the dispute to be resolved	Mandatory
W2.1(2)	submits it statement of case within one week of the notification	Mandatory
W2.1(3)	put into effect the issues agreed by the Senior Representatives	Mandatory
W2.1(4)	do not disclose any evidence, statement or discussion in any subsequent proceedings	Mandatory
W2.2(1)	refer a dispute to the Adjudicator	Discretionary
W2.2(3)	appoint the Adjudicator under the NEC Adjudicator's Contract	Mandatory (joint)

(*Continued*)

Table A2.4 (*Continued*)

Clause	Role	Mandatory or Discretionary
W2.2(5)	if the Adjudicator is not identified in the Contract Data jointly choose a new adjudicator or ask the Adjudicator nominating body to choose an adjudicator	Mandatory
W2.2(5)	if the Adjudicator resigns or is unable to act, jointly choose a new adjudicator or ask the Adjudicator nominating body to choose an adjudicator	Mandatory
W2.2(7)	do not refer a dispute that is substantially the same as one already decided by the Adjudicator	Mandatory
W2.3(1)	give notice to the Client before referring a dispute to the Adjudicator	Mandatory
W2.3(1)	if the Adjudicator is named in the Contract Data send a copy of the notice of adjudication to the Adjudicator when it is issued	Mandatory
W2.3(1)	if a named adjudicator does not notify whether he is able to act or not within three days of the notice act as if the Adjudicator had resigned	Discretionary
W2.3(2)	within seven days of giving a notice of adjudication refer the dispute to the Adjudicator, providing the information relied on including any supporting documents and provide a copy of all such information and documents to the other party	Mandatory
W2.3(2)	provide the Adjudicator with any further information within fourteen days of the referral	Mandatory
W2.3(2)	extend the fourteen day period if agreed with the Adjudicator and other party	Discretionary
W2.3(3)	refer a subcontract dispute to the Adjudicator if the matter disputed is the same	Discretionary
W2.3(4)	provide any further information to the Adjudicator as instructed	Mandatory
W2.3(4)	comply with any instruction of the Adjudicator	Mandatory
W2.3(6)	copy any communication to the Adjudicator to the Client at the same time	Mandatory
W2.3(8)	if the Contractor is the referring party, agree to the extension of the period for the Adjudicator to reach the decision by up to fourteen days	Discretionary
W2.3(8)	if Contractor is the other party, agree to any longer extension of the period for the Adjudicator to reach the decision	Discretionary
W2.3(9)	proceed as if the matter disputed was not disputed	Mandatory
W2.3(10)	if the Adjudicator does not make the decision and notify to the parties in the time provided agree an extension of the period for the Adjudicator to reach the decision	Discretionary
W2.3(10)	if the Adjudicator does not make the decision and notify to the parties in the time provided and an extension to the period is not agreed then act as if the Adjudicator had resigned	Discretionary

Table A2.4 (Continued)

Clause	Role	Mandatory or Discretionary
W2.4(1)	do not refer any dispute to the tribunal unless it has first been decided by the Adjudicator	Mandatory
W2.4(2)	notify the Client that it intends to refer a dispute to the tribunal within four weeks of the Adjudicator's decision	Discretionary
W2.4(5)	do not call the Adjudicator as a witness in tribunal proceedings	Mandatory
W3.1(1)	jointly select the third member of the Dispute Avoidance Board	Mandatory
W3.1(2)	appoint the Dispute Avoidance Board under the NEC Adjudicator's Contract	Mandatory (joint)
W3.1(4)	chose a new member of the Dispute Avoidance Board	Discretionary
W3.1(4)	ask the Dispute Avoidance Board nominating body to nominate a replacement member	Discretionary
W3.1(5)	agree that a site visit by the Dispute Avoidance Board is not required	Discretionary
W3.1(5)	request that the Dispute Avoidance Board makes an extra visit	Discretionary
W3.1(6)	propose the agenda for a site visit by the Dispute Avoidance Board	Mandatory
W3.2(2)	refer a potential dispute to the Dispute Avoidance Board	Mandatory
W3.2(4)	make materials available to the Dispute Avoidance Board	Mandatory
W3.3(1)	do not refer any dispute to the tribunal unless it has first been referred to the Dispute Avoidance Board	Mandatory
W3.3(2)	notify the Client that it intends to refer a dispute to the tribunal within four weeks of the Dispute Avoidance Board's recommendation	Discretionary
W3.3(5)	do not call a member of the Dispute Avoidance Board as a witness in tribunal proceedings	Mandatory
X4.1	give a guarantee from its ultimate holding company to the Client	Mandatory
X4.2	propose an alternative guarantor, also owned by the ultimate holding company	Discretionary
X7.1	pay delay damages if Completion is later than the Completion Date	Mandatory
X8.1	give undertakings to others as stated in the Contract Data	Mandatory
X8.2	arrange for subcontractors to give undertakings to Others as stated in the Contract Data	Mandatory
X8.3	arrange for subcontractors to give undertakings to the Client as stated in the Contract Data	Mandatory

(Continued)

Table A2.4 (Continued)

Clause	Role	Mandatory or Discretionary
X8.5	sign and return any undertakings to Others that it has to give	Mandatory
X8.5	arrange for subcontractors to sign and return any undertakings to Others and the Client that they have to give	Mandatory
X9.1	obtain rights for the Client as stated in the Scope	Mandatory
X9.1	obtain rights from a its subcontractors equivalent to those which it must provide	Mandatory
X9.1	provide documents to the Client that transfer the rights	Mandatory
X10.2	collaborate with other Information Providers as stated in the Information Model Requirements	Mandatory
X10.3	give an early warning of any matter which could adversely affect the creation or use of the Information Model	Mandatory
X10.4(1)	submit the first Information Execution Plan if not identified in the Contract Data	Mandatory
X10.4(2)	issue a notification to the Project Manager reminding the Project Manager that a. Information Execution Plan has not been replied to	Discretionary
X10.4(3)	submit a revised Information Execution Plan as required	Mandatory
X10.4(4)	provide the Project Information in the form stated in the Information Model Requirements and in accordance with the Information Execution Plan	Mandatory
X10.5	if the Information Execution Plan is altered by a compensation event include the alterations to the Information Execution Plan in the quotation for the compensation event	Mandatory
X10.6	obtain rights from a its subcontractors equivalent to those which it must provide	Mandatory
X10.6	provide documents to the Client that transfer the rights	Mandatory
X10.7(3)	provide insurance for claims made against it arising out of its failure to provide the Project Information using the skill and care normally used by professionals providing information similar to the Project Information	Mandatory
X13.1	give a performance bond to the Client for the amount stated in the Contract Data	Mandatory
X14.2	provide the advanced payment bond	Mandatory
X14.3	repay the advanced payment by installments	
X15.1	use reasonable skill and care to ensure that his design complies with the Scope	Mandatory
X15.3	use the material provided by it subject to ownership and the Scope	Mandatory
X15.4	retain copies of drawings, specifications, reports and other documents recording its design for the period of retention	Mandatory

Table A2.4 (Continued)

Clause	Role	Mandatory or Discretionary
X15.5	provides insurance for claims made against it arising out of its failure to use the skill and care normally used by professionals designing works similar to the works	Mandatory
X16.3	give a retention bond if allowed by the Contract Data or agreed to by the Client	Discretionary
X17.1	pay low performance damages if a defect included in the Defects Certificate shows low performance	Mandatory
X20.2	report to the Project Manager his performance against each of the Key Performance Indicators	Mandatory
X20.3	if his forecast final measurement against a Key Performance Indicator will not achieve the target stated in the Incentive Schedule submit his proposals to the Project Manager for improving performance	Mandatory
X21.1	propose to the Project Manager that the Scope is changed in order to reduce the operating and maintenance costs	Discretionary
X21.2	if the Project manager is willing to consider the change submit a quotation for the proposed change	Mandatory
X21.3	consult with the project manager about the proposed change when the Project Manager requires	Mandatory
X22.2(1)	provide detailed forecasts of the total Defined Cost of the work to be done in Stage One at the intervals stated	Mandatory
X22.2(3)	make a revised submission if the Project Manager does not accept a submission	Mandatory
X22.2(5)	prepare forecasts of the Project Cost at the intervals stated and include explanations of any changes	Mandatory
X22.2(5)	consult with the Project Manager about the forecasts being prepared	Mandatory
X22.3(1)	submit its design proposals for Stage Two	Mandatory
X22.3(2)	include a forecast of the effect of the design proposal on Defined Cost and the Accepted Programme with a design proposal submission	Mandatory
X22.3(4)	make a revised submission if the Project Manager does not accept a submission	Mandatory
X22.3(5)	assess the total of the Prices for Stage Two using the Pricing Information	Mandatory
X22.3(6)	obtain approvals and consents from Others as stated in the Contract Data	Mandatory
X22.3(8)	complete any outstanding design during Stage Two	Mandatory
X22.3(9)	if the main Option is C submit the total of the Prices for Stage Two in the form of changes to the Activity Schedule	Mandatory

(Continued)

Table A2.4 (Continued)

Clause	Role	Mandatory or Discretionary
X22.4	do not replace any key person during Stage One unless the Project Manager instructs the replacement or the key person is unable to continue to act	Mandatory
X22.6(1)	discuss with the Project Manager different ways of dealing with changes to the Budget	Mandatory
X22.6(2)	agree changes to the Budget with the Project Manager	Mandatory
Y1.2	establish the project bank account within three weeks of the Contract Date	Mandatory
Y1.3	pay any charges in connection with and receive any interest from the project bank account, unless the Contract Data states otherwise	Mandatory
Y1.4	submit details of the banking arrangements to the Project Manager for acceptance	Mandatory
Y1.4	provide copies of communications with the project bank in connection with the project bank account to the Project Manager	Mandatory
Y1.5	include the arrangements in this contract regarding the project bank account and Trust Deed in its contracts with the Named Suppliers	Mandatory
Y1.5	notify the details of the project bank account and the arrangement for payment to the Named Suppliers	Mandatory
Y1.6	submit proposals for adding a Supplier to the Named Suppliers to the Project Manager	Discretionary
Y1.6	sign the Joining Deed after a Supplier has been accepted as a named Supplier	Mandatory
Y1.7	submit an application for payment showing the amounts due to Named Supplier to the Project Manager before each assessment date	Mandatory
Y1.8	make payment of any amount that the Client has notified to him that it intends to withhold from the certified amount and which is required to pay the Named Suppliers to the project bank account	Mandatory
Y1.9	prepare the Authorisation setting out the sums due to the Named Suppliers as assessed by the Contractor and to the Contractor for the balance of the payment	Mandatory
Y1.9	sign the Authorisation and submit it to the Project Manager no later than four days before the final date for payment	Mandatory
Y1.10	receive payment from the project bank account	Mandatory
Y1.13	sign the Trust Deed before the first assessment date	Mandatory
Y2.3	notify the Client of its intention to pay less by notifying the Client how much it considers is due and the basis on which that sum is calculated no later than seven days before the final date for payment	Discretionary

Table A2.5 Senior Representatives' actions (note that the order of the clauses follows the printed order of the black-covered book).

Clause	Role	Mandatory or Discretionary
W1.1(3)	attend as many meetings as are necessary to try to resolve the dispute	Mandatory
W1.1(3)	at the end of the period produce a list of the issues agreed and the issues not agreed	Mandatory
W2.1(3)	attend as many meetings as are necessary to try to resolve the dispute	Mandatory
W2.1(3)	at the end of the period produce a list of the issues agreed and the issues not agreed	Mandatory

Table A2.6 Adjudicator's actions (note that the order of the clauses follows the printed order of the black-covered book).

Clause	Role	Mandatory or Discretionary
W1.2(1)	accept an appointment under the terms of the NEC Dispute Resolution Service Contract	Mandatory
W1.2(2)	act impartially	Mandatory
W1.2(2)	decide a dispute that is referred	Mandatory
W1.2(2)	decide the dispute as an independent adjudicator and not as an arbitrator	Mandatory
W1.3(3)	extend the period for the submission of additional information with the Parties agreement	Discretionary
W1.3(4)	decide identical disputes under a subcontract at the same time as the main contract dispute	Mandatory
W1.3(5)	review and revise any action or inaction of the Project Manager or Supervisor and alter a quotation that is treated as being accepted	Discretionary
W1.3(5)	take the initiative in ascertaining the facts and the law related to a dispute	Discretionary
W1.3(5)	instruct a party to provide further information within a stated time	Discretionary
W1.3(5)	instruct a party to take any other action considered necessary to reach the decision and to do so within a stated time	Discretionary
W1.3(6)	copy a communication to a party to the other party at the same time	Mandatory
W1.3(7)	make any assessment of additional cost or delay in the same way as a compensation event is assessed	Mandatory

(*Continued*)

Table A2.6 *(Continued)*

Clause	Role	Mandatory or Discretionary
W1.3(8)	decide the dispute and notify the parties within four weeks of the end of the period for receiving further information	Mandatory
W1.3(11)	correct any clerical mistake or ambiguity within the decision within two weeks of giving the decision	Discretionary
W2.1(1)	decide a dispute that is referred to him	Mandatory
W2.2(3)	accept an appointment under the terms of the NEC Dispute Resolution Service Contract	Mandatory
W2.2(4)	act impartially	Mandatory
W2.2(4)	decide the dispute as an independent adjudicator and not as an arbitrator	Mandatory
W2.3(1)	notify the parties within three days of receiving the notice of adjudication that the Adjudicator can or cannot act	Mandatory
W2.3(2)	extend the period for the provision of further information by agreement with the parties	Discretionary
W2.3(3)	decide identical disputes under a subcontract at the same time as the main contract dispute	Mandatory
W2.3(4)	review and revise any action or inaction of the Project Manager or Supervisor and alter a quotation that is treated as being accepted	Discretionary
W2.3(4)	take the initiative in ascertaining the facts and the law related to a dispute	Discretionary
W2.3(4)	instruct a party to provide further information within a stated time	Discretionary
W2.3(4)	instruct a party to take any other action considered necessary to reach the decision and to do so within a stated time	Discretionary
W2.3(5)	continue the adjudication and make a decision based on the information received if a party does not comply with an instruction to provide information within the time stated	Discretionary
W2.3(6)	copy a communication to a party to the other party at the same time	Mandatory
W2.3(7)	make any assessment of additional cost or delay in the same way as a compensation event is assessed	Mandatory
W2.3(8)	decide the dispute and notify the parties within twenty eight days of the dispute being referred or such longer period as the parties have allowed or agreed	Mandatory
W2.3(8)	allocate the Adjudicator's fees and expenses between the Parties in the decision	Discretionary
W2.3(10)	agree to the extension of the time for making the decision if not made it within the time allowed	Discretionary
W2.3(12)	correct any clerical or typographical error arising by accident or omission within the decision within five days of giving the decision	Discretionary

Table A2.7 Tribunal's actions (note that the order of the clauses follows the printed order of the black-covered book).

Clause	Role	Mandatory or Discretionary
W1.4(4)	settles the dispute referred to it	Mandatory
W1.4(4)	reconsider any decision of the Adjudicator	Discretionary
W1.4(4)	review and revise any action or inaction of the Project Manager or the Supervisor related to the dispute	Discretionary
W1.4(5)	if the tribunal is arbitration, use the arbitration procedure stated in the Contract Data	Mandatory
W2.4(3)	settles the dispute referred to it	Mandatory
W2.4(3)	reconsider any decision of the Adjudicator	Discretionary
W2.4(3)	review and revise any action or inaction of the Project Manager or the Supervisor related to the dispute	Discretionary
W2.4(4)	if the tribunal is arbitration, use the arbitration procedure stated in the Contract Data	Mandatory
W3.4(3)	settles the dispute referred to it	Mandatory
W3.4(3)	reconsider any decision of the Dispute Avoidance Board	Discretionary
W3.4(3)	review and revise any action or inaction of the Project Manager or the Supervisor related to the dispute	Discretionary
W3.4(4)	if the tribunal is arbitration, use the arbitration procedure stated in the Contract Data	Mandatory

Table A2.8 Dispute Avoidance Board's actions (including the members thereof) (note that the order of the clauses follows the printed order of the black-covered book).

Clause	Role	Mandatory or Discretionary
W3.1(2)	accept an appointment under the terms of the NEC Dispute Resolution Service Contract	Mandatory
W3.1(3)	acts impartially	Mandatory
W3.1(5)	visit the site at the intervals stated in the Contract Data from the starting date until the defects date unless the Parties agree that a visit is not necessary	Mandatory
W3.1(6)	decide the agenda for the site visit	Mandatory
W3.2(1)	assist the Parties in resolving potential disputes before they become disputes	Mandatory
W3.2(5)	visit the site and inspect the works	Mandatory
W3.2(5)	review all potential disputes and help the Parties settle them before they become disputes	Mandatory
W3.2(5)	prepare a note of their visit	Mandatory
W3.2(5)	provide a recommendation of how to resolve a dispute if the Parties haven't resolved it	Mandatory
W3.2(6)	take the initiative in reviewing potential disputes including asking for further information	Discretionary

The actions of the Promotor, the Partners and the Core Group under secondary Option X12 are not listed in this Appendix – see Section 20.4 for a guide to these actions.
The actions of the Project Bank and Named Suppliers under secondary Option Y(UK)1 are not listed in this Appendix – see Section 13.10.4 for a guide to these actions.

Appendix 3
Tables of Communication Forms and Their Uses

Table A3.1 Table of communication forms in Volume 4 of the User Guide. The following forms are shown in the Guide.

Form
an Instruction
a Notification
a Submission
an Acceptance
a Payment Certificate
a Completion Certificate
a Take Over Certificate
a Defects Certificate

Table A3.2 Table of uses of the communication forms named in Volume 4 of the User Guide.

Communication Form	Clause Number	Used by	For
an Instruction	14.3	Project Manager	To instruct the Contractor to change the Scope
	14.3	Project Manager	To instruct the Contractor to change a Key Date
	15.2	Project Manager	To instruct the Contractor to attend a risk reduction meeting
	15.2	Contractor	To instruct the Project Manager to attend a risk reduction meeting
	15.4	Project Manager	To instruct the Contractor to change the Scope

(*Continued*)

A Practical Guide to the NEC4 Engineering and Construction Contract, First Edition. Michael Rowlinson.
© 2019 John Wiley & Sons Ltd. Published 2019 by John Wiley & Sons Ltd.

Table A3.2 (Continued)

Communication Form	Clause Number	Used by	For
	16.2	Project Manager	To instruct the Contractor to change the Scope
	16.2	Project Manager	To instruct the Contractor to submit a quotation
	17.2	Project Manager	To instruct the Contractor to change the Scope
	19.1	Project Manager	To instruct the Contractor how to deal with a Prevention event
	23.1	Project Manager	To instruct the Contractor to submit particulars of the design of an item of Equipment
	24.2	Project Manager	To instruct the Contractor to remove an employee
	32.2	Project Manager	To instruct the Contractor to submit a revised programme
	34.1	Project Manager	To instruct the Contractor to stop or not start any work
	34.1	Project Manager	To instruct the Contractor to re-start or start any work which had previously been stopped or to remove that work from the Scope
	36.1	Project Manager	To instruct the Contractor to provide a quotation for acceleration
	40.3	Project Manager	To instruct the Contractor to correct a failure to comply with the quality plan
	43.1	Supervisor	To instruct the Contract to search for a Defect
	61.2	Project Manager	To instruct the Contractor to issue a quotation
	62.1	Project Manager	To instruct the Contractor to submit alternative quotations
	62.3	Project Manager	To instruct the Contractor to provide a revised quotation
	65.1	Project Manager	To instruct the Contractor to submit quotations for a proposed instruction or a proposed changed decision
	65.2	Project Manager	To instruct the Contractor to provide a revised quotation
	65.2	Project Manager	To instruct that the proposed instruction is to be implemented

Table A3.2 *(Continued)*

Communication Form	Clause Number	Used by	For
	65.3	Project Manager	To instruct a proposed change to the Scope as a compensation event
	73.1	Project Manager	To instruct the Contractor what to do if an object of value, historical or other interest is found within the Site
	86.1	Contractor	To instruct the submission of the Employer's certificates of insurance
	92.2 (P2)	Client	To instruct the Contractor to leave the Site and clear it.
	60.6 (Opt B & D)	Project Manager	To correct a mistake in the Bill of Quantities
	X10.4(3)	Project Manager	To instruct the Contractor to submit a revised Information Execution Plan
	X12.3(5)	A Partner	To issue instructions to implement a decision of the Core Group
	X12.3(6)	Core Group	To issue an instruction to a Partner
	X22.4	Project Manager	To instruct the Contractor to replace a key person during Stage One
	X22.5(2)	Project Manager	To instruct the removal of the work in Stage Two from the Scope
a Notification	13.2	All	To notify or change the address(es) for the receipt of communications from the Contractor
	13.5	Project Manager	To notify an agreement to the extension of the period for reply to a communication
	14.2	Project Manager or Supervisor	To delegate powers to his delegates
	14.4	Client	To notify the Contractor of the name of the replacement Project Manager or Supervisor
	15.1	Project Manager or Contractor	To notify an early warning matter
	17.1	Project Manager or Contractor	To notify an ambiguity or inconsistency
	17.2	Project Manager or Contractor	To notify an illegal or impossible requirement
	28.1	Client or Contractor	To notify of an intended transfer or any benefit or rights under the contract

(Continued)

Table A3.2 (Continued)

Communication Form	Clause Number	Used by	For
	31.3	Contractor	To notify the Project Manager of a failure to reply to the submission of a programme
	41.3	Contractor or Supervisor	To notify the other in time for a test or inspection
	41.3	Contractor or Supervisor	To notify the other of the result of any test or inspection
	42.1	Supervisor	To notify the Contractor that Plant and Materials which has been the subject of a test or inspection before delivery can be brought to the Working Areas
	43.2	Supervisor or Contractor	To notify a Defect
	44.4	Project Manager	To notify the Contractor that access is available to correct a Defect
	61.1	Project Manager	To notify a compensation event
	61.3	Contractor	To notify a compensation event
	61.4	Project Manager	To notify the Contractor of his decision as to whether a compensation event notified by the Contractor is a compensation event or not
	61.4	Contractor	To notify the Project Manager of a failure to rely to the notification of a compensation event
	61.5	Project Manager	To notify the Contractor that he did not issue an early warning that an experienced contractor could have given
	61.6	Project Manager	To notify a correction to an assumption
	62.3	Project Manager	To notify the Contractor that he will be making his own assessment of a compensation event
	62.5	Project Manager	To notify an agreement to the extension of the time allowed or the submission of or a reply to a quotation for a compensation event
	62.6	Contractor	To notify the Project Manager of a failure to reply to a quotation

Table A3.2 (Continued)

Communication Form	Clause Number	Used by	For
	64.3	Project Manager	To notify the Contractor of his assessment of a compensation event
	64.4	Contractor	To notify the Project Manager of a failure to carry out an assessment
	65.2	Project Manager	To notify that a quotation for a proposed instruction is not accepted
	73.1	Contractor	To notify the Project Manager that an object of value, historical or other interest has been found
	90.1	Client or Contractor	To notify the Project Manager of a wish to terminate
	91.2	Project Manager	To notify that the Contractor has not put right a specified default
	91.3	Project Manager	To notify that the Contractor has not stopped a specified default
	92.2	Project Manager	To notify the Contractor to remove any Equipment from Site
	50.9 (Opt C, D, E or F)	Contractor	To notify the Project Manager when a part of Defined Cost has been finalised
	50.9 (Opt C, D, E or F)	Project Manager	To notify the Contractor that either further records are required or of errors in the final assessment
	W1.1(2)	Client or Contractor	To notify the Senior Representatives, the other Party and the Project Manager of the nature of the dispute to be resolved
	W1.4(2)	Client or Contractor	To notify the other that it is dissatisfied with a decision of the Adjudicator
	W1.4(3)	Client or Contractor	To notify the other that it intends to refer a dispute to the tribunal if the adjudicator has not informed the Parties of the decision in time
	W2.1(2)	Client or Contractor	To notify the Senior Representatives, the other Party and the Project Manager of the nature of the dispute to be resolved
	W2.4(2)	Client or Contractor	To notify the other that it is dissatisfied with a decision of the Adjudicator and state that it intends to refer the matter to the tribunal

(*Continued*)

Table A3.2 (Continued)

Communication Form	Clause Number	Used by	For
	W3.2(3)	Client or Contractor	To notify the Dispute Avoidance Board, the other Party and the Project Manager of a potential dispute
	W3.3(2)	Client or Contractor	To notify the other of a matter which it disputes and which it intends to refer to the tribunal
	X10.3	Contractor or Project Manager	To notify an early warning concerning the Information Model
	X10.4(2)	Contractor	To notify the Project Manager of a failure to reply to the submission of an Information Execution Plan
	X11.1	Client	To notify the Project Manager and the Contractor that it is terminating the Contract for a reason not in the Termination Table
	X12.2(2)	Partner	To nominate a representative to act when dealing with other Partners
	X12.3(2)	Partner	To notify another Partner of the need for information
	X12.3(3)	Partner	To notify an early warning to other Partners
	X12.3(8)	Client	To give advice, information and information to the Core Group
	X12.3(9)	Partner	To notify the Core Group before subcontracting any work
	X18.6	Client	To notify the Contractor of a matter which the Contractor is liable of
	X22.2(2)	Project Manager	To notify the Contractor that a forecast is not accepted
	X22.6(2)	Project Manager	To notify the Contractor of the Project Manager's assessment of the change to the Budget
	Y1.5	Contractor	To notify the Named Suppliers the details of the bank arrangements
	Y2.3	Client or Contractor	To notify the other of an intention to pay less
	Y3.3	Client	To notify the Contractor of the name of a beneficiary once it is known
a Submission	16.1	Contractor	To make a proposal that the Scope should be changed

Table A3.2 (*Continued*)

Communication Form	Clause Number	Used by	For
	16.2	Contractor	To submit a quotation for a proposed instruction to change the Scope
	16.3	Contractor	To submit a proposal for adding an area to the Working Areas
	21.2	Contractor	To submit the Contractors design of the Works for acceptance
	23.1	Contractor	To submit the Contractor's design of an item of Equipment for acceptance
	24.1	Contractor	To submit the details of a replacement key person
	26.2	Contractor	To submit the name of a proposed Subcontractor
	26.3	Contractor	To submit the proposed conditions of contract for a proposed Subcontractor
	31.1	Contractor	To submit a first programme
	32.2	Contractor	To submit a revised programme
	36.1	Contractor	To submit a quotation for acceleration
	40.2	Contractor	To provide a quality policy statement and a quality plan
	40.2	Contractor	To provide a changed quality plan
	45.1	Contractor or Project Manager	To submit a proposal that the Scope should be changed so that a Defect does not have to be corrected
	45.2	Contractor	To submit a quotation to accept a Defect
	50.2	Contractor	To submit a payment application
	62.1	Contractor	To submit a quotation for a compensation event
	62.4	Contractor	To submit a revised quotation for a compensation event
	65.2	Contractor	To submit quotations for a proposed instruction
	84.1	Contractor	To submit insurance certificates for acceptance
	86.1	Project Manager	To submit the Client's insurance certificates for acceptance
	31.4 (Option A)	Contractor	To provide information showing how each activity on the Activity Schedule relates to operations on the programme

(*Continued*)

Table A3.2 (*Continued*)

Communication Form	Clause Number	Used by	For
	55.3 (Options A and C)	Contractor	To submit a revised Activity Schedule as necessary
	20.4 (Options C, D, E and F)	Contractor	To submit forecasts of the total Defined Cost to the Project Manager
	26.4 (Options C, D, E and F)	Contractor	To submit the pricing information for each subcontract
	50.9 (Options C, D, E and F)	Contractor	To provide records to demonstrate the assessment of a finalised part of Defined Cost
	50.9 (Options C, D, E and F)	Contractor	To provide any further records requested
	W1.1(2)	Client and Contractor	To submit their statement of case
	W1.3(3)	Client and Contractor	To provide information to the Adjudicator
	W1.3(5)	Client and Contractor	To provide information to the Adjudicator if instructed to so
	W2.1(2)	Client and Contractor	To submit their statement of case
	W2.3(2)	Client and Contractor	To provide information to the Adjudicator
	W2.3(4)	Client and Contractor	To provide information to the Adjudicator if instructed to so
	W3.2(5)	Dispute Avoidance Board	To provide a recommendation for resolving a dispute
	W3.2(6)	Client and Contractor	To provide additional information requested by the Dispute Avoidance Board
	X4.1	Contractor	To submit an ultimate holding company guarantee to the Client
	X8.1	Contractor	To give an undertaking to others as required
	X8.2	Contractor	To give Subcontractor undertakings to Others to the Client
	X8.3	Contractor	To give Subcontractor undertakings to the Client to the Client
	X9.1	Contractor	To provide documents transferring rights to the Client
	X10.4(1)	Contractor	To submit a first Information Execution Plan

Table A3.2 (Continued)

Communication Form	Clause Number	Used by	For
	X10.4(3)	Contractor	To submit a revised Information Execution Plan
	X10.4(4)	Contractor	To provide the Project Information
	X10.6	Contractor	To provide documents transferring rights to the Client
	X12.3(2)	Partner	To provide another partner with information needed
	X12.3(8)	Contractor	To give advice, information and information to the Core Group
	X13.1	Contractor	To submit the name of a proposed bank or insurer for the performance bond
	X14.2	Contractor	To submit the name of a proposed bank or insurer for the advanced payment bond
	X20.2	Contractor	To submit the Contractor's performance against each Key Performance Indicator to the Project Manager
	X20.3	Contractor	To submit his proposals for improving performance to the Project Manager
	X21.2	Contractor	To submit a quotation for a whole life cost change
	X22.2(1)	Contractor	To provide detail forecasts of the total defined Cost
	X22.2(3)	Contractor	To submit a revised forecast of the total defined Cost
	X22.2(5)	Contractor	To submit forecasts of the Project Cost
	X22.3(1)	Contractor	To submit design proposals for Stage Two
	X22.3(4)	Contractor	To submit revisions to the design proposal for Stage Two
	X22.3(9) (Option C)	Contractor	To submit the total of the Prices for Stage Two
	Y1.4	Contractor	To submit details of the banking arrangements for the project bank account
	Y1.6	Contractor	To submit proposals for adding a supplier to the Named Suppliers

(Continued)

Table A3.2 (*Continued*)

Communication Form	Clause Number	Used by	For
	Y1.9	Contractor	To submit the signed Authorisation to the Project Manager
	Y1.9	Client	To submit the signed Authorisation to the Project Bank
an Acceptance	16.2	Project Manager	Acceptance of a Contractor proposal
	16.3	Project Manager	Acceptance or rejection of a proposal to add an area to the Working Areas
	21.2	Project Manager	Acceptance or rejection of the Contractor's design of the Works
	23.1	Project Manager	Acceptance or rejection of the Contractor's design of an item of Equipment
	24.1	Project Manager	Acceptance or rejection of a proposed replacement for a Contractor's key person
	26.2	Project Manager	Acceptance or rejection of a proposed Subcontractor
	26.3	Project Manager	Acceptance or rejection of the conditions of contract for a proposed Subcontractor
	31.3	Project Manager	Acceptance or rejection of Contractor's programme
	36.1	Project Manager	Acceptance or rejection of a quotation for acceleration
	40.2	Project Manager	Acceptance or rejection of a quality policy statement or a quality plan
	45.2	Project Manager	Acceptance or rejection of a quotation to accept a Defect
	62.3	Project Manager	Acceptance of a quotation for a compensation event
	65.2	Project Manager	Acceptance of a quotation for a proposed instruction
	84.1	Project Manager	Acceptance or rejection of the Contractor's insurance certificates
	86.1	Contractor	Acceptance or rejection of the Client's insurance certificates
	55.3 (Option A)	Project Manager	Acceptance or rejection of a revised Activity Schedule
	50.9 (Options C, D, E and F)	Project Manager	Acceptance the Defined Cost of a finalised part

Table A3.2 (*Continued*)

Communication Form	Clause Number	Used by	For
	X4.2	Project Manager	Acceptance or rejection of an alternative guarantor
	X10.4(2)	Project Manager	Acceptance or rejection of an Information Execution Plan
	X13.1	Project Manager	Acceptance or rejection of a proposed bank or insurer for the performance bond
	X14.2	Project Manager	Acceptance or rejection of a proposed bank or insurer for the advanced payment bond
	X16.3	Project Manager	Acceptance or rejection of a proposed bank or insurer for the retention bond
	X21.3	Project Manager	Acceptance or rejection of a quotation for a whole life cost change
	X22.2(2)	Project Manager	Acceptance or rejection of a forecast of the Stage One costs
	X22.3(3)	Project Manager	Acceptance or rejection of a design proposal for Stage Two
	Y1.4	Project Manager	Acceptance or rejection of the banking arrangements for the project bank account
	Y1.6	Project Manager	Acceptance or rejection of a proposal for adding a supplier to the Named Suppliers
a Payment Certificate	51.1	Project Manager	To certify the amount due at each assessment date
	53.1	Project Manager	To certify a final payment
a Completion Certificate	30.2	Project Manager	To certify the date of Completion
a Take Over Certificate	35.3	Project Manager	To certify the date of takeover of any part of the Works
a Defects Certificate	43.3	Supervisor	To EITHER certify those notified Defects which the Contractor has not corrected OR state that there are no Defects which have not been corrected

Table A3.3 Table of communication forms required but which are not referred to in Volume 4 of the User Guide (in alphabetical order).

Communication Form	Clause Number	Used by	For
Early Warning Register	15.1	Project Manager	To record all early warnings issued by the Contractor or the Project Manager
	15.4	Project Manager	To record the decisions made at an early warning meeting
	15.4	Project Manager	To issue to the Contractor
Termination Certificate	90.1	Project Manager	To certify termination if the reason notified by either party is valid

Printed and bound by CPI Group (UK) Ltd, Croydon, CR0 4YY
19/04/2024

14485888-0001